Studies in Systems, Decision and Control

Volume 327

The series "Studies in Systems, Decision and Control" (SSDC) covers both new developments and advances, as well as the state of the art, in the various areas of broadly perceived systems, decision making and control–quickly, up to date and with a high quality. The intent is to cover the theory, applications, and perspectives on the state of the art and future developments relevant to systems, decision making, control, complex processes and related areas, as embedded in the fields of engineering, computer science, physics, economics, social and life sciences, as well as the paradigms and methodologies behind them. The series contains monographs, textbooks, lecture notes and edited volumes in systems, decision making and control spanning the areas of Cyber-Physical Systems, Autonomous Systems, Sensor Networks, Control Systems, Energy Systems, Automotive Systems, Biological Systems, Vehicular Networking and Connected Vehicles, Aerospace Systems, Automation, Manufacturing, Smart Grids, Nonlinear Systems, Power Systems, Robotics, Social Systems, Economic Systems and other. Of particular value to both the contributors and the readership are the short publication timeframe and the world-wide distribution and exposure which enable both a wide and rapid dissemination of research output.

Indexed by SCOPUS, DBLP, WTI Frankfurt eG, zbMATH, SCImago.

All books published in the series are submitted for consideration in Web of Science.

More information about this series at http://www.springer.com/series/13304

Gorachand Dutta · Arindam Biswas ·
Amlan Chakrabarti
Editors

Modern Techniques in Biosensors

Detection Methods and Commercial Aspects

Springer

Editors
Gorachand Dutta
School of Medical Science and Technology
Indian Institute of Technology Kharagpur
Kharagpur, West Bengal, India

Arindam Biswas
School of Mines and Metallurgy
Kazi Nazrul University
Asansol, West Bengal, India

Amlan Chakrabarti
A. K. Choudhury School of Information Technology
University of Calcutta
Kolkata, West Bengal, India

ISSN 2198-4182 ISSN 2198-4190 (electronic)
Studies in Systems, Decision and Control
ISBN 978-981-15-9611-7 ISBN 978-981-15-9612-4 (eBook)
https://doi.org/10.1007/978-981-15-9612-4

This Springer imprint is published by the registered company Springer Nature Singapore Pte Ltd.
The registered company address is: 152 Beach Road, #21-01/04 Gateway East, Singapore 189721, Singapore

Preface

Biosensors are analytical devices with one or more biologic recognition elements that are specific for a certain target analyte of interest. Biosensor technology offers several benefits over conventional diagnostic analysis including simplicity of use, specificity for the target analyte, capability for continuous monitoring and multiplexing. Major applications for biosensors are clinical diagnosis and disease prevention. Over the years, there are increasing needs for the development of simple, cost-effective, portable, integrated biosensors that can be operated outside the laboratory by untrained personnel. In recent years, many chip-based biosensors have been reported and so far the technology for electrochemical amplification on a chip has always relied on expensive micro- and nanofabrication technologies such as optical and e-beam lithography. This approach has several drawbacks even when leading to reliable results in research laboratories. Most of clinical analysis is carried out in centralized laboratories where high-technology equipment is available, and trained personnel perform the assays under almost ideal conditions. However, a large part of the population in developing countries does not have access to the state-of-the-art diagnostic methods. So, there has been intense demand of various biosensors for portable devices for rapid and sensitive detection.

In this book, we address these challenges for the development of a point-of-care-test platform including printed chip-based assay (laboratory-on-a-chip and laboratory-on-a-PCB) for rapid, inexpensive, multiplex detection of disease biomarkers in real samples. The chip will be able to handle raw samples and to perform on-chip all the processing required from sample-to-result by proposing practical laboratory-on-chip platform that will provide the means for simple, rapid, automated and cost-effective whole blood analysis. This book also addresses how to overcome existing barriers for laboratory-on-a-chip commercialization (lack of cost-effective mass manufacturing methods, self-contained, fully autonomous operation and user-friendliness). Different advanced techniques including electrochemical, optical, mass, colorimetric and signal amplification strategies describe early stage disease diagnosis. This title would focus on recent advances and different research issues in the biosensor technology and would also seek out theoretical, methodological, well-established and validated empirical work dealing with these different topics. The title covers a very vast audience from basic science to engineering and technology experts and learners. This could eventually work as a textbook for engineering and biomedical students or science master's programs and for researchers. This title also serves common public interest by presenting new methods for data evaluation, medical diagnosis, etc., to improve the quality of life in general, with a better integration into society.

Kharagpur, India — Gorachand Dutta
Asansol, India — Arindam Biswas
Kolkata, India — Amlan Chakrabarti

Contents

About the Editors

Dr. Gorachand Dutta is an Assistant Professor at the School of Medical Science and Technology, IIT Kharagpur. He received M.Sc. degree in Chemistry from Indian Institute of Technology, Guwahati, India. His research interests include the design and characterization of portable biosensors, biodevices and sensor interfaces for miniaturized systems and biomedical applications for point-of-care testing. He received his Ph.D. in Biosensor from Pusan National University, South Korea, where he developed different class of electrochemical sensors and studied the electrochemical properties of gold, platinum and palladium-based metal electrodes. He completed his postdoctoral fellowships in Michigan State University, USA, and Pusan National University, South Korea, University of Bath, UK.

Dr. Arindam Biswas received Ph.D. from NIT Durgapur in 2013. He received prestigious BK21PLUS Fellowship, Republic of Korea 2015, Special Appointed Associate Professor (Visiting), Shizuoka University, Japan, in 2017, DST-JSPS Invitation Research Grant Award in 2020, IE(I) Young Engineer Award: 2019–2020, Institute of Engineers, India. Presently, Dr. Biswas is working as an Assistant Professor at Kazi Nazrul University, India. He has 98 publications in in refereed journals and conferences. Dr. Biswas received research grant from SERB, Government of India in 2017, Centre of Biomedical Engineering, Japan, for two consecutive years 2019–2020 and 2020–2021. Presently, Dr. Biswas is serving as an Associate Editor of Cluster Computing, Springer.

Amlan Chakrabarti is a Full Professor of Information Technology in the A.K. Choudhury School of Information Technology at the University of Calcutta. He was a Postdoctoral Fellow at the School of Engineering, Princeton University, USA during 2011–2012. He is the recipient of DST BOYSCAST fellowship award in Engineering Science in 2011, Indian National Science Academy (INSA) Visiting Faculty Fellowship in 2014, JSPS Invitation Research Award in 2016 and Erasmus Mundus Leaders Award from EU in 2017 and Hamied Visiting Fellowship from Cambridge University in 2018. He has been associated as a Visiting Professor at the University of Cambridge, City University of London, University of Oradea

Romania, SUNNY Buffalo USA, GSI Helmholtz Research Laboratory Germany, University of Bremen Germany, CERN Geneva, Kyushu Institute of Technology Japan, etc. He has published around 130 research papers in refereed journals and conferences.

Current Developments in Diagnostic Biosensor Technology: Relevance to Therapeutic Intervention of Infectious and Inflammatory Diseases of Human

Suprabhat Mukherjee and Niladri Mukherjee

Abstract The current scenario of various infectious and inflammatory diseases of humans is indeed alarming. Implication of appropriate therapeutic strategies in many diseases was found to be limited due to the lack/inefficiency of accurate diagnostic approach. In this context, biosensor technology has come out as an efficacious mean to meet the need. Several types of diagnostic biosensors viz. electrochemical, enzymatic, optical, aptamer-based, and immunosensors have been developed and modified throughout the last decade and few of them are currently in use. Since early detection is considered the key behind adopting accurate therapeutic strategies, use of biosensors is growing day by day. Such high demand in the market has come out as a promising prospect in different research institute and industries dedicated to human health care. Major limitations amongst the available biosensing devices are high cost, lower accuracy, and sensitivity. Therefore, the major challenge in modern biosensor research is the development of miniatured biosensors which will facilitate in situ diagnosis and treatment. More collaborative research among different scientific communities are particularly needed to overcome the aforesaid challenge. This chapter has been written with an intension to present a generalized overview on the relevance of the modern diagnostic biosensors in human healthcare especially for diagnosing human health problems with special emphases on infectious and infectious diseases.

Keywords Biosensor technology · Diagnosis · Infectious diseases · Inflammatory diseases · Human health

S. Mukherjee and N. Mukherjee are contributed equally.

S. Mukherjee (✉) · N. Mukherjee
Department of Animal Science, Kazi Nazrul University, Asansol, West Bengal 713340, India
e-mail: suprabhat.mukherjee@knu.ac.in

N. Mukherjee
e-mail: niladrizoology@gmail.com

G. Dutta et al. (eds.), *Modern Techniques in Biosensors*, Studies in Systems, Decision and Control 327, https://doi.org/10.1007/978-981-15-9612-4_1

1 Introduction

Despite having a well-orchestrated immune system, several life-threatening infectious and inflammatory diseases are considered as the major concerns. Infections caused from viruses (Polio, Dengue, Hepatitis B&C, Herpes Simplex virus, Human papilloma virus, Rota virus, etc.) bacteria (*Mycobacterium tuberculosis, Salmonella typhi, Clostridium tetani*), fungus (*Candida* spp., *Aspergillus* spp. etc.) and parasites (*Entamoeba histolytica, Plasmodium falciparum, Leishmania donvani, Fascioa hepatica, Wuchereria bancrofti*) result in huge number of morbidities and mortality worldwide [1]. On the other side diseases caused from the dysregulated immunity viz. diabetes, cancer, rheumatoid arthritis, inflammatory bowel disease (IBD), different types allergies and autoimmune diseases are the reasons behind the death of millions of people each year [1]. In this context, development of new drugs and vaccines as well as advancement of treatment procedures have counteracted to save the affected ones. The World Health Organization initiated and implemented several community-based treatment approaches like mass drug administration (MDA) program and eventually these treatment approaches found to be useful in lowering the prevalence of a number of diseases especially the neglected tropical diseases (e.g. lymphtic filariasis) [1]. However, the current scenario of various infectious and inflammatory diseases of human is indeed alarming. This is due to the dearth of information regarding the reason for disease prior to its pathogenesis and/or treatment as most of the treatment procedures in the developing countries are based on the symptoms. Although several efficacious drugs, vaccines, and probiotics are available, implication of appropriate therapeutic strategies in many diseases have been found to be limited due to the lack/inefficiency of accurate diagnostic tool. In this context, biosensor technology has come out as an efficacious mean to meet the need.

In general, biosensor is an opto-electronic device comprised of a biological sensing element and a transducer which can monitor the presence of an analyte and quantify the same with maximum accuracy in minimum time [2]. Glucose oxidase-coupled amperometric oxygen sensor is considered as the first functional biosensing tool for determining the presence of glucose in blood [3]. Enzyme-coupled sensors were the first-generation biosensors and thereafter several upgraded versions of diagnostic biosensing devices have been developed and modified throughout the last decade and few of them are currently in use. Till date, the available biosensors have been grouped according to the detection system (such as Enzymatic-, immuno-, nucleic acid- and bacterial cell-based biosensors) and the signal transduction methods (electrochemical, optical, thermal, and mass-based sensors) used in the sensing process [4–6].

Assessment of the altered health condition prior to the disease display; continuous monitoring of the parameters of the disease and health status during treatment of a disease as well as during follow up are considered as the major characteristic features of an ideal biosensor [7]. Such kind of features are dependent on three major aspects firstly selection of influencing biomarkers designating normal/healthy state or diseased conditions, second an efficacious instrumentation that enables detection of the biomarker and finally feasibility of distinguishing the biomarkers [7].

In recent years, the major emphasis given on the advancement of the biosensors is 'miniaturization' which seeks to enable the sensors to be used for in vivo real-time purposes and also in tissue engineering applications [8]. Such upgradation in biosensing-devices resulted in ultrasensitive sensing systems which can determine ultra-low quantity signal and differentiate highly overlapping in a cellular system such as DNA, mRNA, protein-like low molecular weight cytokines, etc. [7]. Surface plasmon resonance (SPR), different nanotubes or nanorods, nanowires or nanocantilevers are examples of modern-age biosensors [7, 8]. Furthermore, quantum dots, i.e., fluorescent semiconductor nanocrystals are also emerging as important component in developing biosensor thorough fluorescence resonance energy transfer (FRET) signal transduction system [9]. Researches on the progress in developing biocompatible nano-biosensors with ultra-high sensing efficiency and meant for both diagnosis and treatment are becoming evident in the current literatures [4].

The present chapter aims to present a generalized overview on the biosensing approaches used for diagnosing human health problems with special emphases on infectious and infectious diseases. Use of different biorecognition approaches used in the diagnostic biosensors, selection and use of different biomarkers; current advancements, challenges, and future directions in biosensing technologies have been discussed based on the information available to date.

2 Design and Mechanism of Function of Diagnostic Biosensors

The procedure of designing biosensor comprises a number of critical experimental approaches of engineering, chemistry, and biology that include fabrication of the sensor component, immobilization of the biological component, and most importantly the transducer.

2.1 Design of Biosensing Devices

As discussed in the earlier section, a biosensor is typically designed to convert a biological signal to electronic signal as described by Cammann, the pioneer in biosensor technology. Thus, a biosensing device is constructed by combining a biomolecule/microbe having sensing ability, a bio-transducing element, and an electronic system for processing the physio-chemical signal [10]. A detector is designed to sense the biological molecule to induce a signal which is subjected to transform to an electric signal by a transducer connected to a signal processing system that generates the final output in readable format [4, 7]. The instrumentation and mode of operation of commonly available biosensing devices have been depicted in Fig. 1.

Biosensing tools. Depending on the type of construction used in the biosensing device and working principle, biosensors are divided into four different types viz. enzymatic/biocatalytic biosensors, antigen/antibody/nucleic acid-based/bioaffinity biosensors, microbial biosensors, and nanoparticle/nanocomposite-based biosensors or nano-biosensors [7, 10].

Enzymatic biosensors are comprised of enzymes mostly belonging to oxidoreductases that are immobilized on porous, stable, and biocompatible matrices. Oxidoreductase enzymes [horseradish peroxidase (HRP), Glucose oxidase (GO_x) polyphenol oxidase (PPO) and phosphatase (alkaline phosphatase (ALP))] are commonly used enzymes for developing diagnostic biosensors [4, 7, 10]. Although, each of the mentioned enzymes possesses different mode of catalysis and different substrate specificity but in general the mechanism of their use in the biosensing device is similar. Bio-catalysis results in the uptake or production of detectable molecules (CO_2, O_2, H_2O_2 and NH_3) or ions (H^+); analyte induced stimulation of enzyme or blockage of the enzyme activity [4, 7, 10]. Any of the responses are usually detected by the transducer (discussed in the subsequent section). In the sensing device, an enzyme is usually immobilized on the transducer. Process of immobilization and selection of immobilizing matrix are considered as the pivotal factors in developing enzymatic biosensors [4, 7, 10]. Covalent crosslinking, gel entrapment, and physical adsorption are known as the widely used immobilization procedures [4, 7, 10]. Porosity, stability, and non-interference are considered the key properties of an immobilizing matrix. Several studies on single- and multiwall nanotubes of carbon (SWCNTs and MWCNTs) deciphering high conductivity of electrons, stability, and firmness made these materials as stable network or matrix for immobilizing a biocatalyst [4, 7, 10]. It is noteworthy to mention that biodegradable and biologically benign hydrogels are becoming evident as efficient immobilizing matrix for developing diagnostic biosensor especially for detecting in situ toxic effects of toxic contaminants on living organisms especially humans [11–13]. Utility of polyphenol oxidase (PPO) in biosensing devices is also becoming increasing evident [14]. Excellent substrate specificity, high availability, and relatively low price of PPO are considered as the key advantages of using this enzyme in developing biosensors [14, 15]. Recently, immobilized PPO on gold-nanoparticle-doped graphene oxide nanocomposite has been found to provide excellent sensing of phenolic contaminants from water [16].

Non-enzymatic biosensors are comprised of several modern-age sensors that are designed to meet the increasing need for human health sectors. Microbial biosensors are constructed by fabricating the biosensing scaffold using whole-cell microbe [17]. Microbes metabolize complex organic/inorganic compounds and release several molecules like hydrogen ion, carbon dioxide, ammonia, etc. detected by transducer for tracing the identity of an analyte [17]. Unlike enzymatic sensors, whole-cell microbial sensors are cost-effective as microbes do not require extensive purification [17]. On other hand, microbial sensors can sense a wide array of analytes due to the ability of the microbes to metabolize complex molecules through aerobic as well

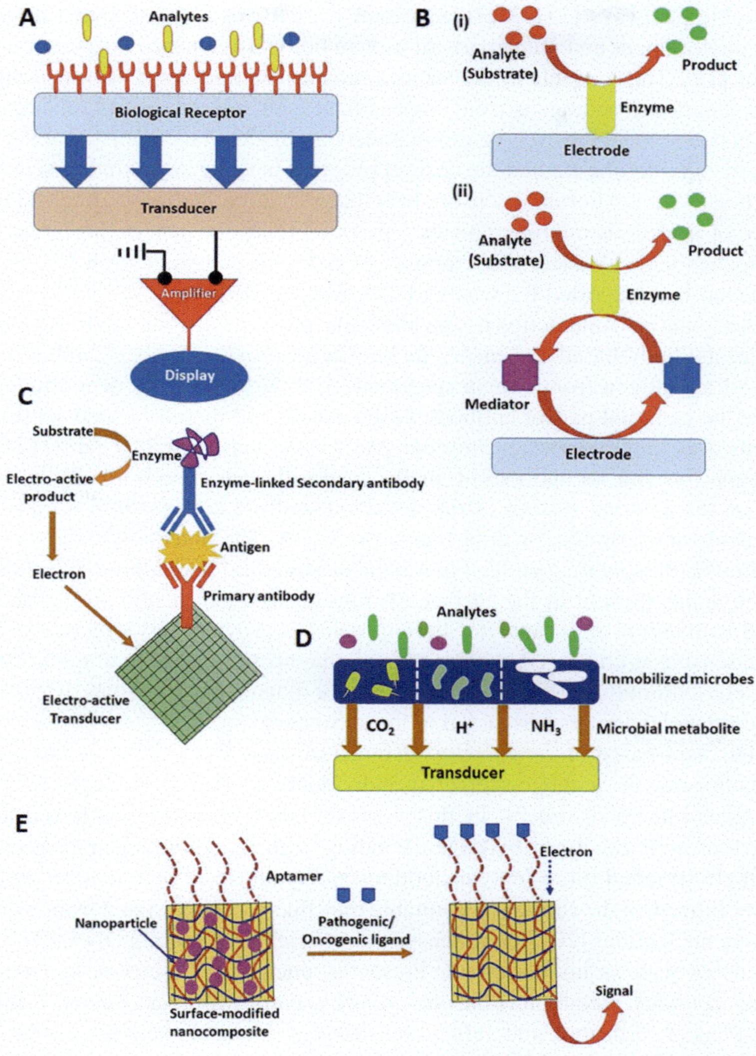

Fig. 1 Design and mechanism of function of different biosensors for diagnosing infectious and inflammatory diseases. a Basic components in a diagnostic biosensor. **b** Design of an enzymatic biosensor for determining a specific analyte from a mixture of compounds. Enzyme is immobilized in the sensing scaffold and the electrochemical signal generated through enzymatic catalysis is transmitted in absence (i) or presence (ii) of a mediator. **c** Mechanism of operation in immunosensor. Antigen-antibody interaction generates electron which in turn received and processed by transducer. **d** Design of microbial biosensor. Microbes are immobilized on porous solid conducting surface that perceives signal from the metabolites generated from microbes and transmits to the transducer. **e** Design and mode of function of nano-biosensor. Specific aptamers corresponding to pathogen-specific nucleic acids are immobilized on nanoparticle-embedded conducting polymers. After binding with specific ligand, electrons are generated and transmitted to the transducer

as anaerobic pathways [17]. Most importantly, microbes are distributed in all the parts of environment and many parts of human body also contain different types of microbes [18]. These enable microbes as a sensing material for detecting contaminants/toxicants and pathogens from environment (soil, water, air, and food) as well as human body [17]. However, dearth in understanding the biochemistry of the individual sensing microbes halted the desired progress in using such kind of sensors in human health care. Microbial sensors have been found effective in diagnosing the presence of infectious microorganisms, microbial toxic substances, hormones, and other biochemical subtance. For example, SOS-EGFP constructed on *E. coli* SOS response has been reported for sensing DNA damage [19].

Antibody-antigen interaction is considered the most specific interaction in biology and hence antibody-based biosensors are known for providing specific and adequate sensing of an antigen from human specimen [4, 7, 10]. Antibody or immunoglobulins are the essential part of antibody-based sensor and therefore such sensors are also termed as immunosensors. Immunosensors are constructed by fabricating the sensing matrix with an antibody. Usually, a specific detection antibody is immobilized on the external surface of the detection scaffold connected to a transducer by covalent esters, amide, or thiol bonds [4, 7, 10]. The detection antibody interacts with its corresponding antigen that induces structural alteration in the chemical groups/moieties present in the surface of transducer which results in the recognition and estimation of an analyte. Immunosensors provide very precise, fast and accurate sensing as compared to other immunodiagnostic approaches (agglutination, ELISA, etc.). Immunosensors are widely used for diagnosing infectious diseases and immune-related disorders (discussed in the subsequent sections).

Nucleic acid-based biosensors are designed to sense DNA, RNA or nucleotides from pathogens or pathogen-infected individuals and/or individuals having a genetic/gene-related disorders. In this context, DNA is usually preferred as the sensing tool as it can form hydrogen bonding with its complementary sequence [20]. Single-stranded oligo-/polynucleotides are synthesized by chemical methods and immobilized on the surface of a suitable transducer to yield signal from a sample containing nucleotides [20]. These sensors are efficient in detecting DNA/RNA from pathogens as well as human blood. Presently, nucleic acid sensors are used for detecting complex metabolic (diabetes, obesity, cardiovascular diseases), inflammatory (cancer, arthritis) and infectious diseases (dengue, hepatitis, pulmonary tuberculosis, cholera, amoebiasis, and salmonellosis), and other genetic diseases [4, 7, 10, 20].

Aptamer-based sensors are the upgraded addition to the list of biosensors that are meant for highly specific sensing in presence of ultralow quantity analyte. Biochemically, aptamers are composed of single-stranded nucleic acid (DNA or RNA) which are synthesized against specific groups of biologically relevant molecules like amino acids, proteins, drugs, etc. [20, 21]. Major characteristic attribute of aptamer technology is that oligonucleotides used in the aptamer are stable and form secondary structure that quickly functions to achieve better sensing with distinct affinity to its target [21]. Moreover, aptamer can distinguish minute change among its targets [20]. Aptamers are now considered as therapeutic agents, specifically for ocular vascular diseases associated with old age, after the approval of USFDA (United State Food and Drug Administration) [20].

2.2 *Signal Transduction Technology for Biosensing Devices*

Developments in the biosensor technologies have been majorly aimed to meet the analytical demands in human health care and other purposes including environmental and research purposes. The technological progress in this field has been found to be dealt with the signal transduction technologies [22]. Based on the difference in the type of transducers used in the biosensor, the currently available sensors are grouped into several types and are discussed in this section.

Electrochemical sensors are the first line of transducers that determine the electrochemical changes that occur between the sensing molecule and analyte at the sensing matrix [22]. Such kind of transducers are linked with the sensing device as electrodes that are categorized into three categories viz. amperometric, conductometric and potentiometric biosensing devices based on the type of electrochemical signal they perceive [22]. Amperometric biosensors are typically designed for sensing the electrochemical signal generated from an enzyme-catalyzed oxidoreductive reaction [22]. In fact, the foremost GO_x-based biosensor was built using this principle and later other enzymatic sensors employing peroxidase, polyphenol oxidase also followed the same [22]. Alike amperometric biosensors, potentiometric sensors also detect small molecules and are comprised of field-effect transistors made up of modern nanomaterials like nanotubes and nanowires of carbon or silicon [22]. The operating mechanisms of these biosensors are almost identical to pH electrodes [22]. Conductometric biosensors are the newest addition to the list of electrochemical sensors which also allow sensing of an analyte/biomolecule without involving an enzyme or nanomaterial [22]. Moreover, high degree specificity is usually found even in the presence of large numbers of analytes with different biochemical properties. Extensive research throughout the last decade has enabled the incorporation of several smart materials likewise graphene, diamond, screen-printed electrodes to improve the efficiency of the electrochemical biosensing devices [4, 7, 10, 22].

Optical sensors exploit both labeled and label-free signal transduction technologies for recognizing analytes ranging from small to large molecules including proteins [7, 10]. Labeling technology involves the use of fluorophores, nanocomposites, nanoparticles, quantum dots, and carbon nanotubes [7, 10, 23]. Such fluorescent or luminescent materials are excited by an evanescent field or electrochemically and finally generate a detection signal. On other side, label-free detection is solely dependent on refraction [24]. Although TIRF is the firstly developed optical sensor, surface plasmon resonance (SPR)-, photonic crystals, optical ring resonators (ORR), and Bragg gratings-based sensors comprise the modern editions of optical sensors [24]. Intriguingly, recent studies using terahertz spectroscopy and surface-enhanced Raman spectroscopy are emphasizing more on developing label-free detection to achieve ultra-low concentration of analytes [24].

Thermal or calorimetric sensors are built on the principle of calorimetry that deals with the change in temperature and therefore temperature-dependent resistor (thermistor) is used as detector [22]. Alike the enzyme electrode of amperometric sensor, thermal biosensors were initially termed as enzyme thermistors. Although rare in biology, thermal biosensors have been reported to provide label-free sensing of small molecules having relevance to human health.

Piezoelectric or acoustic sensors are constructed using gravimetric, i.e., mass-sensitive transducer that detects mechanical acoustic waves generated from the interaction of a definite mass of analyte with the sensing material [4, 7, 10]. The sensing material is usually consisting of vibrating piezoelectric materials quartz crystal [4, 7, 10]. Recent literature have demonstrated the use of nanoparticles for mass labeling that has been found to increase the sensing efficiency. These group of sensors are used mostly for detecting analytes occurring in gas phase [22]. Technological advancements in biosensor technology resulted in several modern version of acoustic sensor like QCN biosensors, cantilever biosensors, and surface acoustic wave (SAW) [4, 7, 10].

Magnetic and radioactive biosensors are the two less popular sensors in human health care. Magnetic or magnetoresistance biosensors are relying on magnetic transduction principles that require magnetic nanoparticles as labeling material for sensing nonmagnetic analytes [4, 7, 10]. These sensors are somewhat less common as compare to electrochemical, optical, or other biosensors as very sophisticated electronic circuits are required. However, these sensors can be used for detecting trace elements in biomolecules. On the other side, radioactive biosensors are developed on the radioactive transduction principle that usually inspired from radioimmunoassay (RIA) [7, 10]. These biosensors are designed to provide excellent signal due to the use of radioactive labeling.

2.3 *Modern-Age Biosensors*

Nanobiosensors constitute the modern or new-age biosensors which are typically meant for sensing wide array of micro- and macro-analytes. In this connection, quantum dots are considered as efficacious addition for the advancement of the sensing efficiency of the optical biosensors [23]. These semiconducting nanocrystals possess unique quantum confinement properties enabling excellent stability, quick excitation as well as slim size-tunable emission bandwidth and minute photobleaching [24, 25]. Similarly, graphene scaffold-based biosensors are also known to exert high-quality sensing [16]. Due to the porous, stable, and conducting supramolecular scaffold of graphene enables a suitable immobilizing matrix for enzymes [16, 26]. Recently, Metal-doped graphene/graphene oxide nanocomposites and graphene-ascorbate quantum dots have been demonstrated to show promise as highly-sensitive biosensors in future [23, 26]. Similarly, single-wall and multi-wall carbon nanotubes derived from cylindrical fabrication of graphene sheet appears to be potential biosensors, especially for diagnostic purposes [23, 26]. Microfluidics-based and microelectromechanical systems-based biosensors have emerged as most promising version of modern diagnostic sensor wherein a biological receptor or antibody is immobilized on the sensing surface that is designed to facilitate the interaction between the sensing element and the analyte present in a sample (viscous liquid) and generate a signal by determining the alteration of mass and/or dielectric behavior [27]. Such sensors have been found useful in diagnosing pathogenic infection and tumor growth [27]. Diagnostic application has been the major emphasis in developing a new technology namely 'lab-on-a-chip'. These sensors are the miniaturized devices which are fabricated with several microchannels made up of polydimethylsiloxane (PDMS) or thermoplastic polymer and conjugated with detection molecules like antigens, antibodies, or oligonucleotides [27]. These sensors are efficient enough to provide intense signal even in presence of numerous analytes from a less amount of sample.

3 Biomarkers

The functional attributes of a biosensor majorly depend on the selection of a 'biomarker' on which the sensing device works. Biomarkers are molecules that can be considered as a selective parameter/indicator for a health state which can be normal or diseased state [28]. Response of a biomarker may be varied with the system using for the measurement [28]. A biomarker essentially acts as an indicator of normal physiological condition or occurrence of a disease/responses to a pharmaceutical agent [28]. In case of cancer, biomarkers are typical growth factors, altered chromosome structure (reciprocally translocated), mutations in tumor suppressor genes, activated proto-oncogenes, abundance of tumor specific antigens (e.g., prostate-specific antigen (PSA) in male prostate cancer) [29]. Similarly, glucose and insulin for diabetes, leptin/fatty acid for obesity, cholesterol for cardiac heart diseases, Lewi Body for Perkinson's disease [30]. Wolbachial component for filariasis, mitochondrial DNA for leishmaniasis, typical glycoprotein (gp120) for HIV are considered as potential biomarkers [31, 32]. Different biomarkers associated with the various infectious and inflammatory diseases of human have been discussed in the subsequent sections.

4 Biosensing for Diagnosing Human Inflammatory Diseases

Inflammation is the cumulative response of human body which is meant for providing defense against invading pathogens or antigens [32]. Such kind of responses also counteract against self-antigens during several disease conditions [33]. Inflammation up to threshold is beneficial to maintain the normal defense function of human body system but dysregulated inflammation is considered as a major cause behind inducing a number of pathophysiological responses and life-threatening diseases [32]. In this connection, diabetes, especially type II diabetes mellitus, currently responsible for the set back of 3% world population [10]. Pathological attributes of diabetes include hypertension, increased heart-related diseases, renal disorders, neurological problems, blindness, obesity, etc. Increased glucose level ($\geq$180 mg/dl) far beyond the normal level (120 mg/dl) is a characteristic biomarker for diabetes [34]. Therefore, quick determination of blood glucose is a critical need for diagnosing and treating the diabetic individuals. One of the fascinating features of glucose is that it is a good substrate for the enzyme namely glucose oxidase (GOD). GOD converts glucuronic acid in a redox/electrochemical reaction. In this context, the very first glucose biosensor [3] was constructed based on the catalytic affinity of GOD towards glucose and subsequent generation of the electrochemical signal. A number of modifications have been introduced to develop several new editions of glucose biosensors. Miniaturization has been the major modification taken into account in modern glucose biosensors for determining minute quantity of glucose from human blood and/or tissue. For example, Cengiz and Tamborlane (2009) reported an efficient in vivo biomonitoring of interstitial glucose in subcutaneous fat tissue by an invasive GOD-based biosensor [35]. Modern age glucose sensors do have several limitations that include drift, calibration error, and signal delay [36]. In spite of these most of the available sensors are practically accurate and are widely used for monitoring blood and tissue glucose. Electrochemical glucose sensors made up of screen-printed disposable electrode is popular choice nowadays for glucose determination.

Non-invasive glucose biosensor has been the ultimate need for meeting need of the physicians and several breakthrough researches have evidenced about the possibility of such kind of sensors in practice [37–39]. Microcarolimetry, enzyme-fabricated electrode, optical sensor, sonophoresis and ionotophoresis, and near-infrared (NIR) spectroscopy-based non-invasive monitoring techniques have been found efficacious in detecting trace quantity glucose from body fluids including saliva, sweat, tears, and urine [37–39]. Although non-invasive techniques are meant for minimum effort and risk of infection, these sensors have been found possess several limitations like low adequacy and sensitivity, requirement of frequent calibration, poor correlation amongst the glucose levels of different sources, etc. [40].

Cancer is one of the major causes of human death worldwide. In spite of the efforts of medical sciences, the risk of this disease is continuously increasing due to the abundance of cancer-favorable environment that principally comprised of different pollutants and carcinogens [41]. Cancer of head and neck, pancreas, lungs, breasts, cervix, prostate, bladder, rectum, and different forms of blood cancer are taking enormous toll from the human. Apart from mortalities, huge socio-economic loss is also another impact of this disease as the treatment cost of the disease is particularly very high [41]. Intriguingly, early detection of tumor following surgical removal and/or chemotherapy found to reduce the mortality rate and, in many cases, complete recovery of the patients as well [42, 43]. Therefore, accurate and early diagnosis of the tumor is the most crucial need to combat this life-threatening disease. Choice of appropriate biosensing approach for cancer is somewhat complex as each type of cancer possesses distinct biomarker [44]. For example, cancers caused by carcinogens (e.g., bladder carcinoma) are different from bacteria-induced cancers (e.g., gastric or stomach) or epigenetic ones [44]. On other side, location of tumors is also different, variable, and/or overlapping in each type of cancer. Therefore, selection of biomarker or a range of biomarkers are particularly recommended to diagnose malignancy [45]. Common biomarkers used for cancer detection comprise specific mutation, presence of protooncogene, chromosomal alteration, DNA modifications, presence of specific transcript (RNA), proteins (enzymes and glycoproteins), specific hormones and growth factors, immune receptors and cytokines, etc. [45–47]. Detection of appropriate biomarker(s) from body fluids (blood, urine) or tissue are aimed through both invasive and non-invasive approaches. Currently used technologies for determining malignancy (onset of cancer) are invasive and principally involve removal of tissue from the suspected organ/part, analysis of the cellular architecture through staining and microscopic study [20, 48]. Although these approaches are cost-effective but such technologies cannot diagnose complex oncogenic transformation. Therefore, efficient biosensing tool is of major need. In this connection, peptide/protein-based biosensing devices for multiple cancer biomarker analyses have shown their potential [20, 48]. In recent years, several types aptamers (nucleotides), synthetic peptides, phage display peptides, nanomaterials, metal oxides (e.g. Zinc oxide) have been implicated for construction detection scaffolds with highest precision [20, 48]. The design of these modern sensors majorly involves use detection antibodies (monoclonal and polyclonal) specifically developed against distinct cancer biomarker and such kind of antibodies can selectively recognize their antigen to produce a detectable signal [25]. Recently these sensors have been modified by incorporating synthetic receptors and biomimetic materials have been incorporated in place of antibodies and improved sensing have been experienced in more specific, robust, and cost-effective way [21]. In addition to these, electrochemical biosensors (amperometric, potentiometric, and impedimetric/conductivity) designed for affinity detection of mutations in cancer critical genes and proteins found to provide highly accurate sensing of cancer biomarkers [49, 50]. Immunodiagnostic sensors for detection of cancer-associated proteins, cytokines, and growth factors are the widely used technologies for cancer diagnosis [49, 50]. Nucleic acids rather DNA-based sensors having immobilized DNA have also been implicated for diagnosing oncogenic mutations [51]. Electrochemical sensors have been found useful in detecting metastatic tumors especially their cell number and tissue distribution [6]. Intriguingly, most of these sensors are made up of impedance transducers capable of providing label/tag-free as well as real-time detection [49, 50]. SPR-based optical sensors [46] and microcantilever-based sensors [52] are examples of modern edition of cancer biosensors which are meant for early-stage detection.

Cardiac heart disease (CHDs) or stroke and cardio-vascular diseases (CVDs) are the global concern as they kill millions of people in each year and the risk of these diseases are increasing day by day. In this context, heart attack, i.e., myocardial infarction (MI), is the principal culprit resulting most of the CVD-associated death cases. MI is caused due to the formation of atherosclerotic plaque owing to the deposition of cholesterol and lipid-containing foam cells and this results in the blockage of blood flow through coronary artery to damage the cardiac muscles [53]. Altered lifestyle, stress, and intake of junk foods are the major reasons behind such diseases. CVD/CHD patients are usually diagnosed by determining serum parameters, typical pain in chest, and abnormal electrocardiogram (ECG) [53]. However, most of the affected and hospitalized patients display history of apparently normal ECG [53]. Rapid and early-stage diagnosis is known as the major factors for successful prognosis and treatment of cardiac problems. These can be resolve by the use of appropriate biosensors and the demand of cost-effective, robust and real-time biosensor is very high. Alike other diseases, selection of appropriate biomarkers is also a crucial aspect in developing biosensors. Humoral factor-like C-reactive protein (CRP), serum cardiac troponin I (cTnI), myoglobin, B-type natriuretic peptide (BNP), and interleukins, interferons have been explored in recently developed biosensors [54]. Electrochemical (amperometric, potentiometric, and impedimetric), optical (colorimetric, luminescence, fluorescence, SPR and fiber optics/bio-optrode), acoustic (CMOS Si chips), and magnetic-based biosensors have been found efficacious in sensing the aforementioned biomarkers with high precision [54, 55]. Immunoarray-based platform comprising fluorescence microsphere has also been developed for detecting CVD-associated inflammatory cytokines as well as chemokines (TNF-α, IL-1, IL-6, and IL-8) [53]. Recently, the existing cardiac biomarkers-based biosensors have been coupled with artificial intelligence (AI) to facilitate point of care (POC) diagnosis [55]. Similarly, non-invasive and ultra-sensitive multiplex system has been reported for real-time diagnosis of multiple biomarkers with minute quantity of samples [56].

Neuroinflammatory and neurodegenerative disease are becoming global concerns due to the increase of affected individuals each year. Currently, Alzheimer's disease (AD), multiple sclerosis (MS), and Parkinson disease (PD) are the major members of neuroinflammatory diseases. All these diseases are primarily characterized by heterogeneous neuroinflammatory consequences resulted from chronic activation of innate immune system that primarily disrupts immunehomeostasis of the affected individual [30]. AD is a typical neurodegenerative disorder primarily diagnosed by a decrease in amyloid-beta (Ab) levels and an elevated tau protein [neurofibrillary tangles (NFTs)] in the cerebrospinal fluid (CSF) which results in the damage of neocortical regions following progressive dementia and/or intermittent loss of memory [57]. NFT formation is known to be associated with complement-driven microglial activation resulting in activation of mitogen-activated protein (MAP) kinase following the release of proinflammatory cytokines [58]. On the other side, steady loss of motor functions are the major clinical manifestations of PD. Intriguingly, several non-motor symptoms in different body organs and systems are considered as associated problems. Degenerative changes in the dopaminergic neurons of substantia nigra inhibiting dopamine release in striatum to impair motor function is the major pathological attribute of PD [59, 60]. Hitherto, several experimental evidences have documented persistent inflammatory responses due to glial cell activation and T cell infiltration as the crucial contributors of PD pathology [59, 60]. The intracellular inclusion protein, α-synuclein (AS) has been demonstrated to be the key mediator of physiological and inflammatory pathology of PD [59, 60]. Chronic activation of Microglia in response to AS is the major inducer of cytokines and reactive oxygen species resulting persistent detrimental insults to the neurons [59, 60]. Microglia can detect oligomeric AS through innate immune receptors viz. toll-like receptor 2 (TLR2) and TLR4 that triggers inflammatory signaling to cause NF-κB activation resulting in the secretion of the pro-inflammatory cytokines [59, 60]. In recent times, there has been growing evidences on the impact of nitric oxide (NO) induced post-translational modification (nitrosylation) of important enzymes (UCHL1) of ubiquitin-proteasome pathway impairing protein turn over and AS aggregation resulting Lewy Body formation [59, 60]. MS is principally diagnosed by continuing demyelination of the central nervous system due to the neuroinflammation and the pathology of MS is known to be dependent on hereditary predisposition and environmental determinants [61]. Assessment of behavioral changes, histology and determination of the amyloid deposition and/or uptake of fluorodeoxyglucose in the temporoparietal cortex through positron-emitting tomography (PET) are considered as the conventional approaches for diagnosing AD [30, 62]. PD diagnosis potentially includes detection of serum/CSF alpha-synuclein meta-iodobenzyl guanidine (MIBG) scintigraphy, olfactory function testing, etc. [59]. However, lack of specificity in detecting a single pathological manifestation is considered as the major drawback of the conventional diagnostic approaches [30]. Recently developed biosensors for sensing distinct biomarkers could be the promising way for diagnosing neuroinflammatory diseases. A list of selective biomarkers for detecting neuroinflammatory diseases has been given in Table 1. Most of the biosensors developed against the neuroinflammatory and neurodegenerative biomarkers involve detection of a critical inflammatory marker from a given body fluid (serum and CSF). In recent years, an invasive electrochemical biosensor has been developed to selectively sense IL-1beta and IL-10 [63]. A similar kind of biosensor has been developed for robust sensing of matrix metalloproteinase (MMP)-9 [64]. However, these sensors are non-specific. Further modification of these biosensors by choosing more selective biomarkers results in development of more accurate sensors. For example, electrochemical sensors for detecting AD-specific antibody and oligomers [65] and impedimetric biosensor for detecting amyloid peptide [66]; photoelectrochemical biosensor for quantitative sensing of alpha-synuclein and electrochemical sensing of thrombin [67] for selectively diagnosing PD patients; SPR-based immunosensor for detecting MS-specific autoantibody from serum [68].

Research and development in biosensing technologies are not only restricted in the aforementioned diseases but also in other diseases having relevance to human health. For example, arthritis and rheumatoid arthritis (RA) are the very common health problems characterized by mild to severe joint pain, mobility problems, and associated health problems. In this connection, graphene-based nano-immunosensor and Zinc oxide nanorod-based sensor having immobilized anti-cyclic citrullinated peptide (a specific biomarker for RA) has been reported as effective approach in diagnosing patients suffering from RA [69, 70]. On the other hand, periodontitis is also a common problem of mankind related to teeth loss with increased list of other associated problems including oral cancer [71]. Recently, an efficient aptamer-based sandwich biosensor has been developed and this sensor has been claimed to

Table 1 Biosensors for diagnosing inflammatory diseases

Disease	Biomarker and source of sample	Biosensor	References
Type II diabetes	Blood and adipose tissue; blood glucose, ketone bodies, free fatty acids	Immunosensor and electrochemical sensors	[3]
Breast cancer	Tissue; BRCA1, BRCA2, HER2/NEU, NY-BR-1, ING-1	Immunosensor and electrochemical sensors	[4]
Liver cancer	Tissue and serum; AFP, CEA	Immunosensor and electrochemical sensors	[4]
Cervical cancer	HPV DNA and abnormal PAP smear	Anti-(epithelial cell adhesion molecule), anti-CD44, and anti-tumor associated calcium signal transducer 2 coated microbeads	[72]
Ovarian cancer	Tissue and blood; HCG, p53, CEA, CA 549, CASA, MCA, MOV-1, TAG72	Immunosensor and electrochemical sensors	[4]
Colon cancer	Tissue; CEA, EGF, p53	Immunosensor and electrochemical sensors	[4]
Prostate cancer	Prostate specific antigens and blood	Citrate ligands-capped gold nanoparticles-based nano-immunosensor	[29]
Melanoma	Tyrosinase, NY-ESO-1	Enzymatic sensors and electrochemical sensors	[4]
Lung cancer	CEA, CA 19-9, SCC, NSE, NY-ESO-1	Immunosensor and electrochemical sensors	[4]
Multiple Sclerosis	CSF and Oligoclonal bands; IgG index, IL-23, IL-17, CXCL10, TNFa, TGF-b, Nfl, NGRN; MMP-9	Immunosensor and electrochemical sensors	[30]
Parkinson diseasea	CSF/serum; alpha-synuclein	Immunosensor and electrochemical sensors	[30]
Alzheimer's diseasea	CSF and serum; amyloid peptide	Optical sensors, Immunosensor, and electrochemical sensors	[30]
Myocardial infarction	Serum/blood; C-reactive protein, IL-6, IL-8, TNF-alpha, IL-1beta	ELISA and RIA-based Immunosensor	[30]

(continued)

Table 1 (continued)

Disease	Biomarker and source of sample	Biosensor	References
Cardio-vascular diseases	Serum/blood; C-reactive protein, IL-6, IL-8, TNF-alpha, IL-1beta	Immunosensor and electrochemical sensors	[69, 70]
Arthritis and rheumatoid arthritis	Serum; having immobilized anti-cyclic citrullinated peptide	Graphene-based nano-immunosensor and Zinc oxide nanorod-based sensor	[69, 70]
Periodontitis	Serum/blood/tissue; odontogenic ameloblast-associated protein (ODAM)	Aptamer-based sandwich biosensor	[71]

detect odontogenic ameloblast-associated protein (ODAM), a selective biomarker for periodontitis [71]. Selection of biomarkers and use of biosensors in different human diseases have been presented in Table 1.

5 Biosensors for Diagnosing Infectious Diseases of Human

Viruses, bacteria, fungi, and parasites majorly constitute causative agents of diverse infectious diseases of varying amplitude with respect to the disease sternness resulted in patient morbidity and death. In recent years due to enhanced and heightened health practices, the mortality allied to infectious diseases is reduced worldwide, however, an estimated causality calculating the present health practices can reach 13 million deaths even in 2050 with added coercion of epidemics and pandemics [73]. Additionally, threats of new emerging and re-emerging ailments instigated by novel, unspecified or importunate infectious agents are overwhelming like always especially in the present scenario of climatic chaos [74].

Focusing on the capable advancement of biosensor technologies, the diagnosis must be regarded as the foremost part of treatment against infectious diseases which essentially needed to be easy, accurate, and immediate. Existing regular diagnostics for infectious diseases comprise various well-established techniques and to name a few most common techniques are ELISA (Enzyme-Linked Immunosorbent Assay), PCR (Polymerase Chain Reaction) which is a nucleic acid-based assays, microscopic observation, and different ex situ culture practices [75]. However, the expansion of novel diagnosis techniques based upon the embellishment of specific and unique biomarkers in the clinical identification of infectious pathogens is very crucial [76]. Various and different diagnostic tests can be unambiguous to the idea that they are techniques derived to distinguish and classify patient ailment and more precisely for infectious diseases, their ability to detect infection presence or absence. In this particular scenario, tools for accurate, affordable, simple, and rapid diagnosis can potentially define life or mortality considering its successive importance on clinical supervision, treatment efficiency, and pathogen transmission and even to circumvent continuing complications [77]. During the development of biosensors for diagnostic purposes especially against infectious diseases, one should take cautious managements towards its specificity, sensitivity, and most importantly the reproducibility of the diagnostic biosensor [77]. Taken into an account, the recent amendments towards diagnosis include Matrix-Assisted Laser Desorption/Ionization Time of Flight Mass Spectrometry (MALDI-TOF-MS) to identify different pathogenic disease-causing agents like bacteria, fungi, and virus through comparing protein fingerprint from the sample with the existing databases [78, 79]. Infectious agents identification timeline showed periodic importance starting from different culture techniques and microscopy followed by antigen detection and polymerase chain reaction (PCR) and recently more relied upon antibodies detection and development of different biosensors for both pathogen identification and related response of the infected hosts especially for the detection of infection status or stage [80]. NGS (Next Generation Sequencing) is potential can evolve into a very useful diagnostic tool against infectious diseases due to its non-predisposition towards pre-established sequence particularly to identify mutating, transforming, or novel pathogens [81, 82] additionally microfluidic [83] as well as nanotechnological [84] devices can also serve as potential biosensors for clinical diagnosis of infectious agents.

Biomarkers can distinct as detailed variables that are measured by the titers of various pathogenic biomolecules, namely metabolites, proteins, and genes which are predisposed with distinct levels in different populations as apparent characteristics [85]. A biomarker can also be defined as a potential marker of basal or normal and distorted biological mechanism due to invading pathogen or metabolic intervention that can be translated into measurable entities for therapeutic intrusion (Biomarkers Definitions Working Group, 2001). A model biomarker generally or ideally represent few different characteristics namely, accessibility for measurement, sensitivity, specificity, and cost-effective as an intact set which can be externally validated in assays [86] and these criteria is equally important in deriving biosensors for identification and progression determination of infectious diseases with additional value towards sample variation stability and reproducibility [87]. Precise detection and monitoring of biomarkers hold a key in clinical science and health practices like drug discovery, as well as during scientific corroboration of prediction, diagnosis, and therapeutics as well as estimation or analyses of infectious disease progression, regression, and outcome [88]. Starting from this millennium scientists elaborated various biomarkers for infectious diseases second to the precision in screening techniques and advancement in the field of bioinformatics, though their rendition towards clinical application needs promotion as the use is still inadequate [86].

Biosensors can be elaborated as diagnostic measures that can appraise a biochemical detection result into a quantifiable indicator [50] with two main components, a probe (biological recognition element) and a detector (physicochemical transducer). The precise aim of a biosensor is to confirm the absence or presence thereafter concentration and activity evaluation of a bio-molecule simultaneously in a solution or matrix [89]. Biosensors can have wide range of functions in different environmental [90], industrial [91] and most importantly medical [92]. Achievements of biosensors are basically due to some of their inherent properties like high selectivity and sensitivity, portability [93], cost-effectivity, real-time quick detection, and requirement of a small sample [93, 94]. Concerning the biophysical attributes of the working principles, biosensors can be differentiated into diverse groups. On the topic of the transducers, optical senses the changes in electromagnetic changes of light waves both the visible and beyond the visible range and allied incidents [95], calorimetric sensors evaluate the alteration of enthalpy [96], the piezoelectric evaluate mass changes [97], and also the electrochemical which is associated with the documentation of the generation or changes of electroactive specimens [98]. Biosensors can also be differentiated based on the biological component of the probe ranging from antibodies [99], enzymes [95] and DNA molecules [99]. The pioneer works of Clark et al. [3] demonstrated use of enzymatic biosensors by coupling glucose oxidase an enzyme to an electrode for the evaluation and measurement of oxygen uptake and enzyme biosensors based on the advantages of the catalytic activity, selectivity and specificity made them one of the majorly successful biosensors [3]. Whereas, genosensors consist of immobilized nucleic acid molecules and can perceive genes are associated with human-related infections and diseases [100] with an additional specialty towards the genetic recognition of pathogens [101, 102]. Various antibodies are used in immunosensors with a specific binding affinity towards an antigen and development of the antibody-antigen complex specific for the recognition of a particular infectious agent [103]. Besides, aptamers [104] different cells and tissues [105] and even microorganisms [106] can also serve as potent biosensors to identify and elaborate an infectious disease condition.

Dengue virus [107, 108], human Papilloma Virus (HPV) [25], and hepatitis virus [101, 109] are among the most widespread infectious agents belongs to the pathogen class virus which is also liable for various other human ailments. Approximately 100 HPV virus genotypes can be found and many of them are connected with cervix and anus cancers [25]. With HPV virus existing methods for the diagnosis are associated with different limitations and most relevant of them is the low specificity of the methods [110]. To address this problem, a highly sensitive electrochemical genosensor was invented having DNA probes for HPV, with a carbon electrode that was functionalized with gold nanorods, graphene, and polymeric film. The basic identification principle based on electrochemical impedance spectroscopy and 1,10-phenanthroline ruthenium dichloride (Ru(phen)32+) performed as the redox indicator with a detection limit of 4.03×10^{-14} M [25]. The protocol of electrochemical impedance spectroscopy which made the equipment simple and portable having a limit of performing 98 simultaneous tests with a very small sample volume [111]. A genosensor based on HPV specific thiolated oligonucleotides was also developed with the use of gold electrodes, which can monitor the interaction of hematoxylin with dsDNA after the hybridization process with the effective principle of cyclic and differential pulse voltammetry; the linear relationship can differentiate with varying DNA concentration and can detect a range of 12.5–350.0 nM in the patient sample nM [92]. A biosensor for HPV was amended on a 4-aminophenol monolayer glued on a surface of poly(methyl methacrylate) with a gold nanolayer and an HPV 16 specific monoclonal antibody (mAb 5051).

Hepatitis B has considered a major infectious disease originated from the infection of a DNA virus that infects hepatocytes. Hepatitis B infected more or less 2 billion people worldwide which made it a global health problem and about one-third of the total world population shows affirmative serology for hepatitis B [112]. Hepatitis B viral infection can damage human liver with reasonable risk of fatality from cirrhosis and cancer [109]. Monitoring is crucial during the chronic phase and thus can minimize further progression of the diseases and thus the scope and need for manufacturing precise and economical biosensors amplifies along with further development of the existing traditional methods. An electrochemical biosensor was developed to perceive an appropriate DNA sequence of the hepatitis B virus by modulating a graphite electrode customized with poly (4-aminophenol) with the detection part which had been made with differential pulse voltammetry and ethidium bromide as hybridization label which has detection limit of 2.61 nM from infected serum [101]. A colorimetric immunosensors have also been developed to sense the surface antigen of the hepatitis B virus through sandwich immunoassay method having gold nanoparticles made efficient with biotin and luminal [112]. A chemiluminescent immunosensor was developed with gold nanoparticles by means of hydrogen peroxide as the oxidant with a detection range of 1.7–1920 pg/mL^{-1}.

Currently, dengue emerged as a prolific public health crisis globally and about 2.5 billion human are at risk of infection in 98 tropical and subtropical countries considered as the endemic zone of the viral pathogenesis. Available diagnostics for the ailment rely on viral RNA confirmation through RT-PCR (Reverse Transcription PCR) or through ELISA. Early diagnosis of dengue is normally performed by techniques associated with expensive, time-taking and relatively quasi-sensitive procedures and documented with false-positive results [96, 107]. These circumstances elaborate on the need and scope of novel biosensor development. In this purpose, a biosensor was developed on gold electrodes activated with gold nanomaterials, polyaniline, and SH-terminal groups (AuNpPANI-SH) with specific oligonucleotides for serotypes 1 (T1), 2 (T2), and 3 (T3) which were immobilized on the surface. The specific, indicator for the biosensor was ferricyanide/ferrocyanide and signal was observed by cyclic voltammetry or electrochemical impedance spectroscopy which can detect genome of dengue virus in the serum of patients with high sensitivity and in minimum concentrations [113]. A dengue non-structural protein (NS1) can also be utilized as a biosensor target. An immunosensor with a self-assembled mercaptoundecanoic acid monolayer which is covalently bonded with the immobilized anti-NS1 antibody having gold electrodes can construct the biosensor. The detection technique was dependent upon electrochemical impedance spectroscopy with a detection limit of 3 ng/mL in PBS and 30 ng/mL in serum [108]. A screen-printed electrode with modified graphite-thiophene ink and a gold nanoparticles layer which previously immobilized anti-NS1 antibodies can detect the amperometric responses of NS1 protein structured the biosensor with monitoring system of cyclic voltammetry coupled with ferrocyanide/ferricyanide with a detection limit of 0.015 μg/ mL [114]. Use of biosensors in detecting viral diseases of human is given in Table 2.

Table 2 Biosensors for the detection of human infectious viral diseases

Organism	Detectable Bio-sensor Class	Combination of Biomarker and probe	References
HPV	**Genosensor**: Electrochemical	Not Specified/DNA oligonucleotide	[92]
HPV	**Genosensor**: Electrochemical	Not Specified/DNA oligonucleotide	[25]
HPV 16	**Biomicrosystem**: Electrochemical	Not Specified/5051 mAb	[111]
Hepatitis B virus	**Genosensor**: Electrochemical	Not Specified/DNA oligonucleotide	[101]
Hepatitis B virus	**Immunosensor**: Chemiluminescence	HBsAg/mAb	[112]
Dengue virus	**Genosensor**: Electrochemical	Not Specified/DNA oligonucleotides	[113]
Dengue virus	**Immunosensor**: Electrochemical	NS1 protein/anti-NS1 antibody	[114]
Dengue virus	**Immunosensor**: Electrochemical	NS1 protein/anti-NS1 antibody	[108]

Pathogenic bacterial detection is especially significant in the fields of food and beverages and also medicine as bacteria are liable for some extremely notorious infections, for instance, tuberculosis, leprosy, and meningitis [115–118]. Pathogenic *Mycobacterium tuberculosis* causes *tuberculosis* and measured for the most number of demises due to any infectious agents and thus represents itself significantly in worldwide communal health precedence [119]. In recent times, various biosensors have been prepared for *M. tuberculosis* recognition by exploring diverse biological essentials with the aid of diverse transducers. An electrochemical genosensor for *M. tuberculosis* had been developed with a response range as low as 1×10^{-15}–1×10^{-9} M by immobilizing a precise sequence of the IS6110 gene on a reduced graphene-oxide gold nanoparticle-modified electrode which acts as the sensing platform and gold nanoparticle-polyaniline as the amplifier [120]. An genosensor was also developed through self-assembled mercaptobenzoic acid monolayers and magnetite nanoparticles (Fe3O4Nps) both enclosed on a bare gold electrode to immobilize a DNA probe having response range as low as 6 ng/μL [121]. CdSe quantum dots were also used as a tag with an amalgamation of MspI as an electrochemical biosensing platform with endonuclease and gold nanoparticles to develop and distinguish the mismatched DNA of *M. tuberculosis* with high selectivity and can magnify the signal with a response range of 1×10^{-14}–1×10^{-9} M [122]. To identify petite amount of pathogen genomic DNA with high specificity, an electrochemical technique had been projected to perceive the existence of *M. tuberculosis* in various samples like sputum, pleural lung lavage, and urine samples of infected persons through entrapment of the augmented single-stranded DNA sequences on magnetic beads and post-amplification was done by hybridization assay [123]. Apart from immunoassays [124–126], different biosensors had also been developed on the principles of Surface Plasmon Resonance (SPR), by optical detection and formulated genosensors for *M. tuberculosis* [127, 128].

A severe chronic disease, leprosy is caused by *Mycobacterium leprae* and according to WHO, in 2018 approximately 208,619 new cases have been reported globally. A correct and precise early diagnosis is very much essential to prevent the transmission as well as can avoid severe damages due to its infection to the patients, who are generally immune-compromised and suffering from malnutrition [129]. An electrochemical genosensors was developed by immobilizing precise single-stranded DNA oligonucleotide on a poly (4-aminophenol) marked graphite electrode which can detect *M. leprae* within a range of 0.35–35.0 ng/μL [130]. A rapid and quantitative biosensor had also been developed to identify *M. leprae* by immobilizing two precise antigens on nitrocellulose membranes which can detect IgM and IgG antibodies [131].

Meningitis is an inflammatory infectious disease of the meninges that can be caused by a different pathogens ranging from viruses, bacteria, fungi, and also parasites, though bacterial meningitis is most common. Among meningitis causing bacteria species, the most likely infections are mediated by *Neisseria meningitidis, Streptococcus pneumoniae, Haemophilus influenzae, Listeria monocytogenes* and *Streptococcus* [132]. Among them, *N. meningitidis* can cause vast outbreaks. To modulate biosensors to detect *N. meningitidis*, Omp85, a virulence gene that code a conserved outer membrane protein as a potential candidate and based on this postulate, antibodies against this outer membrane protein 85 can be used as biological detector with quartz crystal microbalance as the transducer [133]. Besides, an electrochemical genosensor for a virulent gene was formulated through specific oligonucleotides which had been immobilized on screen-printed gold electrodes with a detection range of 2.6 (μA/cm2)/ng [134]. A genosensor had also been formulated by immobilizing the thiol-labeled DNA probe on a gold electrode through hybridization of the ctrA complementary gene sequence measured for a sensitivity range of 115.8 μA/ng [135]. A specific complementary oligonucleotide sequence to the rmpM (reduction-modifiable protein M) virulent gene was used as the detection element with a detection range of 9.5087 μA/ng [136]. Use of different biosensors for determining life-threatening bacterial diseases of human is given in Table 3.

Protozoan parasites are liable for a wide variety of ailments in humans of varying severity, discomfort, and death. For the diagnosis of protozoan parasite caused diseases various approaches have been taken for developing biosensors against ailments namely, malaria, leishmaniasis, trypanosomiasis, Chagas disease (American trypanosomiasis), toxoplasmosis, etc. Protozoan parasites of the genus *Plasmodium* affects malaria and among various species of *Plasmodium* generally five species namely, *P. vivax*, *P. falciparum*, *P. malariae*, *P. ovale*, and *P. knowlesi* can infect humans [139].

There are wide types of biomarkers that have been used to diagnose malaria [139]. During their presence inside the human host, the malaria parasite produces histidine-rich protein-II that has widely been utilized for constructing electrochemical [140–142] and optical immunosensors [143]. Aptamers with elevated attraction for lactate

Table 3 Biosensors for the detection of human infectious bacterial diseases

Organism	Detectable bio-sensor class	Combination of biomarker and probe	References
Mycobacterium tuberculosis	**Immunosensor**: optical	Serum antibodies/Antigens (*W06, W10, W14, W19, W28, W38, W64, W70, W85)*	[125]
Mycobacterium tuberculosis	**Immunosensor**: optical	Ag85, ESAT6 and LAM/specific antibodies	[124]
Mycobacterium tuberculosis	**Immunosensor**: optical	Polyclonal IgY antibodies	[126]
Mycobacterium tuberculosis	**Genosensor**: electrochemical	IS6110 gene/DNA oligonucleotide	[120]
Mycobacterium tuberculosis	**Genosensor**: electrochemical	Genomic DNA/DNA oligonucleotide	[121]
Mycobacterium tuberculosis	**Genosensor**: electrochemical	Not Specified/DNA oligonucleotide	[122]
Mycobacterium tuberculosis	**Genosensor**: electrochemical	IS6110 gene/DNA oligonucleotide	[123]
Mycobacterium tuberculosis, Mycobacterium avium	**Genosensor**: optical	ITS gene/DNA oligonucleotides	[137]
Mycobacterium leprae	**Genosensor**: electrochemical	Not specified/PCR product	[130]
Mycobacterium leprae	**Immunosensor**: lateral flow	IgM and IgG antibodies/ND-O and LID-1 antigens	[131]
Neisseria meningitidis	**Genosensor**: electrochemical	CtrA gene/DNA oligonucleotide	[135]
Neisseria meningitidis	**Genosensor**: electrochemical	CtrA gene/DNA oligonucleotide	[135]
Neisseria meningitidis	**Immunosensor**: piezoelectric	Membrane protein 85 (Omp85)/specific antibody	[133]
Neisseria meningitidis	**Genosensor**: electrochemical	Omp85 gene/DNA oligonucleotide	[134]
Neisseria meningitidis	**Genosensor**: electrochemical	rmpM gene/DNA oligonucleotide	[136]
Neisseria meningitidis	**Genosensor**: electrochemical	CtrA gene/DNA oligonucleotide	[138]

dehydrogenase, another malaria biomarker has profoundly been explored to develop biosensors with the help of electrochemical [144] and colorimetric [145] transducers. Aptamers by SELEX by means of magnetic beads had utilized *Plasmodium* lactate dehydrogenase-specific ssDNA for constructing biosensors [133]. Additionally, these biosensors can also determine species specification of malarial infection, e.g., a genosensor on quartz crystal microbalance can specifically differentiate *P. falciparum* or *P. vivax* infection [146].

Toxoplasma gondii, an obligate intracellular protozoan is the causative agent of the disease called Toxoplasmosis. Though in most cases *Toxoplasma* infection can be asymptomatic or with mild to moderate discomfort associated with fever, malaise and lymphadenopathy, in immune-compromised individuals or congenitally attained infections can cause severe illness and death [147]. Traditional diagnosis of *Toxoplasma gondii* relies on microscopy and molecular identifications [120]. Immunosensors constitute the major sub-class of biosensors reported for the diagnosis of toxoplasmosis and dependent upon anti-*T. gondii* antibodies. Frequency change of a piezoelectric device explored explicit agglutination of gold nanoparticles covered with antigen (10 nm diameter), when corresponding antibody is present it can monitor extremely diluted (1:5500) anti-*T. gondii* antibody [148]. Enzyme-catalyzed amplification based electrochemical biosensor was also created where *T. gondii* antigen had been coated on a gold electrode exterior to bind anti-toxoplasma IgG. Colorimetric identification was followed with horseradish peroxidase and signal had been transduced through quartz crystal microbalance, electrochemical impedance spectroscopy, and cyclic voltammetry that can sense extremely trace amount anti-*T. gondii* antibody (1:9600 in dilution) [149]. Quantum dots-labeled dual aptasensor has also been developed with high affinities to anti-*T. gondii* antibody where the aptamer-protein-aptamer sandwich had been sensed through fluorescence by using quantum dots as a label having the detection level as low as 0.1 IU [150]. Magnetic fluorescent nanoparticles were also utilized to formulate a genosensor that can sense *T. gondii* DNA oligonucleotides, having a detection minimum of 8.339 nM [151].

Leishmania with its 20 odd human intracellular species can cause the tropical ailment Leishmaniasis; vector of which is hematophagous sand fly. According to the clinical disease manifestation leishmaniasis can be differentiated into four groups: cutaneous, diffuse cutaneous, mucocutaneous and visceral; and of them visceral leishmaniasis being the most severe and fatal and caused by *Leishmania donovani* [152]. The Leishmanial clinical manifestation explored the complex interplay of the parasite type and also the delicately intertwined parameters of the host immune system. Anti-*L. infantum* antibodies had been used as immunosensor having SPR technique as the detection system [153]. Magnetic polymer microspheres layered with recombinant antigens has been derived to detect anti-*Leishmania infantum* specific antibodies where the detector was a fluorescence-based immunosensor [154]. A piezoelectric immunosensor was also modified to detect *L. infantum* antigens in infected host tissues with the aid of immobilized antibodies bound on a gold surface which was enclosed by cysteamine and glutaraldehyde as a thin film having detection maxima of 1.8×10^4 amastigotes/g of infected tissue [31]. An electrochemical genosensor was also reported [155], which act by an immobilized DNA sequence that can sense 18S rRNA gene sequences of *Leishmania donovani*.

Trypanosoma cruzi is the causative agent of the ailment American trypanosomiasis or Chagas disease which is endemic in the South American countries vectored by hematophagous *Triatominae* subfamily bugs [156]. Cytoplasmic and flagellar repetitive recombinant antigen of *T. cruzi* had been explored as biosensor element that had been immobilized on gold and platinum electrodes and confirmation was gained through electrochemical impedance spectroscopy [157]. An amperometric immunosensor from *T. cruzi* epimastigote membrane proteins used for the diagnosis of Chagas disease through chronoamperometry using peroxidase-labeled IgG conjugate to detect specific antibodies in the serum of patients [158]. Amperometric bioelectrode can also be used to detect *T. cruzi* antibodies from the patient sera [158]. A screen-printed carbon electrode has also been used to develop an integrated microfluidic system for the quantitative determination of IgG specific antibodies from patient serum [159]. The electrode was customized by the use of gold nanoparticles and can detect *T. cruzi* epimastigote membrane proteins as low as 3.065 ng/mL. Different biosensors for determining fatal diseases of human caused from parasitic infections is given in Table 4.

6 Advantages and Drawbacks

Use of clinical diagnosis has been decreased to a large extent due to several limitations like time and labor-insensitive, deviation in the quantitative value of a parameter, infeasibility of analyzing the sample in non-hospital-based settings. Biosensors can fulfil all these lacunae with elevated accuracy, sensitivity, and specificity. Additionally, a biosensor be capable to provide analysis of multiple analytes at a single time. Efficient detection of bio-micro- and macromolecules such as DNA and RNA molecules/fragments, growth factors, antigens, antibodies, toxins, hormones, protein and nonprotein ligands, organic and inorganic contaminants can be achieved using the aforementioned sensors. Moreover, the design of these biosensors also enables detection of an analyte with highest precision even in altered physicochemical conditions such as pH, temperature, increase abundance of non-specific components that induce noise during sensing [7, 10].

The major advantages of microbial sensor are many. Firstly, microbes are ubiquitously distributed in the nature and therefore availability is major advantage. Secondly, due to the presence of whole-cell microbes, microbial sensors can withstand and adapt to unfavorable environment and can metabolize new molecules accordingly [10, 17, 19]. On the other side, variable action in changeable or complex environment, brief lifetime, very minute signal-to-noise ratio, and genetic instability resulting in variable response are considered as the major drawbacks of microbial sensors [7, 10].

Immunosensors provide better and faster sensing, specificity in comparison to other sensing approaches and currently employed for diagnosing several human

Table 4 Biosensors for the detection of human infectious parasitic diseases

Organism	Detectable bio-sensor class	Combination of biomarker and probe	References
Plasmodium falciparum	**Immunosensor**: electrochemical	*Pf*HRP-2/specific antibody	[140]
Plasmodium falciparum	**Immunosensor**: electrochemical	Anti-PfHRP-2 antibodies/*Pf*HRP-2	[141]
Plasmodium falciparum	**Immunosensor**: piezoelectric	*Pf*HRP-2/specific antibody	[142]
Plasmodium falciparum, *Plasmodium vivax*	**Aptasensor**: electrochemical	pLDH/aptamer	[144]
Plasmodium vivax	**Aptasensor**: colorimetric	pLDH/aptamer	[145]
Plasmodium falciparum, *Plasmodium vivax*	**Genosensor**: piezoelectric	Not specified/DNA oligonucleotide	[146]
Plasmodium falciparum	**Immunosensor**: optical	Anti-PfHRP-2 antibodies/PfHRP-2	[143]
Plasmodium falciparum, *Plasmodium vivax*	**Aptasensor**: colorimetric	pLDH/aptamer	[160]
Trypanosoma cruzi	**Immunosensor**: electrochemical	*T. cruzi* antibodies/CRA and FRA antigens	[157]
Toxoplasma gondii	**Immunosensor**: piezoelectric	IgG antibodies/specific antigen	[148]
Trypanosoma cruzi	**Immunosensor**: electrochemical	*T. cruzi* antibodies/*T. cruzi* antigen	[158]
Toxoplasma gondii	**Immunosensor**: electrochemical	IgG antibodies/specific antigen	[149]
Leishmania donovani	**Genosensor**: electrochemical	18S rRNA gene/DNA oligonucleotide	[155]
Trypanosoma cruzi	**Immunosensor**: electrochemical	IgG antibodies/specific antigens	[159]
Toxoplasma gondii	**Aptasensor**: fluorescence	IgG antibodies/aptamer	[150]
Toxoplasma gondii	**Aptasensor**: fluorescence	Not specified/DNA oligonucleotides	[151]

diseases [70]. Moreover, affinity and avidity of detection antibody can be suitably regulated by biotechnological approaches. However, these sensors are expensive as several purification strategies are required to obtain specific antibodies. Similarly, antigen-antibody complex in many cases are irreversible and therefore can be applied once.

Nucleic acid sensors are advantageous as the probe, i.e., the detection DNA can be used for several times as the interaction (hybridization) between the detection nucleotide chain and the nucleotides of analyte is reversible [161]. The major constraint of nucleic acid sensor is the quantity of DNA in the sample because proper H-bonding between detection nucleic acid and test nucleic acid can only generate a detectable signal. On the other side, PCR had been utilized to increase DNA quantity however, it is a time-consuming and costly technique. In recent times, microRNA (miRNA) based biosensors have emerged as an upgraded and ultrasensitive version of nucleic acid sensors which can detect cancer and other pathological manifestation of human by selectively sensing miRNAs from the serum sample [161]. Aptamers are the newest addition to the nucleic acid-based sensor which provides excellent sensing efficiency and introduced in diagnosing few health disorders of human [161]. However, widespread application in the human health care still needs a lot of clinical trials.

Signal transduction technology has undergone through several phases of evolution and finally a series of new biosensors have eventually launched to serve our purposes. Development of potentiometric biosensor has allowed label-free detection which enables detection of a signal without involving any conjugation with biocatalysts or nanoparticles [17]. These sensors are also less expensive. Newly developed conductometric or impedimetric biosensors can selectively sense various analytes ranging from micromolecules to macromolecules as well as whole cell [108]. Particularly for detecting proteins, recently developed optical sensors have found promising in detecting protein molecules of varying size based on the change in refractive index [4, 108]. Apart from the mentioned sensors, mass-, magnetic- and radioactive biosensors constitute the less popular sensors. Despite of having high sensing efficiency, hazardous effect of the radiation and complex instrumentation have limited the desired popularity of radio-biosensors. Modern graphene-based biosensors enable high stability, thermo-resistance, increase electrical conductance as well as better sensing [70]. Similarly, microfluidic biosensors are very expensive which has subdued its extreme high-quality sensing ability. Antibody-based cancer sensors provide good detection and have been found successful in several trial. Cost for making most of the antibody-based sensors are extremely high and these devices are difficult to maintain. Similarly, for cancer diagnosis, the lack of accuracy and reproducibility as well as high detection cost are the major limitations [10, 162]. Therefore, development of cost-effective technology with proper miniaturization and high sensitivity is the utmost need. Lack of specificity amongst available biosensing devices for neuroinflammatory and neurodegenerative diseases as well as lack of robust non-invasive technologies were the major setback which are currently substituted by the discovery of highly specific and label-free biosensors [30]. It is worthy to mention that the research on the development of stable and efficient nano-scaffold has demonstrated several new composites like biocompatible nanoparticles and nano-spheres which can be explored for biosensor development [163, 164]. Biosensors also possess huge commercial prospects especially in healthcare industries. Currently miniaturization of existing sensors and improvement of biocompatibility are the two most promising research sectors that promise industrial growth to a broader extent. More collaborations between biologists, chemists, and engineers are particularly needed to explore novel way for improving existing biosensors and developing new biosensors.

7 Conclusion and Future Directions

Despite being the most intelligent living organism on the earth orchestrated with multiple facets of immune system, several inflammatory, and infectious diseases are taking enormous toll day by day. The number of life-threatening human diseases are numerous. However, considering global distribution and relative impact on the human health, currently diabetes, cancer, tuberculosis, malaria, and cardiovascular diseases are the top contributors of mortality, morbidity, and socio-economic loss throughout globe. Intriguingly, an accurate early disease diagnosis plays a key role in successful prognosis of any of the life-threatening diseases, adopting suitable therapeutic intervention strategy and can save a lot of lives. Modern intervention strategies demand perfect diagnosis which can only be achieved by means of applying an efficient biosensor. Biosensors are considered to be the most efficacious mean to diagnose human health problems. Biosensor technology is a cumulative result of the interdisciplinary knowledge of biotechnology, nanotechnology, chemistry, engineering, electronics, materials science, and medical science. In recent years, the applications as well as demand of effectual biosensor have been increased tremendously, especially in human health care more expressly due to the emergence of human heath set back. Selection of appropriate biomarker and development of robust methods are the two critical factors behind the success of a biosensing device.

In recent years, toll-like receptors have emerged as frontline research topics in biology as these cellular receptors can recognize a wide array of biomolecules from several pathogens and from diseased individuals [32, 165, 166]. Moreover, these biomarkers will not be limited to human [167] and therefore these receptor-based biosensing could be useful in diagnosing animal health problems. TLR-based biosensors can be a useful and efficacious mean to selectively determine several viral, microbial, parasitic, and auto-immune biomarkers. Neglected tropical diseases like lymphatic filariasis require cost-convenient and robust diagnostic approaches which can only be possible by using a specific biosensing device. Notably, modern age biosensors (abzyme-based biosensors) are now been used for both in situ detection and treatment [168]. Transcription factors-based sensors are currently in developing state but these sensors have shown promises in detecting diseases specific metabolite and expression of genes [169]. In this chapter, we have present an informative overview of the use of various types of biosensors, their application in healthcare and medicine by incorporating pathbreaking research studies available till date. Taken *en massie*, biosensor technology is a growing sector in current age research and therefore more interest and input from different scientific communities are particularly needed to present more efficient sensing devices that will serve for the mankind.

Acknowledgements Space constraint restricted us to incorporate limited and selected publications but we do acknowledge all the uncited related articles/studies which are equally important for the advancement of biosensor technology.

Disclosure Dr. Niladri Mukherjee is working as a guest faculty in the Department of Animal Science, Kazi Nazrul University, Asansol, India.

References

1. WHO Report on different human diseases: World Health Organization, https://www.who.int/. Last accessed 2020/02/21
2. Hall, E.A.H.: Biosensors. Open University Press, Milton Keynes (1990)
3. Clark, L.C., Jr.: Electrode systems for continuous monitoring in cardiovascular surgery. Ann. N. Y. Acad. Sci. **102**, 29–45 (1962)
4. Metkar, S.K., Girigoswami, K.: Diagnostic biosensors in medicine—a review. Biocatal. Agric. Biotechnol. **17**, 271–283 (2019)
5. Spichiger-Keller, U.E.: Chemical Sensors and Biosensors for Medical and Biological Applications. Wiley, New York (2008)
6. Chung, Y.K., Reboud, J., Lee, K.C., Lim, H.M., Lim, P.Y., Wang, K.Y., Tang, K.C., Ji, H., Chen, Y.: An electrical biosensor for the detection of circulating tumor cells. Biosens. Bioelectron. **26**, 2520–2526 (2011). https://doi.org/10.1016/j.bios.2010.10.048https://doi.org/10.1016/j.bios.2010.10.048
7. Patel, S., Nanda, R., Sahoo, S., Mohapatra, E.: Biosensors in health care: the milestones achieved in their development towards lab-on-chip-analysis. Biochem. Res. Int. **2016** (2016)
8. Hasan, A., Memic, A., Annabi, N., Hossain, M., Paul, A., Dokmeci, M.R., Dehghani, F., Khademhosseini, A.: Electrospun scaffolds for tissue engineering of vascular grafts. Acta Biomater. **10**, 11–25 (2014)
9. Boeneman, K., Delehanty, J.B., Susumu, K., Stewart, M.H., Deschamps, J.R., Medintz, I.L.: Quantum dots and fluorescent protein FRET-based biosensors. In: Nano-Biotechnology for Biomedical and Diagnostic Research, pp. 63–74. Springer, Berlin (2012)
10. Gouvea, C.: Biosensors for health applications. In: Biosensors for Health, Environment and Biosecurity. InTech (2011). https://doi.org/10.5772/16983
11. Dey, B., Mondal, R.K., Mukherjee, S., Satpati, B., Mukherjee, N., Mandal, A., Senapati, D., Babu, S.P.S.: A supramolecular hydrogel for generation of a benign DNA-hydrogel. RSC Adv. **5**, 105961–105968 (2015)
12. Dey, B., Mukherjee, S., Mukherjee, N., Mondal, R.K., Satpati, B., Senapati, D., Babu, S.P.S.: Green silver nanoparticles for drug transport, bioactivities and a bacterium (*Bacillus subtilis*)-mediated comparative nano-patterning feature. RSC Adv. **6**, 46573–46581 (2016)
13. Dey, B., Mukherjee, S., Mukherjee, N., Mondal, R.K., Satpati, B., Babu, S.P.S.: Polyphenol oxidase-based luminescent enzyme hydrogel: an efficient redox active immobilized scaffold. Bull. Mater. Sci. **41**, 14 (2018). https://doi.org/10.1007/s12034-017-1529-3https://doi.org/10.1007/s12034-017-1529-3
14. Mukherjee, S., Basak, B., Bhunia, B., Dey, A., Mondal, B.: Potential use of polyphenol oxidases (PPO) in the bioremediation of phenolic contaminants containing industrial wastewater. Rev. Environ. Sci. Bio/Technol. **12**, 61–73 (2013)
15. Mukherjee, S., Bandyopadhayay, B., Basak, B., Mandal, N., Apurba, D.E.Y., Mondal, B.: An improved method of optimizing the extraction of polyphenol oxidase from potato (*Solanum tuberosum* L.) Peel. Not. Sci. Biol. **4**, 98–107 (2012)
16. Mondal, M.K., Mukherjee, S., Saha, S.K., Chowdhury, P., Babu, S.P.S.: Design and synthesis of reduced graphene oxide based supramolecular scaffold: a benign microbial resistant network for enzyme immobilization and cell growth. Mater. Sci. Eng. C **75**, 1168–1177 (2017)

17. Kylilis, N., Riangrungroj, P., Lai, H.E., Salema, V., Fernández, L.A., Stan, G.B.V., Freemont, P.S., Polizzi, K.M.: Whole-cell biosensor with tunable limit of detection enables low-cost agglutination assays for medical diagnostic applications. ACS Sens. **4**, 370–378 (2019)
18. Mukherjee, S., Joardar, N., Sengupta, S., Babu, S.P.S.: Gut microbes as future therapeutics in treating inflammatory and infectious diseases: lessons from recent findings. J. Nutr. Biochem. **61**, 111–128 (2018)
19. Chen, Z., Lu, M., Zou, D., Wang, H.: An *E. coli* SOS-EGFP biosensor for fast and sensitive detection of DNA damaging agents. J. Environ. Sci. **24**, 541–549 (2012)
20. Khati, M.: The future of aptamers in medicine. J. Clin. Pathol. **63**, 480–487 (2010)
21. Liu, G., Mao, X., Phillips, J.A., Xu, H., Tan, W., Zeng, L.: Aptamer-nanoparticle strip biosensor for sensitive detection of cancer cells. Anal. Chem. **81**, 10013–10018 (2009)
22. Gruhl, F.J., Rapp, B.E., Länge, K.: Biosensors for diagnostic applications. In: Molecular Diagnostics, pp. 115–148. Springer, Berlin (2011)
23. Mondal, M.K., Mukherjee, S., Joardar, N., Roy, D., Chowdhury, P., Babu, S.P.S.: Synthesis of smart graphene quantum dots: a benign biomaterial for prominent intracellular imaging and improvement of drug efficacy. Appl. Surf. Sci. **495**, 143562 (2019)
24. Soler, M., Huertas, C.S., Lechuga, L.M.: Label-free plasmonic biosensors for point-of-care diagnostics: a review. Expert Rev. Mol. Diagn. **19**, 71–81 (2019)
25. Huang, H., Bai, W., Dong, C., Guo, R., Liu, Z.: An ultrasensitive electrochemical DNA biosensor based on graphene/Au nanorod/polythionine for human papillomavirus DNA detection. Biosens. Bioelectron. **68**, 442–446 (2015)
26. Ghosh, D., Dhibar, S., Dey, A., Mukherjee, S., Joardar, N., Babu, S.P.S., Dey, B.: Graphene oxide dispersed supramolecular hydrogel capped benign green silver nanoparticles for anti-cancer, antimicrobial, cell attachment and intracellular imaging applications. J. Mol. Liq. **282**, 1–12 (2019). https://doi.org/10.1016/j.molliq.2019.03.010
27. Srinivasan, B., Tung, S.: Development and applications of portable biosensors. J. Lab. Autom. **20**, 365–389 (2015)
28. Rusling, J.F., Kumar, C.V., Gutkind, J.S., Patel, V.: Measurement of biomarker proteins for point-of-care early detection and monitoring of cancer. The Analyst **135**, 2496–2511 (2010)
29. Zheng, T., Pierre-Pierre, N., Yan, X., Huo, Q., Almodovar, A.J.O., Valerio, F., Rivera-Ramirez, I., Griffith, E., Decker, D.D., Chen, S.: Gold nanoparticle-enabled blood test for early stage cancer detection and risk assessment. ACS Appl. Mater. Interfaces **7**, 6819–6827 (2015)
30. Abreu, C.M., Soares-Dos-Reis, R., Melo, P.N., Relvas, J.B., Guimarães, J., Sá, M.J., Cruz, A.P., Mendes Pinto, I.: Emerging biosensing technologies for neuroinflammatory and neurodegenerative disease diagnostics. Front. Mol. Neurosci. **11**, 164 (2018). https://doi.org/10.3389/fnmol.2018.00164https://doi.org/10.3389/fnmol.2018.00164
31. Cabral-Miranda, G., de Jesus, J.R., Oliveira, P.R.S., Britto, G.S.G., Pontes-de-Carvalho, L.C., Dutra, R.F., Alcântara-Neves, N.M.: Detection of parasite antigens in leishmania infantum–infected spleen tissue by monoclonal antibody-piezoelectric-based immunosensors. J. Parasitol. **100**, 73–78 (2014)
32. Mukherjee, S., Karmakar, S., Babu, S.P.S.: TLR2 and TLR4 mediated host immune responses in major infectious diseases: a review. Braz. J. Infect. Dis. **20**, 193–204 (2016)
33. Mukherjee, S., Karnam, A., Das, M., Babu, S.P.S., Bayry, J.: Wuchereria bancrofti filaria activates human dendritic cells and polarizes T helper 1 and regulatory T cells via toll-like receptor 4. Commun. Biol. **2**, 1–11 (2019)
34. Lasker, R.D.: The diabetes control and complications trial. In: Implications for Policy and Practice (1993). https://doi.org/10.1056/NEJM199309303291410
35. Cengiz, E., Tamborlane, W.V.: A tale of two compartments: interstitial versus blood glucose monitoring. Diabetes Technol. Ther. **11**, S-11 (2009)
36. Castle, J.R., Ward, W.K.: Amperometric glucose sensors: sources of error and potential benefit of redundancy. J. Diabetes Sci. Technol. **4**, 221–225 (2010)
37. Beauharnois, M.E., Neelamegham, S., Matta, K.L.: Quantitative measurement of selectin-ligand interactions. In: Glycobiology Protocols, pp. 343–358. Springer, Berlin (2006)

38. Koschwanez, H.E., Reichert, W.M.: In vitro, in vivo and post explantation testing of glucose-detecting biosensors: current methods and recommendations. Biomaterials **28**, 3687–3703 (2007)
39. Chu, M.X., Miyajima, K., Takahashi, D., Arakawa, T., Sano, K., Sawada, S., Kudo, H., Iwasaki, Y., Akiyoshi, K., Mochizuki, M., Mitsubayashi, K.: Soft contact lens biosensor for in situ monitoring of tear glucose as non-invasive blood sugar assessment. Talanta **83**, 960–965 (2011). https://doi.org/10.1016/j.talanta.2010.10.055https://doi.org/10.1016/j.talanta.2010.10.055
40. Pickup, J.C., Hussain, F., Evans, N.D., Sachedina, N.: In vivo glucose monitoring: the clinical reality and the promise. Biosens. Bioelectron. **20**, 1897–1902 (2005)
41. Jemal, A., Bray, F., Center, M.M., Ferlay, J., Ward, E., Forman, D.: Global cancer statistics. CA: Cancer J. Clin. **61**, 69–90 (2011). https://doi.org/10.3322/caac.20107https://doi.org/10.3322/caac.20107
42. Chambers, A.F., Groom, A.C., MacDonald, I.C.: Dissemination and growth of cancer cells in metastatic sites. Nat. Rev. Cancer **2**, 563–572 (2002)
43. Pantel, K., Brakenhoff, R.H.: Dissecting the metastatic cascade. Nat. Rev. Cancer **4**, 448–456 (2004)
44. Vineis, P., Schatzkin, A., Potter, J.D.: Models of carcinogenesis: an overview. Carcinogenesis **31**, 1703–1709 (2010)
45. Robert, J.: Gene polymorphisms. Bull. Cancer **97**, 1253–1264 (2010). https://doi.org/10.1684/bdc.2010.1203https://doi.org/10.1684/bdc.2010.1203
46. Tothill, I.E.: Biosensors for cancer markers diagnosis. In: Seminars in Cell & Developmental Biology, pp. 55–62. Elsevier, Amsterdam (2009)
47. Li, Z., Wang, Y., Wang, J., Tang, Z., Pounds, J.G., Lin, Y.: Rapid and sensitive detection of protein biomarker using a portable fluorescence biosensor based on quantum dots and a lateral flow test strip. Anal. Chem. **82**, 7008–7014 (2010)
48. Sadik, O.A., Aluoch, A.O., Zhou, A.: Status of biomolecular recognition using electrochemical techniques. Biosens. Bioelectron. **24**, 2749–2765 (2009)
49. Lin, J., Ju, H.: Electrochemical and chemiluminescent immunosensors for tumor markers. Biosens. Bioelectron. **20**, 1461–1470 (2005)
50. Wang, J.: Electrochemical biosensors: towards point-of-care cancer diagnostics. Biosens. Bioelectron. **21**, 1887–1892 (2006)
51. Ahmed, F.E.: Mining the oncoproteome and studying molecular interactions for biomarker development by 2DE, ChIP and SPR technologies. Expert Rev. Proteomics **5**, 469–496 (2008)
52. Liu, Y., Li, X., Zhang, Z., Zuo, G., Cheng, Z., Yu, H.: Nanogram per milliliter-level immunologic detection of alpha-fetoprotein with integrated rotating-resonance microcantilevers for early-stage diagnosis of heptocellular carcinoma. Biomed. Microdevices **11**, 183–191 (2009)
53. Pal, M., Khan, R.: Biosensor platforms for detection of cardiovascular disease risk biomarkers. In: Functional Polysaccharides for Biomedical Applications, pp. 397–431. Elsevier, Amsterdam (2019)
54. Qureshi, A., Gurbuz, Y., Niazi, J.H.: Biosensors for cardiac biomarkers detection: a review. Sens. Actuators B: Chem. **171**, 62–76 (2012)
55. Vashistha, R., Dangi, A.K., Kumar, A., Chhabra, D., Shukla, P.: Futuristic biosensors for cardiac health care: an artificial intelligence approach. 3 Biotech. **8**, 358 (2018)
56. Joos, T.O., Stoll, D., Templin, M.F.: Miniaturised multiplexed immunoassays. Curr. Opin. Chem. Biol. **6**, 76–80 (2002)
57. Lane, C.A., Hardy, J., Schott, J.M.: Alzheimer's disease. Eur. J. Neurol. **25**, 59–70 (2018). https://doi.org/10.1111/ene.13439https://doi.org/10.1111/ene.13439
58. Liddelow, S.A., Guttenplan, K.A., Clarke, L.E., Bennett, F.C., Bohlen, C.J., Schirmer, L., Bennett, M.L., Münch, A.E., Chung, W.S., Peterson, T.C.: Neurotoxic reactive astrocytes are induced by activated microglia. Nature **541**, 481–487 (2017)
59. Postuma, R.B., Berg, D., Stern, M., Poewe, W., Olanow, C.W., Oertel, W., Obeso, J., Marek, K., Litvan, I., Lang, A.E.: MDS clinical diagnostic criteria for Parkinson's disease. Mov. Disord. **30**, 1591–1601 (2015)

60. Poewe, W., Seppi, K., Tanner, C.M., Halliday, G.M., Brundin, P., Volkmann, J., Schrag, A.E., Lang, A.E.: Parkinson disease. Nat. Rev. Dis. Prim. **3**, 1–21 (2017)
61. Reich, D.S., Lucchinetti, C.F., Calabresi, P.A.: Multiple sclerosis. N. Engl. J. Med. **378**, 169–180 (2018). https://doi.org/10.1056/NEJMra1401483https://doi.org/10.1056/NEJMra1401483
62. McKhann, G.M., Knopman, D.S., Chertkow, H., Hyman, B.T., Jack, C.R., Jr., Kawas, C.H., Klunk, W.E., Koroshetz, W.J., Manly, J.J., Mayeux, R.: The diagnosis of dementia due to Alzheimer's disease: recommendations from the National Institute on Aging-Alzheimer's Association workgroups on diagnostic guidelines for Alzheimer's disease. Alzheimer's Dement. **7**, 263–269 (2011)
63. Baraket, A., Lee, M., Zine, N., Sigaud, M., Bausells, J., Errachid, A.: A fully integrated electrochemical biosensor platform fabrication process for cytokines detection. Biosens. Bioelectron. **93**, 170–175 (2017)
64. Biela, A., Watkinson, M., Meier, U.C., Baker, D., Giovannoni, G., Becer, C.R., Krause, S.: Disposable MMP-9 sensor based on the degradation of peptide cross-linked hydrogel films using electrochemical impedance spectroscopy. Biosens. Bioelectron. **68**, 660–667 (2015)
65. Carneiro, P., Loureiro, J., Delerue-Matos, C., Morais, S., do Carmo Pereira, M.: Alzheimer's disease: development of a sensitive label-free electrochemical immunosensor for detection of amyloid beta peptide. Sens. Actuators B Chem. **239**, 157–165 (2017). https://doi.org/10.1016/j.snb.2016.07.181
66. Rushworth, J.V., Ahmed, A., Griffiths, H.H., Pollock, N.M., Hooper, N.M., Millner, P.A.: A label-free electrical impedimetric biosensor for the specific detection of Alzheimer's amyloid-beta oligomers. Biosens. Bioelectron. **56**, 83–90 (2014)
67. Heydari-Bafrooei, E., Amini, M., Ardakani, M.H.: An electrochemical aptasensor based on TiO_2/MWCNT and a novel synthesized Schiff base nanocomposite for the ultrasensitive detection of thrombin. Biosens. Bioelectron. **85**, 828–836 (2016)
68. Real-Fernández, F., Passalacqua, I., Peroni, E., Chelli, M., Lolli, F., Papini, A.M., Rovero, P.: Glycopeptide-based antibody detection in multiple sclerosis by surface plasmon resonance. Sensors **12**, 5596–5607 (2012)
69. Ahn, K.Y., Kwon, K., Huh, J., Kim, G.T., Lee, E.B., Park, D., Lee, J.: A sensitive diagnostic assay of rheumatoid arthritis using three-dimensional ZnO nanorod structure. Biosens. Bioelectron. **28**, 378–385 (2011)
70. Islam, S., Shukla, S., Bajpai, V.K., Han, Y.K., Huh, Y.S., Kumar, A., Ghosh, A., Gandhi, S.: A smart nanosensor for the detection of human immunodeficiency virus and associated cardiovascular and arthritis diseases using functionalized graphene-based transistors. Biosens. Bioelectron. **126**, 792–799 (2019)
71. Lee, B.H., Kim, S.H., Ko, Y., Park, J.C., Ji, S., Gu, M.B.: The sensitive detection of ODAM by using sandwich-type biosensors with a cognate pair of aptamers for the early diagnosis of periodontal disease. Biosens. Bioelectron. **126**, 122–128 (2019). https://doi.org/10.1016/j.bios.2018.10.040https://doi.org/10.1016/j.bios.2018.10.040
72. Im, H., Castro, C.M., Shao, H., Liong, M., Song, J., Pathania, D., Fexon, L., Min, C., Avila-Wallace, M., Zurkiya, O.: Digital diffraction analysis enables low-cost molecular diagnostics on a smartphone. Proc. Natl. Acad. Sci. **112**, 5613–5618 (2015)
73. Dye, C.: After 2015: infectious diseases in a new era of health and development. Philos. Trans. R. Soc. London. Ser. B, Biol. Sci. **369**, 20130426 (2014). https://doi.org/10.1098/rstb.2013.0426
74. Almeida, S.L.: Trending now: re-emerging infectious disease update. J. Emerg. Nurs. **41**, 104 (2015)
75. Carinelli, S., Martí, M., Alegret, S., Pividori, M.I.: Biomarker detection of global infectious diseases based on magnetic particles. N. Biotechnol. **32**, 521–532 (2015)
76. Ince, J., McNally, A.: Development of rapid, automated diagnostics for infectious disease: advances and challenges. Expert Rev. Med. Devices **6**, 641–651 (2009)
77. Banoo, S., Bell, D., Bossuyt, P., Herring, A., Mabey, D., Poole, F., Smith, P.G., Sriram, N., Wongsrichanalai, C., Linke, R.: Evaluation of diagnostic tests for infectious diseases: general principles. Nat. Rev. Microbiol. **5**, S21–S31 (2007)

78. Calderaro, A., Arcangeletti, M.C., Rodighiero, I., Buttrini, M., Gorrini, C., Motta, F., Germini, D., Medici, M.C., Chezzi, C., De Conto, F.: Matrix-assisted laser desorption/ionization time-of-flight (MALDI-TOF) mass spectrometry applied to virus identification. Sci. Rep. **4**, 1–10 (2014)
79. Patel, R.: MALDI-TOF MS for the diagnosis of infectious diseases. Clin. Chem. **61**, 100–111 (2015)
80. Ko, E.R., Yang, W.E., McClain, M.T., Woods, C.W., Ginsburg, G.S., Tsalik, E.L.: What was old is new again: using the host response to diagnose infectious disease. Expert Rev. Mol. Diagn. **15**, 1143–1158 (2015)
81. Barzon, L., Lavezzo, E., Costanzi, G., Franchin, E., Toppo, S., Palù, G.: Next-generation sequencing technologies in diagnostic virology. J. Clin. Virol. **58**, 346–350 (2013)
82. Thorburn, F., Bennett, S., Modha, S., Murdoch, D., Gunson, R., Murcia, P.R.: The use of next generation sequencing in the diagnosis and typing of respiratory infections. J. Clin. Virol. **69**, 96–100 (2015)
83. Su, W., Gao, X., Jiang, L., Qin, J.: Microfluidic platform towards point-of-care diagnostics in infectious diseases. J. Chromatogr. A **1377**, 13–26 (2015). https://doi.org/10.1016/j.chroma.2014.12.041https://doi.org/10.1016/j.chroma.2014.12.041
84. Qasim, M., Lim, D.J., Park, H., Na, D.: Nanotechnology for diagnosis and treatment of infectious diseases. J. Nanosci. Nanotechnol. **14**, 7374–7387 (2014)
85. Wehrens, R., Franceschi, P., Vrhovsek, U., Mattivi, F.: Stability-based biomarker selection. Anal. Chim. Acta **705**, 15–23 (2011)
86. Gupta, S., Venkatesh, A., Ray, S., Srivastava, S.: Challenges and prospects for biomarker research: a current perspective from the developing world. Biochim. Biophys. Acta **1844**, 899–908 (2014). https://doi.org/10.1016/j.bbapap.2013.12.020https://doi.org/10.1016/j.bbapap.2013.12.020
87. He, Z., Yu, W.: Stable feature selection for biomarker discovery. Comput. Biol. Chem. **34**, 215–225 (2010)
88. Naylor, S.: Biomarkers: current perspectives and future prospects (2003). https://doi.org/10.1586/14737159.3.5.525
89. Vopálenská, I., Váchová, L., Palková, Z.: New biosensor for detection of copper ions in water based on immobilized genetically modified yeast cells. Biosens. Bioelectron. **72**, 160–167 (2015)
90. Kashem, M.A., Suzuki, M., Kimoto, K., Iribe, Y.: An optical biochemical oxygen demand biosensor chip for environmental monitoring. Sens. Actuators B: Chem. **221**, 1594–1600 (2015)
91. Hughes, G., Pemberton, R.M., Fielden, P.R., Hart, J.P.: Development of a novel reagentless, screen-printed amperometric biosensor based on glutamate dehydrogenase and NAD+, integrated with multi-walled carbon nanotubes for the determination of glutamate in food and clinical applications. Sens. Actuators B: Chem. **216**, 614–621 (2015)
92. Nasirizadeh, N., Zare, H.R., Pournaghi-Azar, M.H., Hejazi, M.S.: Introduction of hematoxylin as an electroactive label for DNA biosensors and its employment in detection of target DNA sequence and single-base mismatch in human papilloma virus corresponding to oligonucleotide. Biosens. Bioelectron. **26**, 2638–2644 (2011)
93. Nambiar, S., Yeow, J.T.W.: Conductive polymer-based sensors for biomedical applications. Biosens. Bioelectron. **26**, 1825–1832 (2011)
94. Chen, J.C., Chung, H.H., Hsu, C.T., Tsai, D.M., Kumar, A.S., Zen, J.M.: A disposable single-use electrochemical sensor for the detection of uric acid in human whole blood. Sens. Actuators B: Chem. **110**, 364–369 (2005)
95. Pahurkar, V.G., Tamgadge, Y.S., Gambhire, A.B., Muley, G.G.: Glucose oxidase immobilized PANI cladding modified fiber optic intrinsic biosensor for detection of glucose. Sens. Actuators B: Chem. **210**, 362–368 (2015)
96. Zhang, Y., Tadigadapa, S.: Calorimetric biosensors with integrated microfluidic channels. Biosens. Bioelectron. **19**, 1733–1743 (2004)

97. Su, L., Zou, L., Fong, C.C., Wong, W.L., Wei, F., Wong, K.Y., Wu, R.S.S., Yang, M.: Detection of cancer biomarkers by piezoelectric biosensor using PZT ceramic resonator as the transducer. Biosens. Bioelectron. **46**, 155–161 (2013)
98. Gerard, M., Chaubey, A., Malhotra, B.D.: Application of conducting polymers to biosensors. Biosens. Bioelectron. **17**, 345–359 (2002). https://doi.org/10.1016/s0956-5663(01)00312-8https://doi.org/10.1016/s0956-5663(01)00312-8
99. Liew, P.S., Lertanantawong, B., Lee, S.Y., Manickam, R., Lee, Y.H., Surareungchai, W.: Electrochemical genosensor assay using lyophilized gold nanoparticles/latex microsphere label for detection of Vibrio cholerae. Talanta **139**, 167–173 (2015)
100. Murphy, L.: Biosensors and bioelectrochemistry. Curr. Opin. Chem. Biol. **10**, 177–184 (2006). https://doi.org/10.1016/j.cbpa.2006.02.023https://doi.org/10.1016/j.cbpa.2006.02.023
101. Castro, A.C.H., França, E.G., de Paula, L.F., Soares, M.M.C.N., Goulart, L.R., Madurro, J.M., Brito-Madurro, A.G.: Preparation of genosensor for detection of specific DNA sequence of the hepatitis B virus. Appl. Surf. Sci. **314**, 273–279 (2014)
102. Ligaj, M., Tichoniuk, M., Gwiazdowska, D., Filipiak, M.: Electrochemical DNA biosensor for the detection of pathogenic bacteria *Aeromonas hydrophila*. Electrochim. Acta. **128**, 67–74 (2014)
103. Dmitriev, D.A., Massino, Y.S., Segal, O.L., Smirnova, M.B., Pavlova, E.V., Gurevich, K.G., Gnedenko, O.V., Ivanov, Y.D., Kolyaskina, G.I., Archakov, A.I., Osipov, A.P., Dmitriev, A.D., Egorov, A.M.: Analysis of the binding of bispecific monoclonal antibodies with immobilized antigens (human IgG and horseradish peroxidase) using a resonant mirror biosensor. J. Immunol. Methods **261**, 103–118 (2002). https://doi.org/10.1016/s0022-1759(01)00558-0https://doi.org/10.1016/s0022-1759(01)00558-0
104. Abe, K., Yoshida, W., Ikebukuro, K.: Electrochemical biosensors using aptamers for theranostics. In: Biosensors Based on Aptamers and Enzymes, pp. 183–202. Springer, Berlin (2013)
105. Sanders, C.A., Rodriguez, M., Jr., Greenbaum, E.: Stand-off tissue-based biosensors for the detection of chemical warfare agents using photosynthetic fluorescence induction. Biosens. Bioelectron. **16**, 439–446 (2001)
106. Park, M., Tsai, S.L., Chen, W.: Microbial biosensors: engineered microorganisms as the sensing machinery. Sensors **13**, 5777–5795 (2013)
107. Nguyen, B.T.T., Peh, A.E.K., Chee, C.Y.L., Fink, K., Chow, V.T.K., Ng, M.M.L., Toh, C.S.: Electrochemical impedance spectroscopy characterization of nanoporous alumina dengue virus biosensor. Bioelectrochemistry **88**, 15–21 (2012)
108. Cecchetto, J., Carvalho, F.C., Santos, A., Fernandes, F.C.B., Bueno, P.R.: An impedimetric biosensor to test neat serum for dengue diagnosis. Sens. Actuators B: Chem. **213**, 150–154 (2015)
109. Yang, L., Du, F., Chen, G., Yasmeen, A., Tang, Z.: A novel colorimetric PCR-based biosensor for detection and quantification of hepatitis B virus. Anal. Chim. Acta **840**, 75–81 (2014)
110. Campos-Ferreira, D.S., Nascimento, G.A., Souza, E.V.M., Souto-Maior, M.A., Arruda, M.S., Zanforlin, D.M.L., Ekert, M.H.F., Bruneska, D., Lima-Filho, J.L.: Electrochemical DNA biosensor for human papillomavirus 16 detection in real samples. Anal. Chim. Acta **804**, 258–263 (2013)
111. Urrego, L.F., Lopez, D.I., Ramirez, K.A., Ramirez, C., Osma, J.F.: Biomicrosystem design and fabrication for the human papilloma virus 16 detection. Sens. Actuators B: Chem. **207**, 97–104 (2015)
112. Shourian, M., Ghourchian, H., Boutorabi, M.: Ultra-sensitive immunosensor for detection of hepatitis B surface antigen using multi-functionalized gold nanoparticles. Anal. Chim. Acta **895**, 1–11 (2015)
113. Nascimento, H.P.O., Oliveira, M.D.L., de Melo, C.P., Silva, G.J.L., Cordeiro, M.T., Andrade, C.A.S.: An impedimetric biosensor for detection of dengue serotype at picomolar concentration based on gold nanoparticles-polyaniline hybrid composites. Colloids Surfaces B Biointerfaces **86**, 414–419 (2011)

114. Silva, M.M.S., Dias, A., Cordeiro, M.T., Marques, E., Jr., Goulart, M.O.F., Dutra, R.F.: A thiophene-modified screen printed electrode for detection of dengue virus NS1 protein. Talanta **128**, 505–510 (2014)
115. Sanvicens, N., Pastells, C., Pascual, N., Marco, M.P.: Nanoparticle-based biosensors for detection of pathogenic bacteria. TrAC Trends Anal. Chem. **28**, 1243–1252 (2009)
116. Wang, Y., Ye, Z., Ying, Y.: New trends in impedimetric biosensors for the detection of foodborne pathogenic bacteria. Sensors **12**, 3449–3471 (2012)
117. Singh, R., Mukherjee, M.D., Sumana, G., Gupta, R.K., Sood, S., Malhotra, B.D.: Biosensors for pathogen detection: a smart approach towards clinical diagnosis. Sens. Actuators B: Chem. **197**, 385–404 (2014)
118. Ibrahim, F., Thio, T.H.G., Faisal, T., Neuman, M.: The application of biomedical engineering techniques to the diagnosis and management of tropical diseases: a review. Sensors **15**, 6947–6995 (2015)
119. Garcia-Monco, J.C.: Tuberculosis. Handb. Clin. Neurol. **121**, 1485–1499 (2014). https://doi.org/10.1016/B978-0-7020-4088-7.00100-0https://doi.org/10.1016/B978-0-7020-4088-7.00100-0
120. Liu, C., Jiang, D., Xiang, G., Liu, L., Liu, F., Pu, X.: An electrochemical DNA biosensor for the detection of *Mycobacterium tuberculosis*, based on signal amplification of graphene and a gold nanoparticle–polyaniline nanocomposite. Analyst **139**, 5460–5465 (2014)
121. Costa, M.P., Andrade, C.A.S., Montenegro, R.A., Melo, F.L., Oliveira, M.D.L.: Self-assembled monolayers of mercaptobenzoic acid and magnetite nanoparticles as an efficient support for development of tuberculosis genosensor. J. Colloid Interface Sci. **433**, 141–148 (2014)
122. Zhang, C., Lou, J., Tu, W., Bao, J., Dai, Z.: Ultrasensitive electrochemical biosensing for DNA using quantum dots combined with restriction endonuclease. The Analyst **140**, 506–511 (2015)
123. Barreda-García, S., González-Álvarez, M.J., de-los-Santos-Álvarez, N., Palacios-Gutiérrez, J.J., Miranda-Ordieres, A.J., Lobo-Castañón, M.J.: Attomolar quantitation of Mycobacterium tuberculosis by asymmetric helicase-dependent isothermal DNA-amplification and electrochemical detection. Biosens. Bioelectron. **68**, 122–128 (2015)
124. Mukundan, H., Kumar, S., Price, D.N., Ray, S.M., Lee, Y.J., Min, S., Eum, S., Kubicek-Sutherland, J., Resnick, J.M., Grace, W.K.: Rapid detection of Mycobacterium tuberculosis biomarkers in a sandwich immunoassay format using a waveguide-based optical biosensor. Tuberculosis **92**, 407–416 (2012)
125. Hsieh, S.C., Chang, C.C., Lu, C.C., Wei, C.F., Lin, C.S., Lai, H.C., Lin, C.W.: Rapid identification of Mycobacterium tuberculosis infection by a new array format-based surface plasmon resonance method. Nanoscale Res. Lett. **7**, 180 (2012)
126. Kim, J.H., Yeo, W.H., Shu, Z., Soelberg, S.D., Inoue, S., Kalyanasundaram, D., Ludwig, J., Furlong, C.E., Riley, J.J., Weigel, K.M.: Immunosensor towards low-cost, rapid diagnosis of tuberculosis. Lab Chip. **12**, 1437–1440 (2012)
127. Rachkov, A., Patskovsky, S., Soldatkin, A., Meunier, M.: Discrimination of single base mismatched oligonucleotides related to the rpoB gene of *Mycobacterium tuberculosis* using a surface plasmon resonance biosensor. Biotechnol. Appl. Biochem. **60**, 453–458 (2013)
128. Hsu, S.H., Lin, Y.Y., Lu, S.H., Tsai, I., Lu, Y.T., Ho, H.T.: Mycobacterium tuberculosis DNA detection using surface plasmon resonance modulated by telecommunication wavelength. Sensors **14**, 458–467 (2014)
129. Goulart, I.M.B., Goulart, L.R.: Leprosy: diagnostic and control challenges for a worldwide disease. Arch. Dermatol. Res. **300**, 269–290 (2008). https://doi.org/10.1007/s00403-008-0857-yhttps://doi.org/10.1007/s00403-008-0857-y
130. Afonso, A.S., Goulart, L.R., Goulart, I.M.B., Machado, A.E.H., Madurro, J.M., Brito-Madurro, A.G.: A promising bioelectrode based on gene of *Mycobacterium leprae* immobilized onto poly (4-aminophenol). J. Appl. Polym. Sci. **118**, 2921–2928 (2010)
131. Cardoso, L.P.V., Dias, R.F., Freitas, A.A., Hungria, E.M., Oliveira, R.M., Collovati, M., Reed, S.G., Duthie, M.S., Stefani, M.M.A.: Development of a quantitative rapid diagnostic test for multibacillary leprosy using smart phone technology. BMC Infect. Dis. **13**, 1–10 (2013)

132. Abio, A., Neal, K.R., Beck, C.R.: An epidemiological review of changes in meningococcal biology during the last 100 years (2013)
133. Reddy, S.B., Mainwaring, D.E., Al Kobaisi, M., Zeephongsekul, P., Fecondo, J.V.: Acoustic wave immunosensing of a meningococcal antigen using gold nanoparticle-enhanced mass sensitivity. Biosens. Bioelectron. **31**, 382–387 (2012)
134. Dash, S.K., Sharma, M., Khare, S., Kumar, A.: Omp85 genosensor for detection of human brain bacterial meningitis. Biotechnol. Lett. **35**, 929–935 (2013a)
135. Patel, M.K., Solanki, P.R., Seth, S., Gupta, S., Khare, S., Kumar, A., Malhotra, B.D.: CtrA gene based electrochemical DNA sensor for detection of meningitis. Electrochem. Commun. **11**, 969–973 (2009)
136. Dash, S.K., Sharma, M., Khare, S., Kumar, A.: rmpM genosensor for detection of human brain bacterial meningitis in cerebrospinal fluid. Appl. Biochem. Biotechnol. **171**, 198–208 (2013b)
137. Xiang, Y., Zhu, X., Huang, Q., Zheng, J., Fu, W.: Real-time monitoring of mycobacterium genomic DNA with target-primed rolling circle amplification by a Au nanoparticle-embedded SPR biosensor. Biosens. Bioelectron. **66**, 512–519 (2015)
138. Tak, M., Gupta, V., Tomar, M.: Flower-like ZnO nanostructure based electrochemical DNA biosensor for bacterial meningitis detection. Biosens. Bioelectron. **59**, 200–207 (2014)
139. Jain, P., Chakma, B., Patra, S., Goswami, P.: Potential biomarkers and their applications for rapid and reliable detection of malaria. Biomed. Res. Int. **2014** (2014)
140. Sharma, M.K., Rao, V.K., Agarwal, G.S., Rai, G.P., Gopalan, N., Prakash, S., Sharma, S.K., Vijayaraghavan, R.: Highly sensitive amperometric immunosensor for detection of *Plasmodium falciparum* histidine-rich protein 2 in serum of humans with malaria: comparison with a commercial kit. J. Clin. Microbiol. **46**, 3759–3765 (2008)
141. Sharma, M.K., Agarwal, G.S., Rao, V.K., Upadhyay, S., Merwyn, S., Gopalan, N., Rai, G.P., Vijayaraghavan, R., Prakash, S.: Amperometric immunosensor based on gold nanoparticles/alumina sol–gel modified screen-printed electrodes for antibodies to Plasmodium falciparum histidine rich protein-2. The Analyst **135**, 608–614 (2010)
142. Sharma, M.K., Rao, V.K., Merwyn, S., Agarwal, G.S., Upadhyay, S., Vijayaraghavan, R.: A novel piezoelectric immunosensor for the detection of malarial Plasmodium falciparum histidine rich protein-2 antigen. Talanta **85**, 1812–1817 (2011)
143. Sikarwar, B., Sharma, P.K., Srivastava, A., Agarwal, G.S., Boopathi, M., Singh, B., Jaiswal, Y.K.: Surface plasmon resonance characterization of monoclonal and polyclonal antibodies of malaria for biosensor applications. Biosens. Bioelectron. **60**, 201–209 (2014)
144. Lee, S., Song, K.M., Jeon, W., Jo, H., Shim, Y.B., Ban, C.: A highly sensitive aptasensor towards Plasmodium lactate dehydrogenase for the diagnosis of malaria. Biosens. Bioelectron. **35**, 291–296 (2012)
145. Jeon, W., Lee, S., Manjunatha, D.H., Ban, C.: A colorimetric aptasensor for the diagnosis of malaria based on cationic polymers and gold nanoparticles. Anal. Biochem. **439**, 11–16 (2013)
146. Ittarat, W., Chomean, S., Sanchomphu, C., Wangmaung, N., Promptmas, C., Ngrenngarmlert, W.: Biosensor as a molecular malaria differential diagnosis. Clin. Chim. Acta **419**, 47–51 (2013)
147. Xiao, J., Yolken, R.H.: Strain hypothesis of Toxoplasma gondii infection on the outcome of human diseases. Acta Physiol. **213**, 828–845 (2015)
148. Wang, H., Lei, C., Li, J., Wu, Z., Shen, G., Yu, R.: A piezoelectric immunoagglutination assay for *Toxoplasma gondii* antibodies using gold nanoparticles. Biosens. Bioelectron. **19**, 701–709 (2004)
149. Ding, Y., Wang, H., Shen, G., Yu, R.: Enzyme-catalyzed amplified immunoassay for the detection of Toxoplasma gondii-specific IgG using Faradaic impedance spectroscopy, CV and QCM. Anal. Bioanal. Chem. **382**, 1491–1499 (2005). https://doi.org/10.1007/s00216-005-3350-xhttps://doi.org/10.1007/s00216-005-3350-x
150. Luo, Y., Liu, X., Jiang, T., Liao, P., Fu, W.: Dual-aptamer-based biosensing of toxoplasma antibody. Anal. Chem. **85**, 8354–8360 (2013)

151. He, L., Ni, L., Zhang, X., Zhang, C., Li, R., Xu, S.: Fluorescent detection of specific DNA sequences related to *Toxoplasma gondii* based on magnetic fluorescent nanoparticles Fe_3O_4/CdTe biosensor. Int. J. Biochem. Res. Rev. **6**, 130 (2015)
152. Kevric, I., Cappel, M.A., Keeling, J.H.: New world and old world Leishmania infections: a practical review. Dermatol. Clin. **33**, 579–593 (2015)
153. Souto, D.E.P., Silva, J.V., Martins, H.R., Reis, A.B., Luz, R.C.S., Kubota, L.T., Damos, F.S.: Development of a label-free immunosensor based on surface plasmon resonance technique for the detection of anti-Leishmania infantum antibodies in canine serum. Biosens. Bioelectron. **46**, 22–29 (2013)
154. Sousa, S., Cardoso, L., Reed, S.G., Reis, A.B., Martins-Filho, O.A., Silvestre, R., da Silva, A.C.: Development of a fluorescent based immunosensor for the serodiagnosis of canine leishmaniasis combining immunomagnetic separation and flow cytometry. PLoS Negl. Trop. Dis. **7**, e2371 (2013)
155. Mohan, S., Srivastava, P., Maheshwari, S.N., Sundar, S., Prakash, R.: Nano-structured nickel oxide based DNA biosensor for detection of visceral leishmaniasis (Kala-azar). The Analyst **136**, 2845–2851 (2011)
156. Nouvellet, P., Cucunubá, Z.M., Gourbière, S.: Ecology, evolution and control of Chagas disease: a century of neglected modelling and a promising future. In: Advances in Parasitology, pp. 135–191. Elsevier, Amsterdam (2015)
157. Diniz, F.B., Ueta, R.R., Pedrosa, A.M.D.C., Areias, M.D.C., Pereira, V.R.A., Silva, E.D., da Silva, J.G.J., Ferreira, A.G.P., Gomes, Y.M.: Impedimetric evaluation for diagnosis of Chagas' disease: antigen-antibody interactions on metallic electrodes. Biosens. Bioelectron. **19**, 79–84 (2003). https://doi.org/10.1016/s0956-5663(03)00213-6
158. Ferreira, A.A.P., Colli, W., da Costa, P.I., Yamanaka, H.: Immunosensor for the diagnosis of Chagas' disease. Biosens. Bioelectron. **21**, 175–181 (2005). https://doi.org/10.1016/j.bios.2004.08.001https://doi.org/10.1016/j.bios.2004.08.001
159. Pereira, S.V., Bertolino, F.A., Fernández-Baldo, M.A., Messina, G.A., Salinas, E., Sanz, M.I., Raba, J.: A microfluidic device based on a screen-printed carbon electrode with electrodeposited gold nanoparticles for the detection of IgG anti-Trypanosoma cruzi antibodies. The Analyst **136**, 4745–4751 (2011)
160. Lee, S., Manjunatha, D.H., Jeon, W., Ban, C.: Cationic surfactant-based colorimetric detection of *Plasmodium* lactate dehydrogenase, a biomarker for malaria, using the specific DNA aptamer. PLoS ONE **9**, e100847 (2014)
161. Zhu, Y., Chandra, P., Shim, Y.B.: Ultrasensitive and selective electrochemical diagnosis of breast cancer based on a hydrazine–Au nanoparticle–aptamer bioconjugate. Anal. Chem. **85**, 1058–1064 (2013)
162. Rasooly, A., Jacobson, J.: Development of biosensors for cancer clinical testing. Biosens. Bioelectron. **21**, 1851–1858 (2006)
163. Roy, B., Mukherjee, S., Mukherjee, N., Chowdhury, P., Babu, S.P.S.: Design and green synthesis of polymer inspired nanoparticles for the evaluation of their antimicrobial and antifilarial efficiency. RSC Adv. **4**, 34487–34499 (2014)
164. Chakraborty, I., Saha, U., Mandal, D., Mukherjee, S., Joardar, N., Sinha Babu, S.P., Suresh Kumar, G., Mandal, K.: Effect of bovine serum albumin on tartrate-modified manganese ferrite nano hollow spheres: spectroscopic and toxicity study. Phys. Chem. Chem. Phys. **21**, 10726–10737 (2019). https://doi.org/10.1039/C9CP01877Hhttps://doi.org/10.1039/C9CP01877H
165. Mukherjee, S., Mukherjee, S., Bhattacharya, S., Sinha Babu, S.P.: Surface proteins of Setaria cervi induce inflammation in macrophage through Toll-like receptor 4 (TLR4)-mediated signalling pathway. Parasite Immunol. **39** (2017). https://doi.org/10.1111/pim.12389
166. Mukherjee, S., Huda, S., Sinha Babu, S.P.: Toll-like receptor polymorphism in host immune response to infectious diseases: a review. Scand. J. Immunol. **90**, e12771 (2019). https://doi.org/10.1111/sji.12771https://doi.org/10.1111/sji.12771
167. Mukherjee, S., Mukherjee, S., Maiti, T.K., Bhattacharya, S., Sinha Babu, S.P.: A novel ligand of Toll-like receptor 4 from the sheath of *Wuchereria bancrofti* microfilaria induces proinflammatory response in macrophages. J. Infect. Dis. **215**, 954–965 (2017)

168. Cavallo, M.F., Kats, A.M., Chen, R., Hartmann, J.X., Pavlovic, M.: A novel method for real-time, continuous, fluorescence-based analysis of anti-DNA abzyme activity in systemic lupus. Autoimmune Dis. **2012** (2012)
169. Carpenter, A.C., Paulsen, I.T., Williams, T.C.: Blueprints for biosensors: design, limitations, and applications. Genes (Basel). **9**, 375 (2018)

Design and Analysis of a Capacitive MEMS Accelerometer as a Wearable Sensor in Identifying Low-Frequency Vibration Profiles

M. Preeti, Koushik Guha, K. L. Baishnab, and A. S. C. S. Sastry

Abstract The presented MEMS Sensor uses Accelerometer technology to capture human motion and vibration data helping in clinical diagnosis and medical monitoring. The chapter elucidates upon analyzing the response of a Proof mass on changing different parameters in the process of designing a capacitive accelerometer for Human motion Monitoring. Modelling the structures of different proofmass and varying the geometric structure and measurements yields in different values of Effective Spring Constant which in turn impacts operating frequency. Judicious selection of the design parameters and their optimization plays a crucial role in modelling a Proof mass whose resonant frequency is the function of parameters identified thereby. The Proof mass responds to low-frequency inputs less than 10 Hz and has fixed bandwidth of (2–8) Hz. The proofmass could be used as the sensing element in a wearable sensor that diagnoses Parkinson's disease in early stages.

Keywords Wearable sensor · Point of care diagnostic testing · Accelerometer · MEMS · Parkinson's disease

1 Introduction

Multifariousness of the field of Micro Electro Mechanical Systems (MEMS) makes its technology burst into the unreached fields of Science, Engineering, and Medicine. MEMS have diverse applications including giving various profitable products to the field of medicine. Their journey into the field of medicine became possible because of its possibility of using variant materials. The MEMS component materials are both bio-compatible like, gold, silicon nitride, silicon dioxide, SU-8(TM), and silicon metallic gold, and also those which are not bio-compatible [1]. Invasive devices are

M. Preeti · K. Guha (✉) · K. L. Baishnab
Electronics and Communication Engineering Department, NIT Silchar, Silchar 788010, India
e-mail: koushikguha2009@gmail.com

M. Preeti · A. S. C. S. Sastry
Department of Electronics and Communication Engineering, Koneru Lakshamaiah Education Foundation, Vaddeswaram, A.P., India

G. Dutta et al. (eds.), *Modern Techniques in Biosensors*, Studies in Systems, Decision and Control 327, https://doi.org/10.1007/978-981-15-9612-4_2

made up of biocompatible materials and noninvasive devices can be made up of any material. The material properties that are of importance are Young's modulus, Poisson's ratio, density, fracture strength. MEMS Devices are used in various ways to assess and monitor health of a machine or human body in different ways. One of the main criterion of assessment of the strength and working characteristic of a machinery or Mechanical structure is the Vibration profile. Changes in the vibration observed by certain human activities like tremors usually in hands and fingers, Speech or writing problems dent the day to day life of the person. These vibrations fall in the range of few tens of Hz. Vibrations produced by Mechanical instruments, devices, transport vehicles, and seismic Micro tremors are in the range of few 10 s of Hz and also less than that [2]. Assessing Human Movements and monitoring Motor abilities in Humans have shown to be useful in anticipating health risk and also diagnosing some symptomatic Chronic diseases in early stages. Modern medicine does not currently have the technology to make a definitive diagnosis of certain disease due to lack of objective medical tests. One important parameter based on which different Health conditions can be studied is by measuring the vibration of Human body by means of wearable sensors. In the cases where there is a need for improved and constant clinical care, wearable sensors can be implanted and the movement and vibration profile of the patient, Gait analysis can be acquired. Neurological diseases like Parkinson's disease [PD] is one such disease which is progressive and degenerative [3, 4]. The onset of Parkinson's disease shows early symptoms in the extremities of the patient's body parts like legs, hands, and head in the form of involuntary shaking. The disease is accompanied by deterioration of neuro transmitting cells in the part of the brain by decrease in dopamine (Figs. 1 and 2).

Chronic diseases like heart disease, stroke, Alzheimer's disease, chronic obstructive pulmonary disease and Parkinson's disease are by far the leading cause of decrease in quality of life in the patients [5]. Such diseases call for high quality

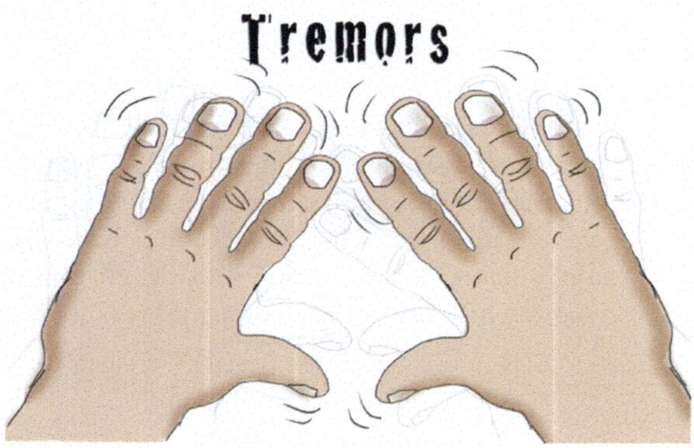

Fig. 1 Tremors in the hands of a patient with PD

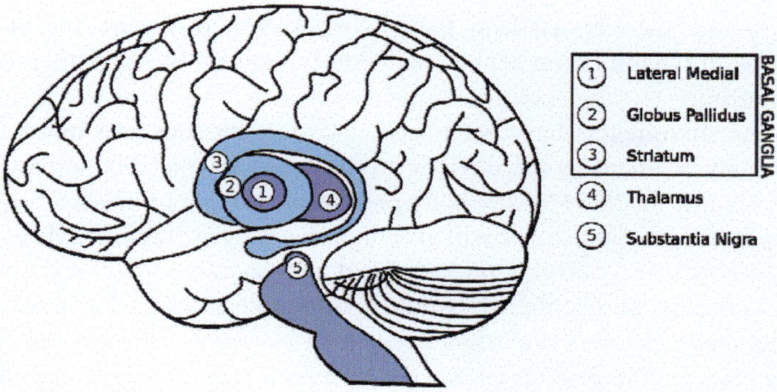

Fig. 2 Illustrates structures of the brain related to the basal ganglia circuit and *Substantia nigra*

of medical service, that is to say that the patient needs long term treatment, monitoring the ups-and-downs of the symptoms, and investigation of vital signals and physiological parameters (Fig. 3).

That's where technology can budge in and create alternate tools for early diagnosis, continuous and objective monitoring and compliment clinical assessment and patient-reported outcomes. Integrating MEMS sensor technology into miniaturized and wearable devices will attend to the present and growing challenges the Medical field is facing. These wearable sensors can be invasive and sometimes non-invasive devices too. MEMS Smart sensors technologies ranges from measuring physiological parameters like heart rate, blood analysis, fall risk assessment, giving continuous

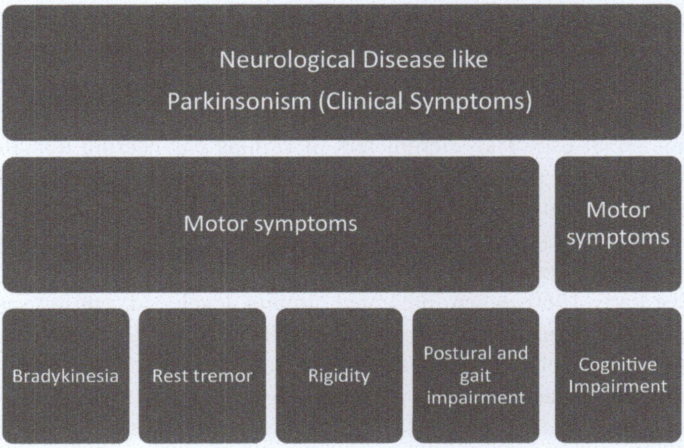

Fig. 3 Symptoms of Parkinson's disease [6]

health assessment there by assisting the personnel in choosing best health management regimes in rehabilitation centres and elderly patients by collecting movement data in daily life.

MEMS Accelerometers have been used since few decades to acquire data from respective body segment as the devices are portable, having biocompatible materials and so can be installed on any body part invasively or noninvasively. Adopting technology for medical assistance will give high accuracy in predicting disease symptoms and cuts back the control of measurement and diagnosis to clinical observation alone. Reducing the subjection of symptom measurements to clinical observation and patients inputs become important because of its dependence on human assessment alone. As technology intervenes, it complements clinical diagnosis and does not restrict itself to measurements that are observant dependent making them biased upon the person diagnosing. The disease biomarkers obtained thus have high precision and therefore high perception of the state of patients. The quality of signals acquired from the movement or vibration measurement as symptomatic of the disease deeply relies on the response of the sensor. MEMS sensors are of various kinds like MEMS Accelerometers, pressure sensors, Temperature Sensors, Flow sensors, and Force sensors. MEMS Accelerometers have seen vast applications in various Engineering fields ever since they got commercialized and officially registered in 1936 [7].

MEMS accelerometers are inertial sensors which can detect accelerations in a single axis. They are very noted and successful MEMS devices which are able to measure wide range of parameters like Vibration, tilt, shock, and acceleration. Measuring acceleration in all three dimensions employs three accelerometers oriented orthogonal to each other as three different devices or as a single device having three sensing outputs. This design task could be challenging because of the manufacturing limitations in the microscale.

Accelerometers by themselves are having their own applications and are also used in conjunction with other devices and while being integrated make powerful devices like gyroscopes, inertial measurement units. This versatile quality has given MEMS Accelerometers a great portion of MEMS Market.

The commercial use of accelerometers in various engineering domains drastically increased since 1930s. Various vendors like Analog Devices, Hitachi Metals, Bosch, Dytrain, Matsushita EG&G IC Sensors, Honeywell, Endevco, Freescale, MEMSIc, Summit Instruments, Entran Devices Indsensors have come up with their Accelerometers into the Market (Fig. 4).

It was the early 1960s when integrated circuits developed, putting sensors and chips together on same chip. Soon by the 1970s developed selective anisotropic etching, Impurity and electrochemically-based etch-stops which lead to the development of "bulk micromachining" By the early 1980s, inkjet print heads, 3D printing, fabricating accelerometers came into existence with the practice of bulk micromachining. In mid-80s came up the technique of Surface Micromachining that lead to new applications of Accelerometers, pressure sensors and new structures were realized in micro-electromechanical industry. Microsystem design has now taken cutting-edge design and fabrication technologies that their expansion and utilization in MEMS industry foresees and promises the possibility of miniaturization.

Fig. 4 Accelerometer manufacturers

MEMS is not restricted with the material aspect as it is in IC industry. Therefore a variety of materials, fabrication techniques, and packaging is possible which gave accelerometers a scope of being realized with new working principles.

Accelerometers are used in the field of Military and defence for test and evaluation, aerospace machine monitoring and maintenance applications, automobile industry with many important sensors, Bio-Medical field to specify many disease bio-markers, thereby helping in detection, monitoring, and controlling various acute and chronic diseases. Choice of using a particular Accelerometer out of the many depends upon the particular application how the specifications of the accelerometer meet that need. The specifications one has to look into are Sensitivity, Resolution, Bandwidth, Gravity, Linearity and Noise. While the Industry looks into aspects like Low cost, high reliability and ability to be interfaced with external electric circuitry for presentable output, they have designed and brought into market many Accelerometers. ADXL50 was the first commercial Accelerometer made for airbag deployment by Analog devices had a footprint of 5 mm^2 and was produced in large number. Accelerometers in combination with gyroscopes make the important device known as Inertial Measurement Unit (IMU) which is used for motion applications, space orientation analysis, in aeroplanes, and also for accessing the unknown trajectories, device stability and control, anticollision assistance and lead leveling. Not only for such major applications as in land, water, and air, but accelerometers become famous in Consumer products also as in mobile phones, game controller, hand-held gadgets, and robotics (Fig. 5).

Seeing that they have covered such a large expanse of industry and day to day life, it is not surprising that this versatile device known as accelerometer has brought in new dimensions in the field of Medicine as BIO-MEMS devices. Focusing on Biological and biomedical sectors, one can find tremendous scope for accelerometers. Wearable sensors networks have used wearable embedded system technology to gain remote access to the patient's physiological data, helping with remote health monitoring and activity monitoring [8]. Wearable sensor technology supports to enhance the

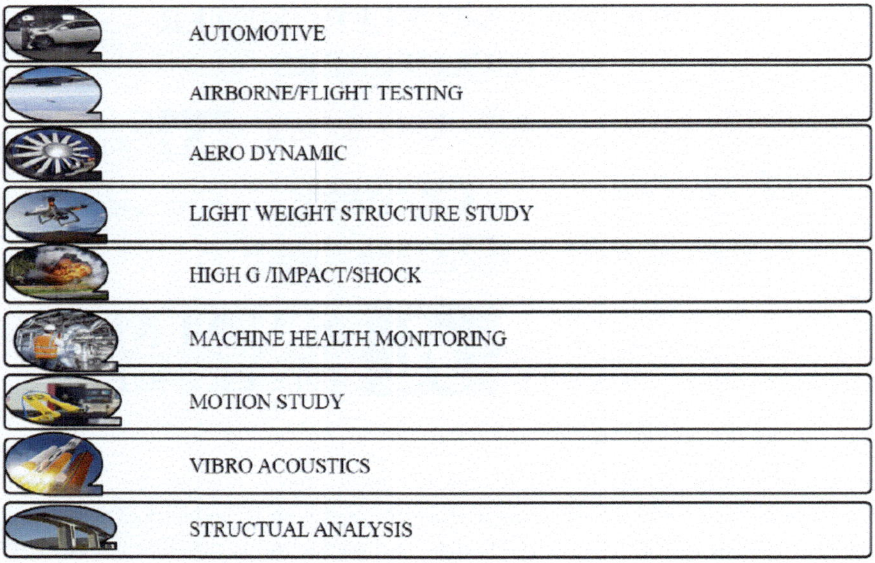

Fig. 5 Applications of MEMS accelerometers

quality of life of a patient suffering with chronic diseases by allowing independent living by continuous assessment and monitoring. Patients with Chronic diseases like leg ailments, Parkinson's disease Alzheimer's disease, suffer with difficulty in movement and therefore have to depend on others continuously. This dependence for a long time for care and monitoring could make the life of the patient void of privacy and independence and may not be flavoured after a while. Assessment of tiny movements using sensors at different locations of the patient's body would bring back that independence and improve the patient's quality of life and confidence.

The data acquired from monitoring these mechanical movements of the patient could be sent to doctors and assessment of the patient's condition can be done. Sometimes, these devices also help in early detection of some chronic diseases like Parkinson's disease, cardiovascular disease.

2 Literature Survey

Literature reports the usage of Accelerometers in biomedical field through a wide range of disease diagnosis, monitoring and controlling by collecting the accelerometer data and processing it through many signal processing units and different integrated modules. Hongzhi Sun et.al, developed a MEMS accelerometer motion sensing module by using multiple MEMS accelerometers to aid the diagnosis and thus helping in treatment of stressed urinary incontinence. Technology is used here to track the position of the bladder in real time thereby helping the surgeons to restore the bladder's neck by giving accurate position [9]. Apart from using an accelerometer for motion data, it is important to know the body location where it must be used. Inaccurate body localization of wearable sensors would lead to data that is incorrect, inaccurate and thereby reduces the usefulness of the data and the adaptability of the sensor. The work conducted in [8] uses accelerometers at different parts of the body to monitor the locations on the body which gives away maximum movement data in a walking activity. The pattern of data received in the study conducted in [8] shows that the shin area of the leg is impacted significantly higher than other locations of the leg.

Risk of being affected by Parkinson's disease is high in older people of age over 65 years. The disease has affected 3% of population of that age. It is true that the patients are affected by Parkinsonism, a condition where the central nervous system gets affected and there by the motor abilities of the patient are hindered, even before they show out the visible symptoms significantly. Wearable sensors having accelerometers are used in different investigations by using them over extended periods of time on the patients and monitoring the motor fluctuations like slowness of movement, bradykinesia, and other disabilities. The data collected thus will help in assessing the patient's condition and monitor the drug dosages [10, 11]. Carlos et.al, performed multidimensional continuous monitoring of Parkinson's disease using wearable biosensors. A study was done on 12 patients with PD and 12 Control patients by placing a small watch-like device on the less affected hand on PD patients and non-dominant hand in controls. The movement data, including sleep and wake periods are subjected to different statistical analysis and machine-Learning analysis. The characteristics of the patients are thus determined and disease duration was found to be 12 ± 1.8. The corresponding (Unified Parkinson's disease rating scale) UPDRS ranges are also given which indicate the severity of symptoms in the patients [11]. Tremor is defined as the involuntary movements of the body parts which can occur symmetrically on both sides of the body or can be unilateral. But this poses serious inconveniences in daily life even though it is not life-threatening in all cases. Literature shows that more than 4% of elderly people above the age of 65 years are affected by tremor. Daily activities like writing, eating or picking an object could be inconvenient and if left unaddressed, the symptoms would worsen by time (Fig. 6).

Neurological patients show this symptom of tremor which is the rhythmic shaking of body part at rest or during a posture. The mode of tremor depends on its clinical presentation. Rest tremor being in the range of 3–6 Hz of frequency range [12].

Table 1 showing the comparison of commonly used sensors for tremor monitoring [13]. In a study conducted by Sigcha et al. [13] it has been identified that accelerometers that are built into other smart devices are giving inaccurate data during acquisition. The data acquired is uncertain as it is observed that the accelerometer in one smartwatch and another smartwatch of the same brand have given distinct values of frequency while analyzed [13]. Hence off the shelf accelerometers, even if they have broad specifications would not be useful for precise measurements and application-oriented specific medical diagnosis. Parkinson's disease is identified with motor and

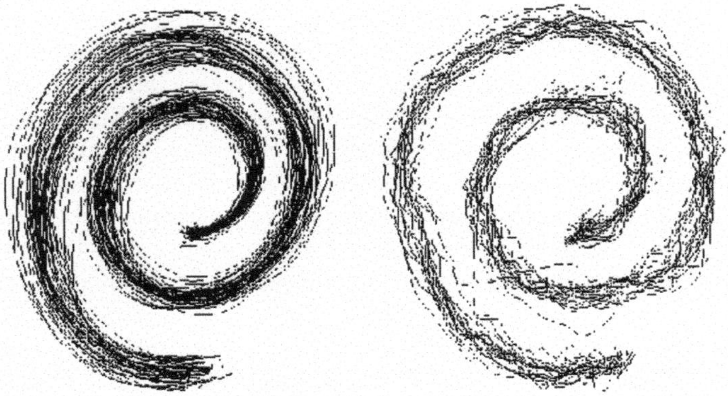

Fig. 6 Spirals drawn by control subject on left and patient with tremor on right [11]

Table 1 Commonly used Sensors in observing tremor [13]

	Surface EMG (SEMG)	Force transducers	Video	Accelerometer
Gravity effect	No	No	No	Yes
Accuracy of reading	Good	Good	Can be low	Good
Signal to noise ratio	High	High	Depends on device	Low to high, depending on design
Electrical contact with subjects	Yes	Yes	No	No
Size	Small	Large	Relatively Large	Small
Cost	Depends on device	Expensive	Cheap to expensive	Cheap

non-motor symptoms that make the patient vulnerable to perform their daily activities and thus reduces their quality of life. Origin of the disease is not yet identified but is characterized by loss of transmission of information from brain cells to the parts of the body due to loss of dopamine in the neurons of the *Substantia nigra* part of the brain [10, 11]. The motor symptoms include Tremors in the extremities of the body parts like shaking of hands, bradykinesia-slowness of movement, freezing, and gait disturbance. The early symptom of the disease is the resting tremor that is observed in the finger tips which is also called pill-rolling effect.

Several researches were conducted in the medical streams to analyze these symptoms through oral interviews, video recordings of the symptoms, and similar studies. The severity of the disease symptoms is marked by a neurologist dependent evaluation scale called Unified Parkinson's disease rating scale [14]. Several sensors were also tried over patients to monitor their daily activities by using sensors integrated into smart phones, Accelerometers, Gyroscopes, and computer vision [15]. MEMS devices have aided in condition monitoring from past few years, and the main device among them is an accelerometer. Measuring the vibration was the important criterion to find any deviation from normal operations, either in a machine or a human body [16] Accelerometer was used in [17] to monitor the subacute stroke patients who are discharged immediately from hospital. They were advised to wear a unilateral accelerometer on the wrist to assess the usage of upper extremity of the body. The continuous monitoring was intended to help them in rehabilitation.

Luca Lonini et al. worked on understanding how wearable sensors can be used on a patient with PD, how many sensors have to be used and how much data is needed to acquire sufficient inputs regarding the status of the disease. They have studied on real subjects and concluded that a single sensor could detect tremor in upper extremities [18]. From the literature, it could be observed that Parkinson's disease is a disease showing symptoms of motor impairment leading to reduced quality of life in terms of performing daily and normal activities. The patient shows visible symptoms and faces problem with daily life only after a while but the early symptom of the disease is the resting tremors that occur at the extremities of the hands. These tremors are of a very less frequency range of (3–7) Hz reported in literature. It is worth detecting the tremors in early stage of the disease so that the symptoms can be used as early sign of warning of the onset of PD. Early diagnosis is crucial in PD as the disease itself doesn't have a permanent cure but only the symptoms can be treated with dosages like levodopa or carbidopa. These drugs are helpful in improving the condition of the patient but do not cure the disease completely because PD is a neurodegenerative disease. While the patients are left untreated, PD can become life-threatening in cases where the patient may face sudden falls due to gait disturbances in major cases. PD drastically reduces the health related quality of life of the patient and further can lead to dementia, cognitive loss and loss of memory. Further stages of PD are treated with Deep brain stimulation (DBS) which is claimed to have fewer risks while some literature points that it may lead to suicidal tendency in the patient. These issues arise only if the disease progresses to bad stages. But if PD is identified in early stages, it could be a life-saving diagnosis for the patient. Hence it is proposed that a sensor be designed to accurately detect the low-frequency PD resting tremors with high sensitivity. It has been a practice of few decades to use wearable devices and sensors for biomedical applications to monitor vital signals of the body. As MEMS gives the advantage of having small size, portable and low-cost devices, also possessing the ability of batch processing, the design of a MEMS sensor for early detection of Parkinson's disease would help the patients greatly.

3 MEMS Wearable Sensor to Detect Low Frequency Tremors

Sensors are at the core of MEMS systems. Whenever there is a need of analysing mechanical movements and output it in the form of a presentable electrical signal, MEMS devices come into picture, so does an accelerometer. There is a huge unexplored area of BIO MEMS wither to the study has to stretch making human-centric applications akin and par the present research. MEMS have novel fabrication methods and they use the conventional Silicon as in integrated circuits also use non-microelectronic materials like polymers, plastic, Quartz, Ceramics, Biocompatible materials also to build mechanical device structures, biomedical and optical devices [19]. Microsensors can be termed as transducers because they convert one form of energy into another. It detects the physical signals from the surroundings and reacts in discernment to the information detected and produces a response in the form of a signal. The manner in which the sensor reacts solely depends upon the principle of transduction used in the sensor. The kind of physical signal that they can sense include light, heat, temperature, gas, vibration, tilt, shock, force, acceleration, etc., MEMS structures consist of both in-plane and out-of-plane designs, with a compilation of movable and immovable members and sometimes thin membranes and immobilization layers. There are many transduction principles making Mechanical Transducers, Thermal Transducers, Radiation Transducers, Magnetic Transducers, Chemical, and Biological Transducers.

3.1 Overview of Different Principles of Mechanical Transduction

Imagine any physical signal and a sensing mechanism can be applied to it to sense it and present it into a presentable and readable form. There are many such microsensors and there includes a mechanical action in them while they operate to transduce the signal. That is called the sensing mechanism. The general Mechanical sensing mechanisms are piezoresistive, piezoelectric, and capacitive and Resonant Sensors.

1. **Piezoresistive Sensors**

These sensors operate on the principle of piezoresistive effect which is the change in the resistivity of the material with change in its deformation due to strain. Change in resistance is sensed by a wheat stone bridge [20].

The deforming part of the structure is made of piezoresistive material or is generally coated with the piezoresistive coating at the most responsive part of the structure to acquire maximum of the signal sensed. Silicon is an excellent sensor of strain.

2. **Piezoelectric Sensor**

Piezoelectric sensors use the piezoelectric effect, i.e., when a piezoelectric crystal is deformed, it produces a corresponding potential across the crystal, and the reverse action also is effective. When a potential is applied across a piezoelectric crystal, it undergoes deformation which is inverse piezoelectric effect generally used in actuation techniques. The potential against the deforming structure gives the amount of strain the material has undergone which becomes the principle of its operation. The common materials used in MEMS devices are Quarts, ZnO, lead zirconate titanate (PZT), polyvinylidene fluoride (PVDF). These materials are generally coated on silicon base to extract the piezo electric effect while the silicon structure undergoes mechanical deformation.

3. **Capacitive Sensors**

These are one of the widely used sensors because of their high sensitivity. They are designed against the simple principle of a parallel plate capacitor. So, the structure essentially consists of at least one or more fixed plates and moving plates. They are placed parallel to each other and the movement of the moving plate relative to the fixed plate changes the distance between the plates thus producing change in capacitance according to the equation

$$c = \frac{\varepsilon_o \varepsilon_r A}{d}. \tag{1}$$

where

$\varepsilon_o = 8.85 \times 10^{-12}$ permittivity of free space.

ε_r = relative permittivity.

A = Overlapping area between the plates (m).

d = distance between the plates (m).

4. **Resonant Sensors**

Resonant sensors are those sensors which are meant to vibrate at resonant frequency. This advantage of this phenomenon of resonance can be extracted in different ways in biosensors. When the device is attached to a member of body part, or when the vibrating layers are immobilized, the movement of the layers or maximum build-up of the binding substance on the immobilization layer would affect the resonant frequency and can be acquired using any transduction principle. Such sensors include the following

(1) **Strain gauge**

The working principle of a vibration sensor or accelerometer depends upon the kind of deformation or mechanical displacement it is undergoing. This deformation is due to the mechanical force that is applied on it in the form of acceleration or force or a distinct movement. The corresponding response of the sensor to the external input which will be in the form of deformation is translated into corresponding electric signal in a presentable form. The deformation is related to the transduction principle according to its principle of operation. The transduction principle could be Piezoresistive, Capacitive, Magnetic, Piezo electric, optical etc.

The general Micro Accelerometer structure consists of a proofmass suspended by complaint beams attached to anchors or fixed frames. The displacement of the proofmass proportionally gives the acceleration depending upon its transduction principle. Piezo resistive and Capacitive transduction principles have become famous. Dynamic acceleration cannot be measured properly with Piezo resistive; hence Capacitive in preferred where ever there is a measurement involving change of acceleration like vibrations [21, 22].

(2) **Gyroscope**

Gyroscope is a device which detects inertial angular motion. For this reason, it is used in navigation systems, missile applications and in transport systems. Fabrication of micro rotating parts is a challenging task due to which MEMS gyroscopes uses vibrating structures in its design. An Accelerometer and a Gyroscope in combination make the very useful inertial measurement unit.

(3) **Pressure Sensor**

MEMS pressure sensors are also very famous sensors having thin membranes with a sealed cavity on one side and open to physical pressure signal on other side. The membranes deflect in response to the pressure it senses and uses piezo resistive or capacitive principles for transduction.

(4) **Accelerometer**

Accelerometer is a sensor which senses and measures the physical acceleration due to inertial forces or mechanical excitation. Acceleration is sensed using a seismic mass or a suspended proof mass that takes the inertial input. A force (F) is experienced by the proofmass when an excitation in the form of movement, shock, tilt, vibration or force is applied on it resulting in proportional displacement. This displacement is measured using the piezo resistive or capacitive principles.

An accelerometer is generally described using a simple spring mass damper composition. Whether it be a piezo resistive or a capacitive accelerometer, the basic description or the analogy of this spring mass damper holds good for both of them. The sensor essentially consists of a proof mass (m) which is meant to receive the external excitation. This proofmass is supported by suspended beams with spring stiffness (k) and a damping effect is also considered with a damping coefficient (c) (Fig. 7).

Owing to the structure and sensing axes of the accelerometer, the corresponding proof mass deformations will happen. If the accelerometer is a single axis sensitive sensor, there exists a sensitive axis for it against which the sensitivity will be

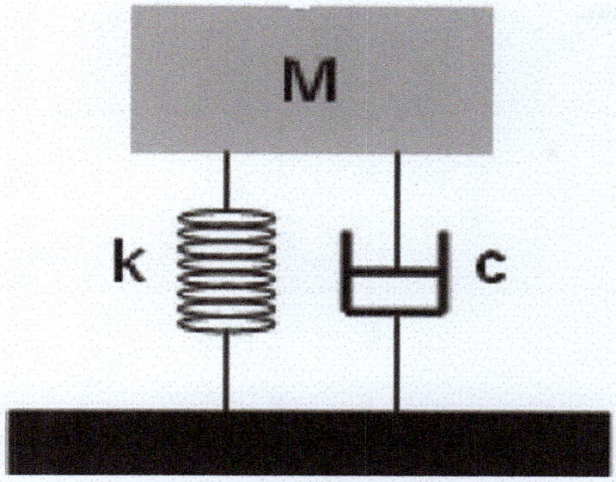

Fig. 7 Spring mass damper analogy of an accelerometer

maximum and the other axis signals would only be noise. If the device is intended to sense 3 axes signals, it will be designed accordingly. Figure 8 shows that the proofmass shown as a solid block here is capable of sensing acceleration in all three axes. The proofmass block is only for the representation where as its actual design includes a lot more details.

5. **Piezo resistive detection**

As described in the piezo resistive principle of transduction above, the materials used in this sensor have the property of converting the mechanical stress into change in resistance. They have high linearity and low output impedance. The change in resistance is proportional to the applied stress and is measured using a wheat stone bridge (Fig. 9).

The modelling of electrical part of accelerometer gives an output voltage Vout

$$V_{\text{out}} = \left[V_{\text{DD}}(\Delta R_Z + \Delta R_Y + \Delta R_{Y-})\right]/2R_0$$

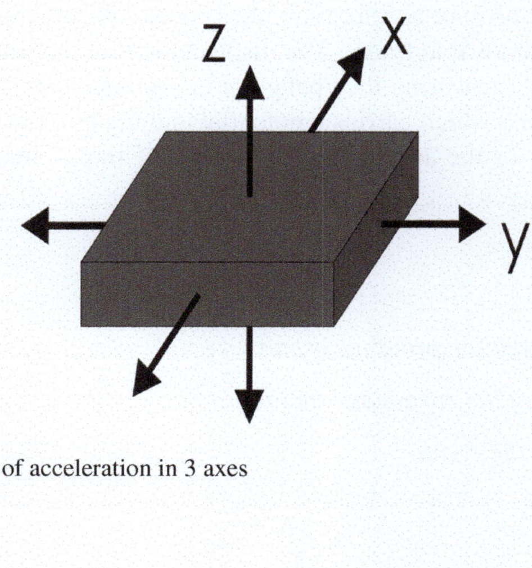

Fig. 8 Measurement of acceleration in 3 axes

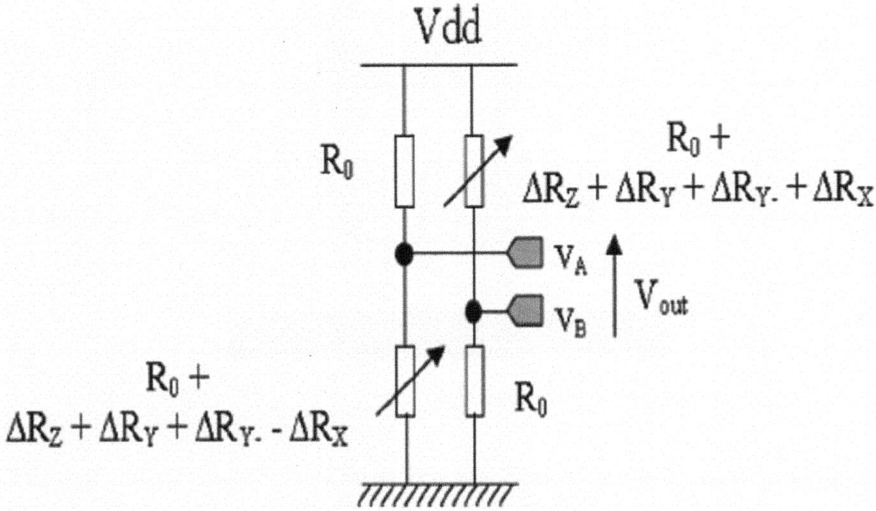

Fig. 9 Electrical model of a piezo resistive accelerometer [23]

The equation of Vout here doesn't have the term ΔR_X in it which means that the output voltage is independent of variation in resistance due to acceleration sensed along *x*-direction.

A good piezo resistive accelerometer must have good stability, accuracy and minimum response time.

6. **Capacitive**

The schematic of a Capacitive Accelerometer is of the form shown in Fig. 4. The primary part of the accelerometer is the suspended proofmass. Since the principle of operation is Capacitive, the structure of the device essentially has a fixed frame or the reference frame and a movable proofmass in parallel to it. When an external force is applied in the form of acceleration, the proofmass deflects. The deflection is with respect to the immovable frame. The displacement of the proofmass is the one which is measured by measuring the change of capacitance between the proofmass and the reference frame which carries a potential and the other plate being ground.

The specifications of the accelerometer that are of importance while designing are

Area of the Proof mass.

Mass of the Proofmass.

The maximum force that the proof mass can take.

Maximum deflection of the suspended beams.

Moment of Inertia of the beam.

Maximum bending stress in the beam.

Natural frequency.

Full scale range of gravity to be measured.

Sensitivity of the accelerometer.

Resolution

Bandwidth

Cross Axis Sensitivity.

To model for all these specifications, the governing equations of the Accelerometer have to be known. As the Accelerometer has the schematic of a spring-mass-damper, its modelling starts with first analyzing its movement.

The equation of motion of a spring mass damper is given by its second-order equation, according to the D'Almebert's Principle,

$$F_{\text{applied}} = ma_{\text{applied}}. \tag{2}$$

$$F_{\text{spring}} = kx. \tag{3}$$

$$F_{\text{damping}} = b\frac{\mathrm{d}x}{\mathrm{d}t}. \tag{4}$$

From Newton's second law, we have

$$F_{\text{applied}} - F_{\text{spring}} - F_{\text{damping}} = m\frac{\mathrm{d}^2x}{\mathrm{d}t^2}. \tag{5}$$

$$m\frac{\mathrm{d}^2x}{\mathrm{d}t^2} + b\frac{\mathrm{d}x}{\mathrm{d}t} + kx = F_{\text{applied}} = ma_{\text{applied}}. \tag{6}$$

So the equation of motion is

$$mx + cx + k\ddot{\dot{x}} = F(t). \tag{7}$$

Solution is given by

$$x = Ae^{\lambda t}. \tag{8}$$

Characteristic Equation is

$$\lambda^2 + 2\frac{c}{2m}\lambda + \omega_n^2 = 0. \tag{9}$$

Solution to the Characteristic equation is

$$\lambda_{1,2} = -\left(\frac{c}{2m}\right) \pm \sqrt{\left(\frac{c}{2m}\right)^2 - \omega_n^2}. \tag{10}$$

Displacement x is given by the general solution

$$x = e^{-\omega_n \zeta t}\left[x_0 \cos \omega_0 t + \frac{\dot{x}_0 + \omega_n \zeta x_0}{\omega_0} \sin \omega_0 t\right]. \tag{11}$$

The Amplitude A is given by

$$A = \sqrt{x_0^2 + \left(\frac{\dot{x}_0 + \omega_n \zeta x_0}{\omega_0}\right)^2}. \tag{12}$$

where ζ is the damping ratio

$$\zeta = \frac{c}{2m\omega_n}. \tag{13}$$

The suspended proofmass is supported by supporting beams in different kinds of ways depending on the requirement and the mandate of meeting the specifications for which the supporting beams with spring constant (k) are sometime in a combination of series and sometimes in parallel giving an equivalent spring constant (k_e) (Figs. 10 and 11).

The equivalent spring constant of springs in series is given by

$$\frac{1}{k_e} = \frac{1}{k_1} + \frac{1}{k_2}. \tag{14}$$

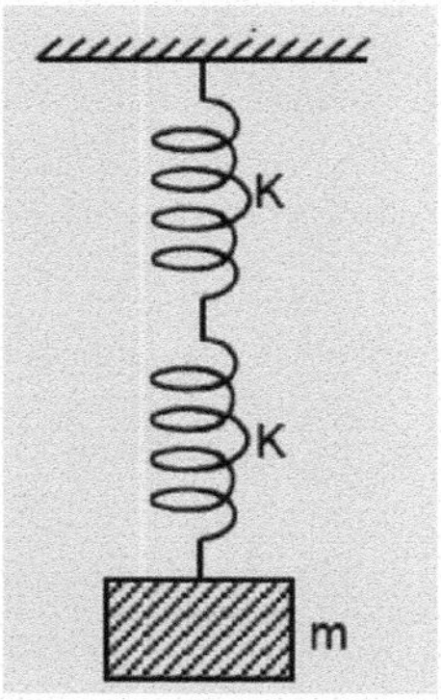

Fig. 10 Springs in series

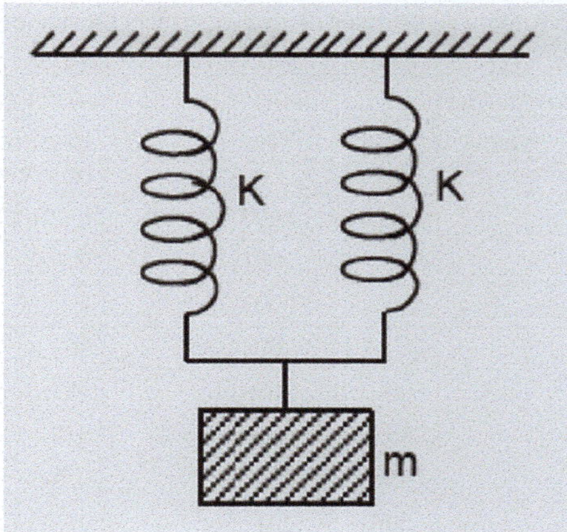

Fig. 11 Springs in parallel

The equivalent spring constant of springs in Parallel is given by

$$k_e = k_1 + k_2. \tag{15}$$

When an accelerometer is designed, all the goodness of a perfect design may not be accomplished in a single design. There will be trade off between parameters like sensitivity, area of the device, noise levels, etc. The proofmass, which is responsible for the reception of excitement as signal, needs to be modelled carefully to meet much of the desired specifications. A Combdrive design having fingers of movable structures and fingers of immovable structure is also one kind of design that improves sensitivity. Table 2 shows different Accelerometer's sensitivity with respect to kind of proofmass that is used in design.

The sensitivity (v/g) of the accelerometer is an important parameter and is the range of gravity for which the accelerometer gives a constant output, measured for a capacitive accelerometer as Change in capacitance to that of Applied Acceleration.

As the accelerometers design progresses, considering its material properties becomes important to attain the constraints kept on the specifications needed, thus the material properties and the quality of such data becomes crucial [25]. These devices have wide ranging structural designs and different principles of working. Hence they are worth a study before one picks an accelerometer for any application. Figure 12 shows the comparison of different accelerometers based on their principle of working.

In view of designing a low-frequency accelerometer for biomedical application of sensing low-frequency body tremors, the major design challenge would be suppression of noise levels and gaining high sensitivity. Therefore a capacitive Accelerometer deems to be the right choice for such a design.

Table 2 Table showing sensitivity comparison of accelerometers for single proofmass and multiple proof mass [24]

Multiple proof-mass accelerometers			Single proof mass accelerometers		
Year	Sensitivity/area (mV/mm^2)	Noise × area	Year	Sensitivity/area (mV/mm^2)	Noise × area
1999	–	15,840	1997	0.0513	12,160
1999	3.42	–	2001	9.259	–
2000	2.59	–	2003	–	50
2005	0.56	–	2008	20	1760
2005	157	100	2009	18	4800
2013	21.35	2553	2010	0.06	1,131,118
2015	–	1058	2010	0.745	264
2016	0.381	1008	2012	91.25	336
–	–	–	2013	1.204	–
–	–	–	2014	24	6

4 Proofmass Design of a Low-Frequency MEMS Capacitive Accelerometer

As it is specified in the previous sections about the importance of moving mechanical parts of the Proofmass, it is noteworthy to enhance the role of the geometrical structure and the corresponding measurements of such moving structures. The moving mechanical structure consists of a proof mass supported by beams of a particular spring constant. Therefore the mass of the moving structure is given by adding the mass of the proof mass and the supporting beams.

The wearable sensor that has to identify the particular disease Biomarkers needs to be designed with the specifications that essentially associates with the significant symptom of the disease. The diseases that show off movement disorders, involuntary tremors as marking symptoms are related to neurological disorders in which the frequency of the physiological signals emitted from human body are of very low frequency. Resting tremors of Parkinson's disease which falls into a frequency range of [3–7] Hz. When these tremors at the extremities of hands that is at the fingertips is identified clinically at a very early stage, the patient can be further screened for the symptoms of Parkinson's disease. The clinical analysis is completely based on the symptoms specified by the patient and is subjected to the observers discretion. Intervention of technology to identify the frequency of resting tremors which are the early symptom of the Parkinson's would help in early detection of the disease and also reduces subjection of diagnosis to the observer's analysis and the interpretation of the patient.

Hence a MEMS Sensor designed to identify low-frequency human tremors has to be designed. An accelerometer with the capability of sensing low frequency with high sensitivity can be designed with capacitive principle. The resonant frequency

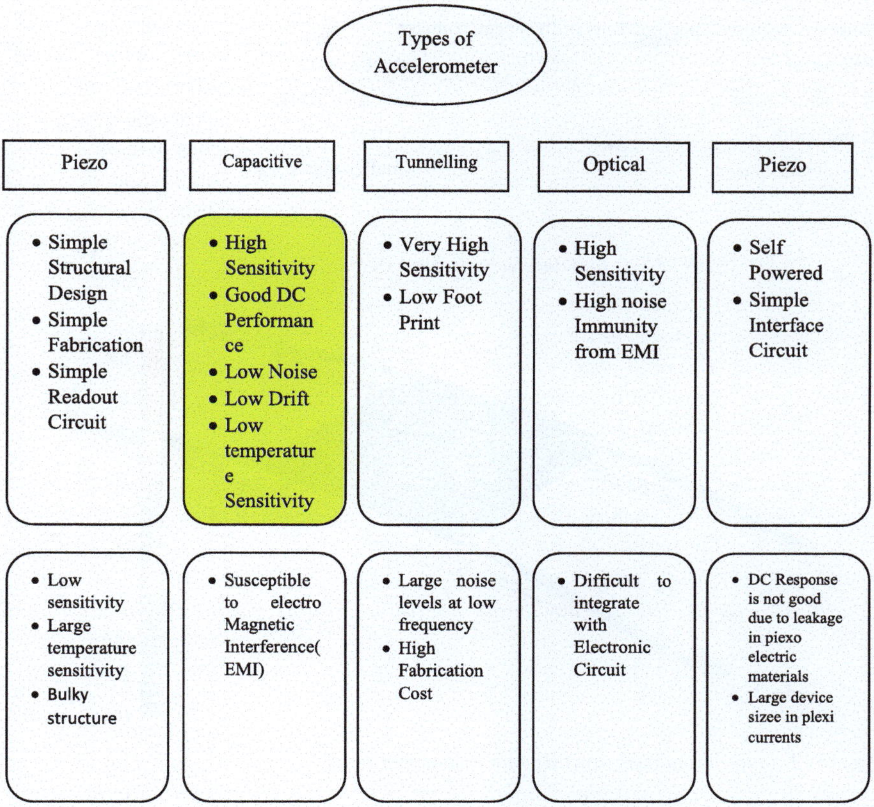

Fig. 12 Comparison of different accelerometers [24]

of such a sensor depends upon the mass of the proof mass and the spring constant of the supporting beams given by

$$f = \frac{1}{2\pi}\sqrt{\frac{k}{m}}. \tag{16}$$

Mass (m) of the proof mass constitutes of its Area × Density

$$m_{\mathrm{PM}} = \text{Area} \times \text{Density}. \tag{17}$$

$$m_{\mathrm{PM}} = (l_{\mathrm{PM}} + w_{\mathrm{PM}} + t_{\mathrm{PM}}) \times \rho. \tag{18}$$

where l_{PM} is the length of the proof mass, W_{PM} and t_{PM} are the width and thickness respectively (for a simple rectangular structure).

Table 3 Required specifications of the accelerometer

Parameter	Specification
Resonant frequency, f	Less than 10 Hz
Band width, ω_c	2–10 Hz
Acceleration, a	Smaller than 0.1 g

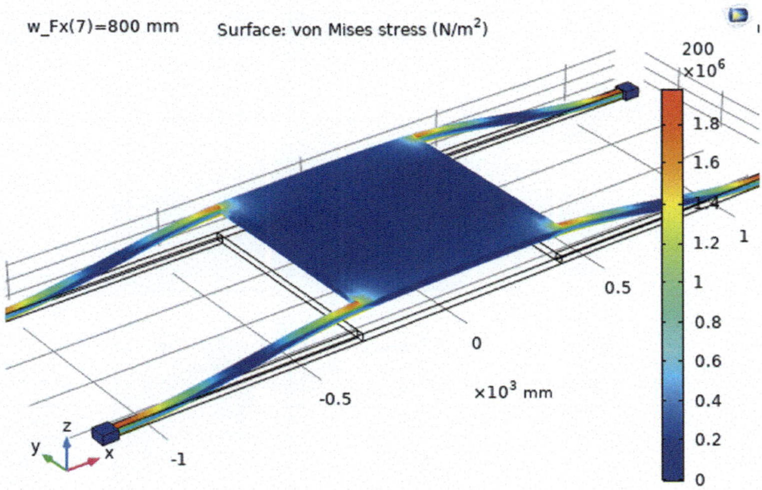

Fig. 13 A simple proofmass structure with four supports

Required Specifications of the Accelerometer are (Table 3).

From Fig. 14 it could be observed that the frequency range over 50 Hz was giving a constant displacement of 0 mm for the structure of proof mass in Fig. 13.

It is important here to note that the measurements of the proofmass and the beams play a vital role in determining the range of operation.

When these measurements are changed and outputs analyzed using FEM tool, we arrive at certain numbers where the proofmass vibrates at fixed and low frequency (Fig. 15).

5 Discussion

For The specified parameters and design structure, with Silicon as the material, the frequency of operation has reached the required low-frequency range of nearly 4 Hz for a beam length of 2000 mm. Though the length of the beams seems to be comparatively long for a wearable sensor, the graphs definitely show that for a length of 2000 mm beam, the required frequency range could be achieved. The

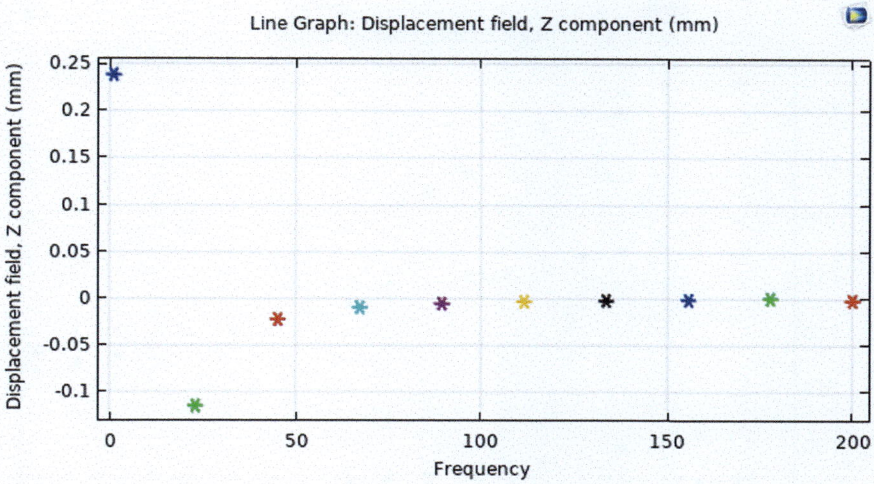

Fig. 14 Frequency versus displacement of the proof mass

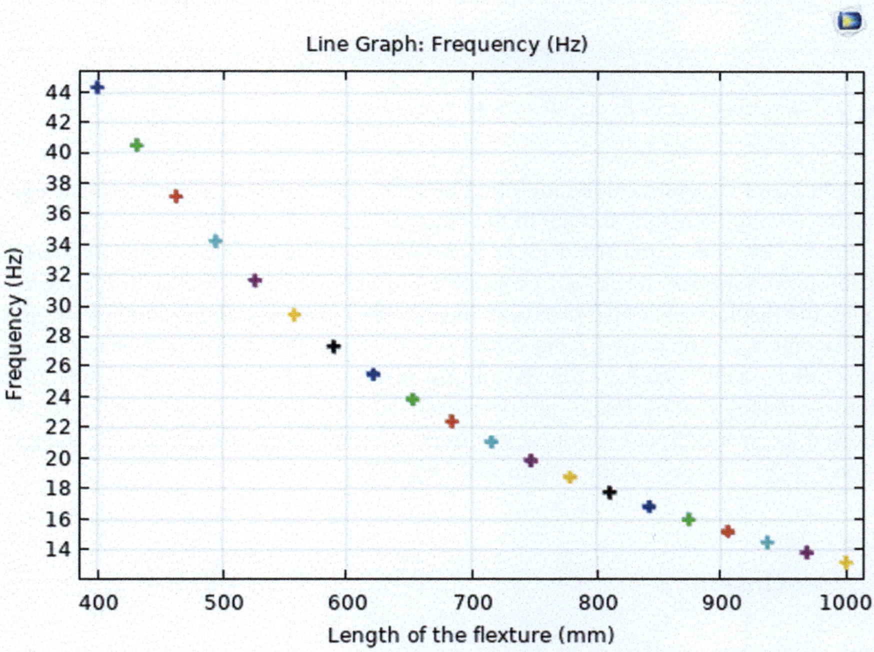

Fig. 15 Decrease in frequency of operation with increase in length of the beams

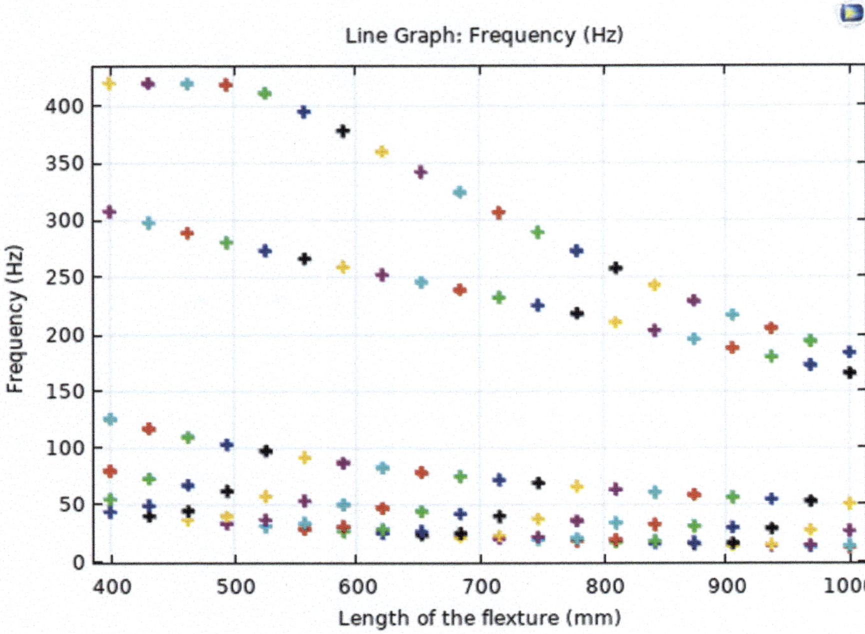

Fig. 16 Different Eigen mode operations with change in length

straight beam structure can be modified by realizing the beam as a meander whose lengths combined would amount to the required measure of 2000 mm.

Figure 16 shows that for a given constant length of the beam, the frequency of the proofmass has decreased proportionally. It is observed that with increasing force, the displacement also increases linearly, and also with increasing length, the displacement has increased. This shows that the length of the support beams has a great effect on the response of the proofmass in terms of displacement. Similarly, the increase in length of the beams has decreased the stiffness of the beams and there by allowed the proofmass to vibrate at low frequency. All these Simulations are done at a force of 1 N applied on the proofmass (Fig. 17).

6 Conclusion

It has been elaborated through the chapter about the importance of MEMS in the present Medical diagnostic procedures. Even though MEMS has been developed in recent decades, their wings have spread broadly into the Medical diagnostic field. Parkinson's disease can be detected early in its stage using MEMS sensor. This sensor is an accelerometer that reacts to a very low frequency by picking up the Parkinson's disease biomarker which is the resting tremor in the frequency range of

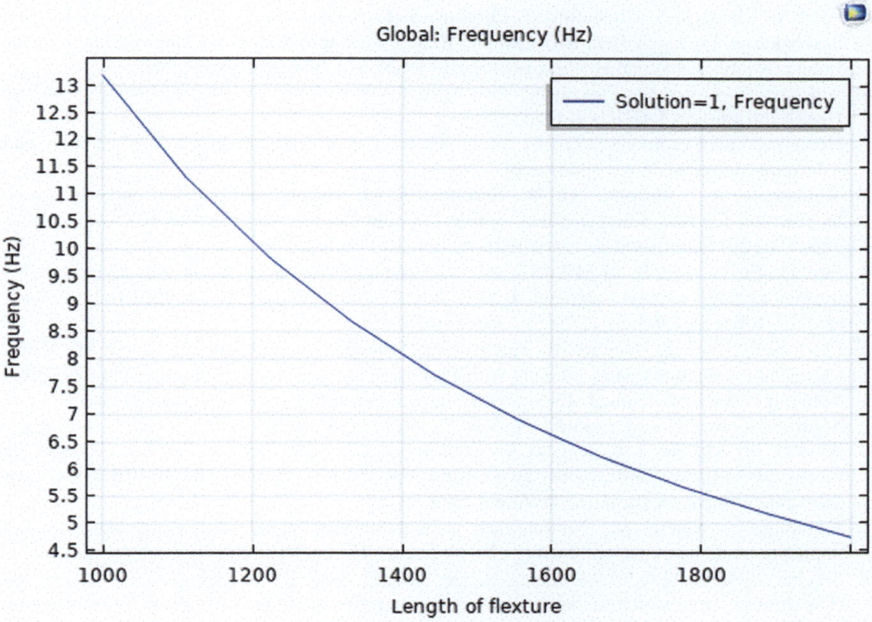

Fig. 17 Decrease in frequency of operation with increase in length

[3–7] Hz. Different MEMS sensors are presented and the different principle of operation. Literature has specified many accelerometers applications in medical assistance and monitoring but a application-specific disease diagnostic sensor is not proposed. The results shown in the chapter are the simulation results of structural analysis of a particular low frequency less than 10 Hz. The structure and measurements can be improvised for high sensitive and low noise performance.

References

1. Voskerician, G. et al.: Bio compatibility and bio fouling of MEMS Drug delivery devices. Biomaterials **24**(11), 1959–1967 (2003)
2. Li, G., Wang, J., Chen, D., Chen, J., Chen, L., Xu, C.: An electrochemical, low-frequency seismic micro-sensor based on MEMS with a force-balanced feedback system. Sensors **17**, 2103 (2017). https://doi.org/10.3390/s17092103
3. Lones, M., Smith, S., Alty, J., Lacy, S., Possin, K., Stuart Jamieson, D., Tyrrell, A.: Evolving classifiers to recognise the movement characteristics of Parkinson's disease patients. IEEE Trans. Evol. Comput. **18** (2013)
4. Tien, I., Glaser, S.D., Bajcsy, R., Goodin, D.S., Aminoff, M.: Results of using a wireless inertial measuring system to quantify gait motions in control subjects. IEEE Trans. Inf. Technol. Biomed. **14**(4) (2010)

5. Patel, V., Chatterji, S., Chisholm, D., Ebrahim, S., Gopalakrishna, G., Mathers, C., Mohan, V., Prabhakaran, D., Ravindran, R.D., Srinath Reddy, K.: Chronic diseases and injuries in India, India: Towards Universal Health Coverage 3, Series, vol. 377 January 29, 2011
6. Ahlrichs, C., Lawo, M.: Parkinson's disease motor symptoms in machine learning: a review. Health Inform. Int. J. (HIIJ) **2**(4) (2013)
7. History of accelerometer 1920 to 1996, Sound and Vibration/January 2007, Sound and Vibration Magazine, www.sandv.com
8. Vahdatpour, A., Amini, N., Xu, W., Sarrafzadeh, M.: Accelerometer-based on-body sensor localization for health and medical monitoring applications. Pervasive Mob. Comput. **7**(6), 746–760 (2011). https://doi.org/10.1016/j.pmcj.2011.09.002
9. Sun, H., Fu, G., Xie, H.: A MEMS accelerometer-based real-time motion-sensing module for urological diagnosis and treatment. J. Med. Eng. Technol. **37**(2), 127–134 (2013). https://doi.org/10.3109/03091902.2012.753127. ISSN: 0309-1902
10. Patel, S., Lorincz, K., Hughes, R., Huggins, N., Growdon, J., Standaert, D., Akay, M.: Monitoring motor fluctuations in patients with Parkinson's disease using wearable sensors. IEEE Trans. Inf. Technol. Biomed. **13**(6) (2009). https://doi.org/10.1109/Titb.2009.2033471
11. Madrid-Navarro, C.J., Escamilla- Sevilla, F., Mínguez-Castellanos, A., Campos, M., Ruiz-Abellán, F., Madrid, J.A., Rol, M.A.: Multidimensional circadian monitoring by wearable biosensors in Parkinson's disease. Front. Neurol. **9**, 157 (2018). https://doi.org/10.3389/fneur.2018.00157https://doi.org/10.3389/fneur.2018.00157
12. Grimaldi, G., Manto, M.: Neurological tremor: sensors, signal processing and emerging applications. Sensors **10**, 1399–1422 (2010). https://doi.org/10.3390/s100201399
13. Sigcha, L., Pavón, I., Arezes, P., Costa, N., De Arcas, G., López, J.M.: Occupational risk prevention through smartwatches: precision and uncertainty effects of the built-in accelerometer. Sensors **18**, 3805 (2018). https://doi.org/10.3390/s18113805
14. Ramaker, C., Marinus, J., Stiggelbout, A.M., van Hilten, B.J.: Systematic evaluation of rating scales for impairment and disability in Parkinson's disease. Mov. Disord. **17**(5), 867–876 (2002)
15. Torres, R., Huerta, M., González, R., Clotet, R., Bermeo, J., Vayas, G.: Sensors for Parkinson's Disease Evaluation. 978-1-5386-1962-9/17/$31.00 ©2017 IEEE
16. Albarbar, A., Mekid, S., Starr, A., Pietruszkiewicz, R.: Suitability of MEMS accelerometers for condition monitoring: an experimental study. Sensors **8**, 784–799 (2008). ISSN: 1424-8220
17. Wei, W.X.J., Fong, K.N.K., Chung, R.C.K., Myint, J.M.W.W., Cheung, H.K.Y., Chow, E.S.L.: Utility of a unilateral accelerometer for monitoring upper extremity use in subacute stroke patients after discharge from hospital. Assistive Technol. (2017). https://doi.org/10.1080/10400435.2017.1414085https://doi.org/10.1080/10400435.2017.1414085
18. Lonini, L., Dai, A., Shawen, N., Simuni, T., Poon, C., Shimanovich, L., Daeschler, M., Ghaffari, R., Rogers, J.A., Jayaraman, A.: Wearable sensors for Parkinson's disease: which data are worth collecting for training symptom detection models. NPJ Dig. Med. **1**, 64 (2018). https://doi.org/10.1038/s41746-018-0071-z
19. Li, R.J., Lei, Y.J., Chang, Z.X., Zhang, L.S., Fan, K.C.: Development of a high-sensitivity optical accelerometer for low-frequency vibration measurement. Article Sens. (2018). https://doi.org/10.3390/s18092910https://doi.org/10.3390/s18092910
20. Zhang, L., Jian, Lu., Takagi, H., Maeda, R.: Frontside-micromachined planar piezoresistive vibration sensor: evaluating performance in the low frequency test range. AIP Adv. **4**, 017112 (2014). https://doi.org/10.1063/1.4862253https://doi.org/10.1063/1.4862253
21. Pandey, K.P., Kumar, A.: Design and analysis of dual axis MEMS capacitive accelerometer. Int. J. Electron. Eng. Res. **9**(5), 779–790 (2017). ISSN: 0975-6450
22. Beliveau, A., et al.: Evaluation of MEMS capacitive accelerometer, 0740-7475189/$10.00 0 1999 IEEE
23. Ghemari, Z.: Study and analysis of the piezoresistive accelerometer stability and improvement of their performances. Int. J. Syst. Assur. Eng. Manage. (2017). https://doi.org/10.1007/s13198-017-0622-8

24. Mohammed, Z., et al.: Monolithic multi degree of freedom (MDoF) capacitive MEMS accelerometers. Micromachines **9**, 602 (2018). https://doi.org/10.3390/mi9110602https://doi.org/10.3390/mi9110602
25. Ghodssi, R., Lin, P. (eds.): MEMS Materials and Processes Handbook, pp. 22–28. Springer Publications, Berlin (2011)

Strategies for Multiplexed Electrochemical Sensor Development

Uroš Zupančič, Joshua Rainbow, Connor Flynn, Josephine Aidoo-Brown, Pedro Estrela, and Despina Moschou

Abstract Detection of multiple biomarkers for disease diagnosis or treatment monitoring has received a lot of attention due to their potential impact on clinical decision making. Electrochemical biosensors have become one of the preferred detection approaches, due to the simplicity of the accompanying instrumentation. This chapter will explore how electrochemical sensors can be utilized for detection of multiple analytes by integration of sensors into microfluidic microsystems. Some key fabrication technologies for such devices will be presented utilizing polymer microfabrication, paper-based approaches, and the use of printed circuit boards. Next, the use of electrode arrays will be presented along with some commercial platforms, outlining plausible paths towards a successful electrochemical multiplexed sensor. Novel approaches based on microbeads and various labels will then be introduced along with various strategies and technologies utilized to achieve ultrasensitive multiplexed detection.

Keywords Electrochemical biosensors · Multiplex immunoassays · Microfluidics integration · Printed circuit boards · Electrode arrays

1 Introduction

Electrochemical sensors for ultrasensitive multiplexed diagnostics of biologically relevant health and environmental markers are an important area of scientific interest. The importance of developing such devices has never been more topical than now, with global disease prevalence increasing, associated with a growing global population and worsening environmental pollution levels. Thus, the next generation of medical and environmental diagnostic and monitoring technologies needs to be robust

U. Zupančič · J. Rainbow · C. Flynn · J. Aidoo-Brown · P. Estrela · D. Moschou (✉)
Centre for Biosensors, Bioelectronics and Biodevices (C3Bio) and Department of Electronic and Electrical Engineering, University of Bath, Bath, UK
e-mail: D.Moschou@bath.ac.uk

C. Flynn
Department of Chemistry, University of New Brunswick, Fredericton, Canada

G. Dutta et al. (eds.), *Modern Techniques in Biosensors*, Studies in Systems, Decision and Control 327, https://doi.org/10.1007/978-981-15-9612-4_3

enough to address current and future unmet needs. To develop robust and truly integrated electrochemical sensors for multiplexed and ultrasensitive devices, a number of strategies have been developed and utilized.

This chapter discusses the importance of microfluidics and fabrication techniques associated with the development of electrochemical sensor devices. It looks at the various techniques implemented and how they may be applied to the fabrication of important components for sample transport, preparation, and analysis. It takes a specific look at polymer-based, paper-based, and printed circuit board (PCB)-based devices for this effort of developing truly integrated microsystems.

This chapter has further been divided into the three main routes for multiplexed device development including microfluidic microsystems, multi-electrode arrays, and the use of microbeads and labels. Microfluidic microsystems are discussed encompassing the use of fabricated and polymer-based systems as mentioned above. The Multi-electrode arrays section explores the use of sequencing and detection of nucleic acids using electrochemical devices alongside detection of proteins, whole cell pathogens, and small molecules. Finally, we discuss the important role of microbeads and labels for multiplexing. In this section we cover the many ways in which this approach can be implemented such as microbeads, enzyme labels, using multiple labels on single transducer surfaces (barcoding), and charged nanoparticles. Throughout this chapter, specific examples are detailed of devices and technologies that have been developed using these various approaches. Also discussed are some of the key commercial devices that have made it to market using these techniques and strategies.

2 Microfluidic Microsystems for Multiplexed Analysis

Microfluidics enables spatial separation of multiple biosensing reactions or analyte recognition sites, allowing detection of multiple analytes in a single system. A multitude of techniques have been developed for the fabrication of channels, chambers, mixers, and filters ranging from the macro- to the nanoscale. Using these techniques to fabricate such miniaturized structures allows scientists and engineers to develop integrated devices. Integrated devices can include sensing electrodes, fluidic channels, sample chambers, micromixers, pumps, and filters. All of these components can be fabricated using such materials as glass, silicone, metals, quartz, hydrogels, paper alongside a multitude of polymers [1]. All built into one, this allows for the realization of a truly portable 'lab-on-chip' (LoC) device. The first LoC device was created by S.C. Terry at Stanford University in 1979 for gas chromatography analysis [2]. Lab-on-Chip devices have now gained much interest since the early 1990s for the integration of multiple laboratory-based procedures onto miniaturized platforms [3].

Several fabrication techniques for sensing microsystems have been established. Some techniques for construction of signal transduction elements, e.g., electrodes include physical vapor deposition (PVD) (i.e., e-beam, resistive thermal evaporation and sputtering), etching, inkjet printing, screen-printing and electrodeposition. Each approach comes with a multitude of advantages for various applications, some of which are described throughout this chapter. For fabrication of microfluidic components, e.g., channels, mixers, filters, etc. there exist various polymer-based techniques such as photolithography, soft lithography, PDMS casting/molding, superimposed acrylic layers, 3D resin printing, and injection molding.

2.1 Microfabricated Microsystems with Polymer-Based Microfluidics

Microfabrication of transducer electrodes in combination with polymer chemistry techniques is commonly used for construction of diagnostic microsystems. After significant growth of the field, it has become clear that microfabrication is vital as an enabler for the growth of already existing technologies [4]. One way in which microfabrication has enabled existing diagnostic approaches is the miniaturization of electrochemical sensors. This has enabled electrochemical sensing devices capable of rapid detection, reduced reagent cost, high sensitivity, and high-throughput capability [5].

For example, Hwang et al. utilized several microfabrication techniques for construction of electrodes and microchannels using polymer materials to fabricate an integrated biosensor chip detecting matrix metalloproteinases (MMPs) for cancer diagnostics [6]. Electrodes were fabricated by first spin coating positive photoresist onto glass wafers and patterned using photolithography. This was followed by separately evaporating Au and Pt using E-beam PVD and then wet etching to reveal the electrode designs (see Fig. 1d). The microchannels were then created through spin coating and photolithography of negative photoresist (SU-8) for chamber I, II, and III (see Fig. 1a). The final top layer consists of three gates and a pectinate micromixer (see Fig. 1b) fabricated by patterning PDMS over an SU-8 mold. This was then bonded to the chip through oxygen plasma treatment heated in an oven at 100 °C for an hour.

A similarly fabricated device by Lee et al. showed the high-detail extent to which these kinds of fabrication techniques can be used [7]. In the developed device, a three-layered PDMS microfluidic system was developed, superimposed over evaporated Au electrodes on a glass wafer. The bottom layer of the PDMS systems consisted of the microchannels and chambers for reagent transport and reactions. The middle layer of PDMS acted as a membrane layer and finally the top layer of PDMS hosted numerous pneumatic valves, three inlets and fourteen outlets (see Fig. 2).

A study in 2017 shows the simple use of two double-sided tape layers with various laser cut designs followed by a third laser cut acrylic polymer layer [8]. Figure 3 shows the design of a multiplexed electrode layer superimposed by a three-layer microfluidic system for simple sample flow over separated channel wells for each set of three-electrode sensors. First, a positive photoresist layer was photopatterned

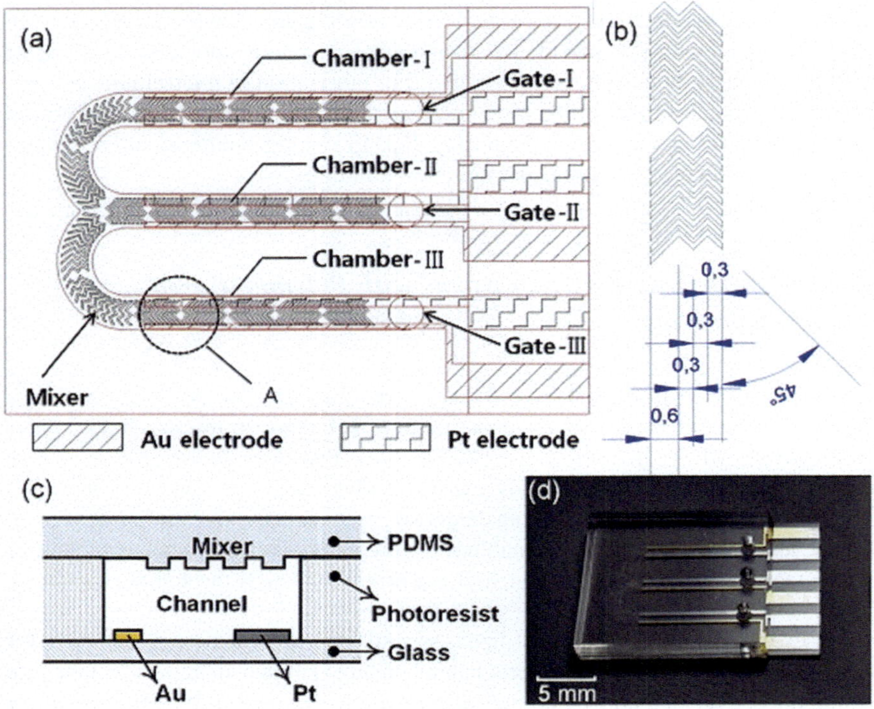

Fig. 1 Fabricated multiplex electrochemical biochip. Shown are the 2-D drawings of **a** the design **b** the chaotic passive mixer (in mm) and **c** a schematic sectional view of 'A'. Also shown is **d** a photograph of the fabricated microchip on glass substrate. Reprinted from [6], with permission from Elsevier

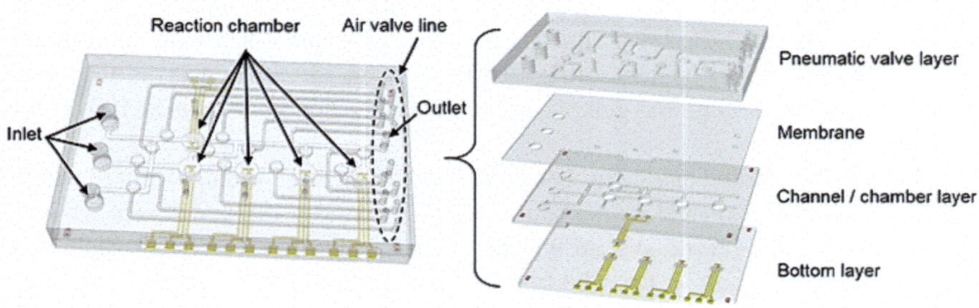

Fig. 2 Schematic illustration of the microfluidic system for electrochemical analysis in a single device. Shown above (left) is the schematic description of the microfluidic system integrated electrochemical sensor. Also shown (right) is an exploded model of the microfluidic electrochemical biosensor composed of a glass bottom layer with electrodes, PDMS channel/chamber layer, PDMS membrane, and PDMS pneumatic valve layer Reprinted from [7], with permission from Springer Nature

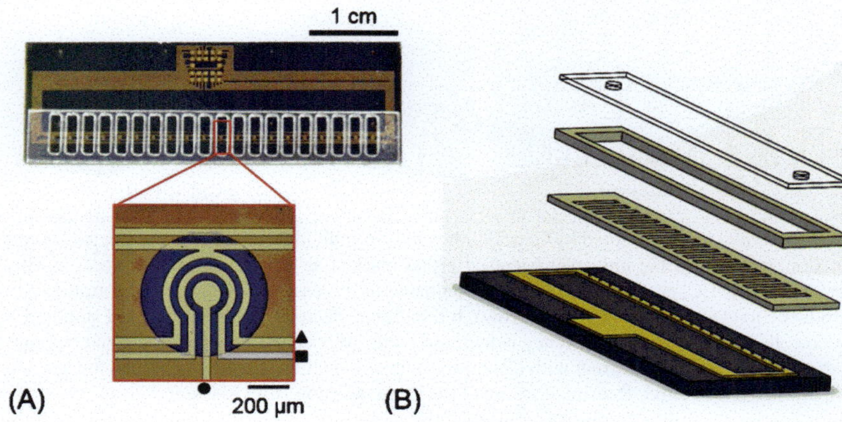

Fig. 3 Simplistic design of the microfluidic microsystem for detection of biological samples. Shown above is the photograph of **a** the device was used with an array of 20 sensors including a magnified image of an individual sensor with working electrode (scale bars indicate the length scale). Also shown is the exploded 3D model for **b** the schematic of the electrode device and the microfluidic channel design including assembly parts Reprinted from [8], with permission from Elsevier

onto a SiO_2 wafer, followed by thermal evaporation of Au and wet etching to create the electrode layer (see Fig. 3a). Second, two layers of double-sided tape were laser cut. The bottom layer consisting of a 20× array of separated wells to cover each set of electrodes, while the middle layer acted as a chamber spacer between the bottom and top layers. Finally, a plastic layer was laser cut with two holes at either end of the sensor array, acting as an inlet and outlet fit with microfluidic tubing for sample flow (see Fig. 3b). The developed sensor utilized DNA nucleic acids for detection of bladder cancer-specific DNA samples down to a sensitive detection limit of 250 fM within only 20 min. The developed sensor demonstrates the vast potential of simplistically designed microsystems for the detection of clinically relevant analytes.

An example of pathogen identification from whole blood with minimal sample pre-treatment and no amplification steps has been demonstrated by GeneFluidics Inc. Their detection chip was based on plastic substrate and sputtered gold electrodes. In total 16 sets of working, reference, and counter electrodes were presented on the chip where each working electrode can be functionalized by individual thiolated DNA capturing probes, see Fig. 4.

GeneFluidics are focusing on detection of urinary tract infections along with antibiotic susceptibility testing. Pathogen identification is achieved by pathogen-specific probes for 16S rRNA, immobilized on the chip. Sample is enzymatically lysed and hybridization of DNA probe to rRNA is facilitated using electrokinetic hybridization based on AC current generating localized heating. The assay is completed by a detection probe-HRP conjugate and incubation with TMB [9, 10]. The assay takes 50 min to complete and GeneFluidics have used it to commercialize

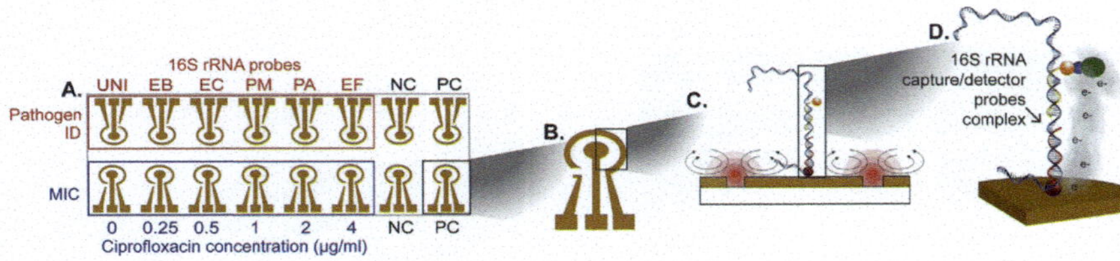

Fig. 4 **a** The microchip presented by GeneFluidics Inc. 16 individually addressable electrodes can be functionalized (top row; universal bacterial probe marked as UNI, Enterobacteriaceae is EB, Escherichia coli is EC, etc. while sample pre-incubated in increasing antibiotic concentration can be introduced to electrodes in the bottom row). **b** Individual electrode set comprising of standard 3 electrode cell. **c** Hybridization efficiency is increased with AC current mediated localized heating. **d** A sandwich ELISA assay format with DNA probes is utilized to bring HRP closer to the surface for amperometric detection. Reprinted from [11], with permission from Elsevier

urinary tract infection diagnostic system UtiMax™. However, to perform an antibiotic susceptibility test, a sample first needs to be incubated in increasing antibiotic concentrations before analysis. 2.5 h incubation is sufficient, but this increases the total analysis time. One of the advantages of 16S rRNA detection is that multiple copies of rRNA are present in a single pathogen, hence amplification steps are not required when a sensitive sensor is utilized [11].

More examples of microfluidic devices fabricated using lithographic technologies are presented in Table 1.

Although such devices can be fabricated with high precision, the manufacturing process requires cleanrooms adding to the cost of the final LoC device. Furthermore, multi-step assays are usually based on pressure or vacuum-driven microfluidic flow requiring a pump for the chip to be operational. This increases the cost of the instrumentation needed, which makes such devices best suitable for applications where accuracy and precision are the main requirements. On the other hand, low-cost devices with the need for minimal or no external instrumentation are best utilized using paper-based systems, which are presented in the next section.

2.2 *Paper-Based Systems*

Since its invention, paper has provided humanity with a simple, multipurpose substrate that has proven useful in countless applications. In sensing, paper is most well-known for its use in lateral flow pregnancy tests; however, it has also been particularly helpful in the fabrication of microfluidic paper-based analytical devices (μPADs)—microfluidic sensors that substitute traditional substrates (e.g. plastic, glass) for paper alternatives. Paper offers several significant advantages given its low cost, lightweight nature, flexibility, biodegradability, and chemical compatibility. Most importantly, the fibrous nature of paper allows liquid to flow via capillary action, with individual fibers acting as capillaries, without the need for external driving

Table 1 Examples of electrochemical biosensor microsystems utilizing microfabrication techniques and various polymer materials

Electrode type	Microfluidic components	Analyte	Bio-recognition element	Detection technique	References
Au and Ag thermally evaporated electrodes	PDMS microfluidic layer	Cocaine, ATP	Aptamers	CV, eSPR and CC	[12]
Au microneedles sputtered after MRDL on flexi-PCB	In-house fabricated multi-channel portable electrochemical analyzer	Glucose, uric acid & cholesterol	GO_x, uricase & CHOx	CV & CA	[13]
Inserted Ag/AgCl macro electrodes	PDMS capture chamber & two sensing channels using double-layer SU-8 mold	Human ferritin, mouse anti-rabbit IgG	Antibody-functionalized polystyrene NPs	Resistive pulse sensing	[14]
Au & Pt electrodes using E-beam evaporation and photolithography patterning	One layer of negative photoresist, superimposed by PDMS layer	Matrix metalloprotease (MMP) peptides	Peptides	CV & EIS	[6]
Au & Ag electrodes using E-beam and stencil lithography utilizing shadow mask	Three-layer PDMS microfluidic layers containing; channels, chambers, membranes, and pneumatic valves	Fibrinogen, adiponectin, low-density lipoprotein, 8-isoprostane	Antibodies	CV & SWV	[7]

forces [15]. The elimination of an external pump not only saves space and energy, but also makes μPADs more portable and easier to implement in non-laboratory environments.

The basic μPAD design consists of hydrophilic channels that are isolated through either precision cutting of the paper or the creation of hydrophobic boundaries. Hydrophobic regions are created through various methods including photolithography, inkjet printing, wax printing, and screen printing [16]. There are several types of μPADs, classified by the detection technique they use; here, we focus specifically on electrochemical paper-based analytical devices (ePADs)—devices that employ electrochemical techniques to detect and quantify analytes. Electrochemistry offers several advantages in microfluidic systems as electrodes are easily miniaturized, potentiostats can be small and portable, and previously developed sensors have shown both high sensitivity and selectivity [17].

In recent years, there has been increased interest in the development of ePADs capable of simultaneously detecting multiple analytes. These multiplexed sensors offer a new level of complexity, without necessarily increasing the device size or cost. Several methods have been employed to construct multiplexed ePADs, which typically depend on both the nature of the analyte/biomarker and the number of electrodes (one or multiple) in the device [18].

When analytes are not redox-active, or their redox activity is irrelevant to the mechanism of detection, steps must be taken to ensure the analyte is detectable at the electrode interface. This can be achieved through the addition of biorecognition elements (e.g. antibodies, aptamers, enzymes) to the electrode. Wang et al. reported an ePAD for the simultaneous detection of two cancer biomarkers—carcinoembryonic antigen (CEA) and neuron-specific enolase (NSE)—using nanocomposite gold nanoparticle-bound aptamers on separate working electrodes [19]. With the aid of the nanocomposite particles to enhance electron transfer, the ePAD achieved limits of detection (LODs) of 2 pg/mL (CEA) and 10 pg/mL (NSE). To achieve multiplexed detection without the need for multiple working electrodes, it is also possible to add distinguishable labels so that one electrode is sufficient to measure multiple analytes. Li et al. developed a foldable origami ePAD with a nanoporous silver (NPS)-modified electrode for the simultaneous detection of CEA and alpha-fetoprotein (AFP) [20]. By printing the working electrode and reference/counter electrodes on separate sides of paper, the device was easily folded to bring all electrodes together for sampling (Fig. 5). As the biorecognition element, antibodies specific to CEA and AFP were bound to the electrode. After exposure to the analytes, CEA and AFP were sandwiched with a second antibody bound to nanoporous gold-chitosan, complexed with either Cu^{2+} or Pb^{2+}, respectively. Given the disparity in redox potentials between these ions, voltammetry was able to measure both simultaneously and obtain LODs of 0.06 pg/mL (CEA) and 0.08 pg/mL (AFP).

When the analytes of interest are redox-active, and have sufficiently different oxidation/reduction potentials, then multiplexed detection can be carried out like the above Cu^{2+}/Pb^{2+} label example. However, analytes with similar redox potentials, and thus overlapping curves, require either additional electrode modification, to alter the kinetics of the redox reaction, or separate quantification on different electrodes

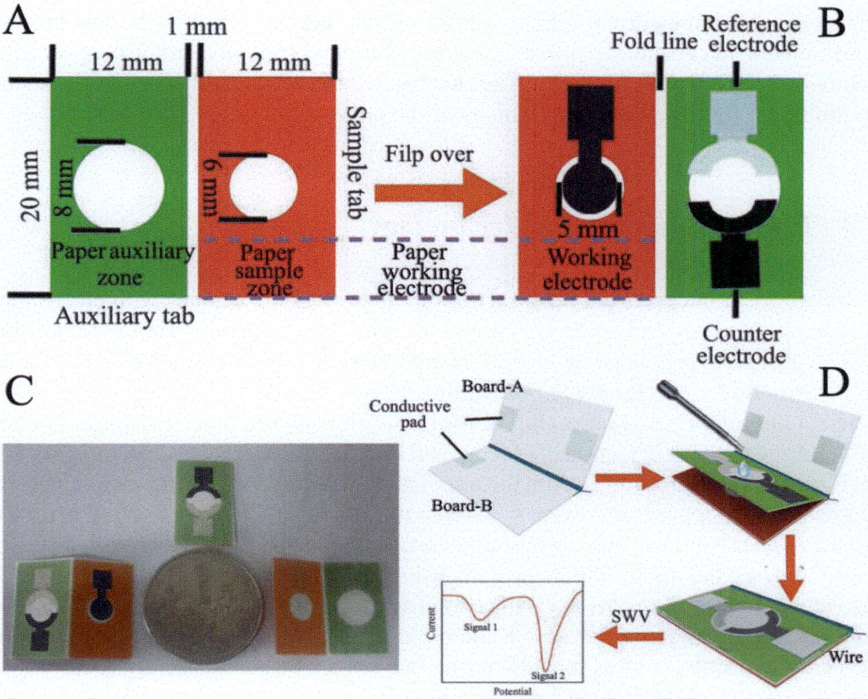

Fig. 5 Foldable ePAD constructed using origami paper showing **a** the back of the sample and auxiliary tabs of the device **b** the front of the tabs with the working electrode (sample tab) and reference and counter electrodes (auxiliary tab) **c** several views of the device relative to a yuan coin **d** the insertion of the device into a transparent device holder, with subsequent sample addition and electrochemical measurement. Reprinted from [20], with permission from the Royal Society of Chemistry

[18]. Separate working electrodes can also be used to maintain unique chemical environments that are optimal for detection of specific analytes. Janus ePADs, aptly named after the two-faced roman god of duality, store differing reagents in paper near the working electrodes to tailor the pH for each target [21]. Nantaphol et al. describe a Janus ePAD for the simultaneous detection of two neurotransmitters, norepinephrine (NE) and serotonin (5-HT), whose electrochemical behavior is ideal under different pH conditions. By impregnating the paper with H_3PO_4 and NaOH, the solution reaching the working electrodes measured pH 6 and pH 8, respectively. Under their optimal pH, NE and 5-HT were measured with LODs of 0.71 μM and 0.38 μM, respectively.

While there are many benefits to using paper in place of other materials, there are still challenges that have yet to be overcome. The largest issue with μPADs is the lack of flow control within the capillary-driven system; the wicking of water into all available fibers can disrupt the desired flow path and lead to inconsistencies between trials [15]. Also, the current scarcity of out-of-lab, real-world μPAD studies makes development and eventual commercialization difficult [16]. An alternative approach to low-cost LoC devices for multiplexed analysis is the use of readily available printed circuit board (PCB) technology, which will be presented next.

2.3 Printed Circuit Board Technology

In the effort to integrate the necessary components of LoC devices with electronic readers, printed circuit boards are commonly used as a chip-to-world interfaces. In 1997, Jobst et al. presented an idea to incorporate LoC components within the PCB by constructing a PCB interface, with outlined microchannel and a counter electrode, where a microfabricated glass chip bearing working and reference electrode can be inserted and used for detection of glucose and lactate [22]. Merkel et al. demonstrated PCB integrated microfluidics with the use of photopatterned copper traces as channel borders [23]. Copper traces were covered with thin layer (4 μm) of epoxy resin to isolate copper from the fluid, preventing corrosion and enable bonding with a cover board, closing the microfluidic channels using a hot press. Using this strategy they also report heating elements that can be used as temperature sensors and microfluidic valves and actuators using flexible Kapton® foil. The advantages of this approach lie in the accessibility of PCB manufacturing, which has been extensively used in the last decades to fulfil the need for booming consumer electronics market. Materials used in PCB construction have excellent thermal and mechanical stability, excellent dielectric properties, and are resilient to organic solvents and acid/base solutions, with available recycling strategies already standardized and established.

One strategy for multiplexed detection was demonstrated by Kling et al. where a dry photoresist was used to fabricate a device with microfluidic channels and platinum electrodes were evaporated by physical vapor deposition [24]. Microchannels were pre-treated with anti-fluorescein antibodies so any DNA sequence labeled with fluorescein can be used as a capturing probe. Detection of tetracycline and pristinamycin was achieved using tetO or Pir operator fluorescein tagged DNA sequences, specific to TetR or PIP repressor proteins, which were biotinylated to enable labeling with streptavidin-glucose oxidase (GO_x) conjugate. Eight separate channels were constructed and although reference and counter electrode were short circuited to connect in two individual pads, it was the integration of microfluidics that enabled multiplexed detection, see Fig. 6.

Panneer Selvam and Prasad demonstrated a label free multiplexed detection of procalcitonin, lipoteichoic acid and lipopolysaccharide on PCB electrodes using non-faradaic electrochemical impedance spectroscopy [25]. The PCB electrodes were covered with microporous membrane and encapsulated with PDMS to achieve a nanoconfinement stimulating macromolecular crowding, contributing to increased

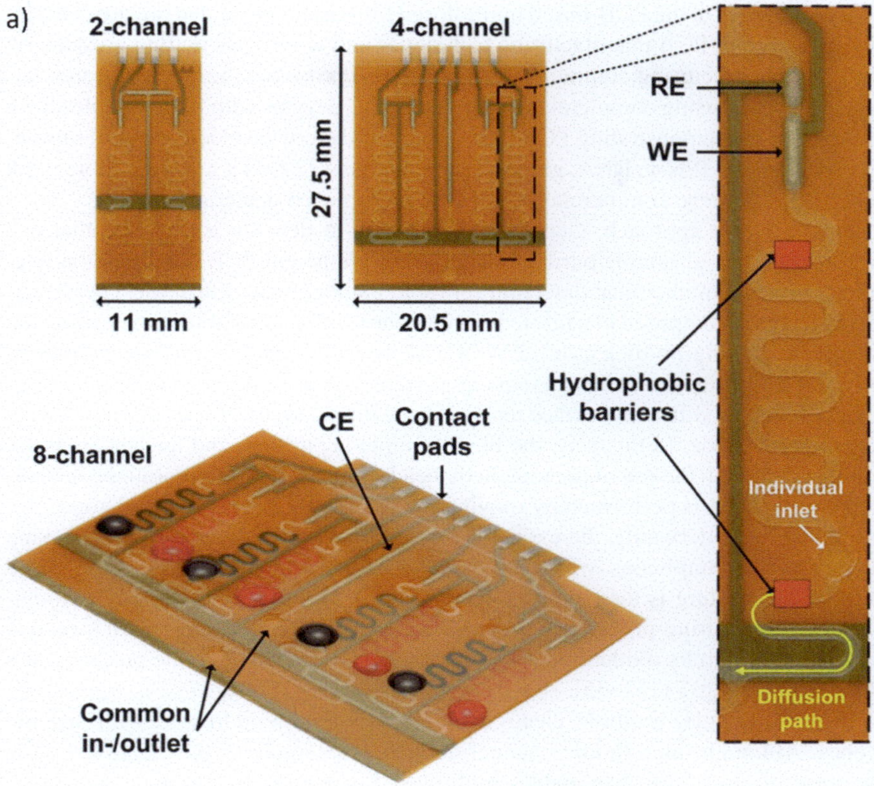

Fig. 6 Tetracycline and pristinamycin detection based on dry photoresist technology. 2-, 4-, and 8-channel system is presented with 8 individual measuring chambers incorporating reference and working electrode, while counter electrode is positioned downstream form the measuring chamber. Reprinted from [24], with permission from American Chemical Society

binding affinity of the analytes. Spiked human serum samples were tested and clinically relevant dynamic ranges were achieved. PCBs have also been used for detection of multiple mRNAs [26], but in a microarray-based approach, described in the following section.

The potential of integrating all necessary lab-on-chip components in a PCB gave rise to the Lab-on-PCB approach [27, 28]. Overall, PCB technology offers extensive commercial opportunities. The materials used are inexpensive leading to potentially extremely low-cost devices, especially when manufactured at scale. As the manufacturing facilities are widely available, product upscaling can be performed with ease and at minimal costs. LoC components can be standardized, in a similar way compared to electronics components PCB footprints, to promote system-level integration.

To date, most of the PCB-based multiplexed biosensors presented are not compatible with current PCB manufacturing techniques and workflows putting pressure on the PCB manufacturing community to uptake new processes. Limited research has been performed using commercially fabricated LoC devices with few examples from Moschou et al. demonstrating PCB based Ag/AgCl reference electrodes and biosensors for cytokine IFN-γ, DNA, glucose, and pathogens [29–31]; however, more work is needed to achieve commercial PCB based multiplexed devices.

In a different approach, GenMark used PCBs to develop a cartridge that can perform automated sample purification, concentration, and PCR amplification integrated with electrochemical detection for rapid pathogen identification. The single-use cartridge is composed of a printed circuit board with hydrophobic cover, allowing for electrowetting-on-dielectric (EWOD) technology for sample pre-treatment. Droplets can be moved in 2D by sequential voltage application in neighboring electrodes, which modifies the surface tension in the drop and the droplet aligns with the activated electrode. In this way, the blood sample is purified, and nucleic acids are concentrated with the use of paramagnetic beads, magnetically immobilized in dedicated areas. PCR is performed by moving the droplet through the heated zones, for amplification. Afterwards, the exonuclease digestion of DNA is performed leaving single-stranded amplicons, which are then hybridized with a ferrocene (Fc) labeled probe. This mixture is then exposed to PCB electrodes with individually immobilized DNA capturing probes and when a target is recognized, a potential sweep is performed and an Fc oxidation peak can be observed, identifying the presence of a pathogen (see Fig. 7) [32–35].

In this example, multiple electrodes are individually addressed with specific probes. Although microfluidic channel delivers the sample to the electrodes, it is the array of electrodes that enables multiplexed detection. Electrode arrays and it uses will be covered in more detail in the next section.

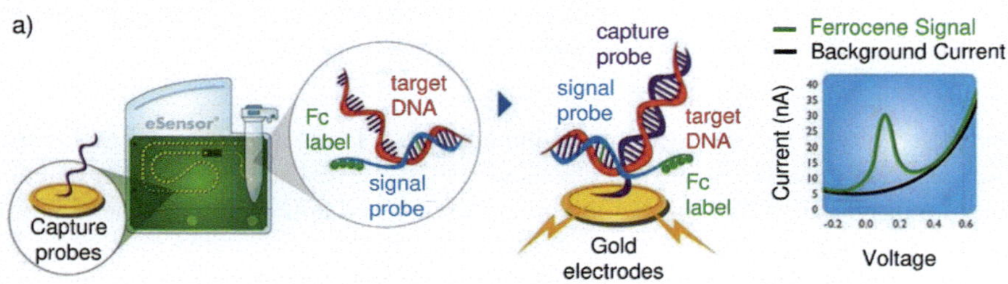

Fig. 7 **a** GenMark's first version of eSensor technology, employing PCB gold electrode with immobilized capture DNA binding to target DNA, which is labeled with ferrocene conjugated signal DNA strand enabling electrochemical detection. Figure modified from [35], with permission from John Wiley and Sons

3 Electrode Arrays

Microarrays on planar surfaces are a popular method for multiplexed analysis enabled by precise inkjet printing of microspots with volumes as low as picoliters [36]. Microarrays are mostly used in combination with optical detection methods, but here we will focus solely on electrochemical approaches. This includes arrays of individually addressable electrodes as well as arrays of ion-sensitive field-effect transistors (ISFETs) and nanopores used for electrochemical sequencing of nucleic acids.

3.1 Electrochemical Sequencing and Detection of Nucleic Acids

Nucleic acids comprise an important group of macromolecules that are highly specific, ubiquitous to all life, and can be amplified to detectable quantities. For these reasons, nucleic acids are often the target of sensors, both for the sequencing of genetic material and the identification of known sequences. Given the immense diversity within many nucleic acid samples, multiplexed devices that can simultaneously measure multiple sequences are of great interest to scientists targeting DNA and RNA. To achieve this level of detection, devices commonly employ arrays—networks of repeating units, either identical or variable—which can provide thousands of individual recognition sites. Some current array technologies include microwell- and nanopore-based sequencing, as well as diagnostics that employ DNA/RNA microarrays [37]. Though these methods differ in their approach to nucleic acid analysis, all allow for the simultaneous measurement of complex mixtures of sequences.

Since the development of Sanger sequencing in 1977, there have been substantial efforts to improve the speed and ease with which genetic code is sequenced. Next-generation sequencing (NGS) methods aim to process genetic material in a massively parallel manner, allowing thousands to billions of sequences to be read simultaneously [38]. While there are many types of NGS available, we will focus on two specific methods employing electrochemical detection. Ion semiconductor and nanopore sequencing both provide rapid, simultaneous sequencing, while also eliminating the optical component found in many other NGS approaches [39].

Ion semiconductor sequencing is a relatively new and rapid method of DNA sequencing that exploits natural nucleotide chemistry to identify DNA bases. This technology, first released as the Ion Torrent™ line by Life Technologies and later acquired by ThermoFisher, consists of millions of microwells, each of which contains an ion-sensitive field-effect transistor, embedded in a semiconductor chip (Fig. 8) [40]. In preparation for sequencing, carrier beads, each bound with clones of a single template, are deposited in the microwells such that each well contains a single bead. Sequencing is achieved through the repeated sequential addition of the four nucleotides (A, C, G, T); when the next complementary nucleotide is added, DNA

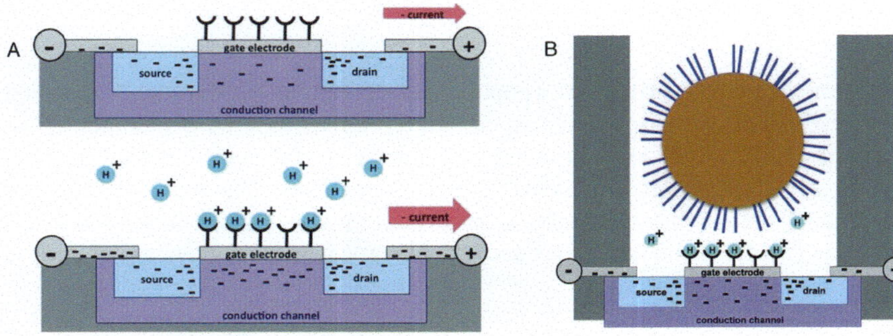

Fig. 8 Ion Torrent sequencing setup showing **a** the field-effect transistor-based sensor with electrons (-) flowing from source to drain through the conduction channel (top) and increased electron flow upon collection of hydrogen ions in proximity of the gate. **b** A carrier bead-bound with template clones inside of a well. Reprinted from, with permission from John Wiley and Sons

polymerase catalyzes the addition, and hydrogen ions (H^+) are released as a byproduct of the reaction. This decreases the pH in the well which is detected by the ISFET in real-time. If consecutive identical bases are present in the sequence, the pH lowers accordingly. Sequential flow of reagents with different nucleotides causes proportional changes in the electrical signal. As of 2020, the Ion GeneStudio™ S5 Prime System is capable of sequencing 100–130 million reads (200 bp per read) in just 8.5 h [41].

A different approach is based on nanopore sequencing, an NGS technique that determines nucleotide identity as DNA or RNA is transported through pores in a membrane. Nanopore sensors consist of large arrays of nanopores, inserted into membranes either through the incorporation of transmembrane proteins (biological pores) [42] or, more recently, the fabrication of synthetic nanopores (solid-state nanopores) [43]. MspA and α-hemolysin are two bacterial pore-forming proteins commonly used to create biological nanopores, owing partly to their optimal channel diameters [42]. Figure 10 demonstrates how a single strand of DNA, moving from the cis side of the membrane to the trans side, leads to changes in current across an MspA pore. These alterations in current are unique for each respective nucleotide, and thus allow for accurate sequencing of the entire strand. The pore depicted in Fig. 9 also contains an assistive enzyme, whose helicase activity helps to both separate the DNA strand and control the speed of strand passage. Oxford Nanopore's MinION™, released in 2014, is one example of a successful commercial nanopore-based sequencer [43]. As of 2020, the MinION™ Mk1C can read up to 30 Gbp in less than 48 h with a single flow cell [44].

When arrays of electrodes can be individually addressed with specific probes, they can be used for detection of multiple analytes simultaneously. An example of a PCB-based electrode array was presented by Sánchez et al. who demonstrated the detection of seven breast cancer microRNA sequences using an asymmetric multiplex ligation-dependent probe amplification and electrochemical detection using an HRP

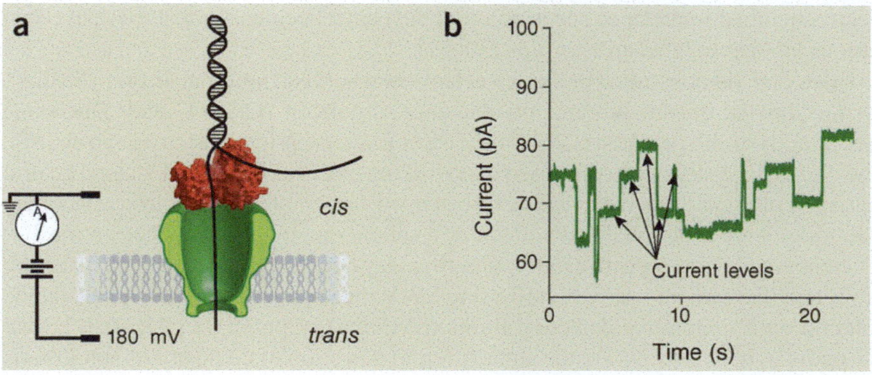

Fig. 9 Nanopore sequencing setup showing **a** a helicase enzyme (red) unwinding double-stranded DNA as the single strand passes through the MspA nanopore (green) in the membrane **b** changes in current that is observed across the nanopore as different nucleotides pass through the channel. Reprinted from [42], with permission from Springer Nature

label. The authors used precipitating TMB to detect hybridization leading to minimal cross-reactivity [26].

A similar setup was presented by GenScript® utilizing a CMOS chip named the CustomArray 12 K™, containing 12,000 individually addressable microelectrodes, each of which can capture a different oligonucleotide sequence simultaneously [45]. Once functionalized, these microarrays bind sample DNA or RNA and are processed with the ElectraSense® Reader—a device capable of electrochemically measuring up to 12,000 probes in less than 45 s. This device has been employed for numerous applications, including influenza A subtype analysis, where it exhibited 100% specificity and 95.2% sensitivity [46]. The use of electrode arrays is by far not limited to DNA and RNA based applications but can be used with varying analytes from proteins to pathogens and small molecules, described in the next section.

3.2 Detection of Proteins, Pathogens, and Small Molecules

Detection of multiple analytes can be performed in parallel with a multi-channel potentiostat or using sequential measurements. Due to the ability to detecting a large number of analytes, electrode arrays can be cost-effective in terms of cost per single test and also require low sample volumes for multiplexed detection due to the miniaturization achieved via microfabrication [47]. Rapid, high-throughput, reproducible, stable, and sensitive biosensors, exhibiting wide dynamic ranges have been demonstrated using multi-electrode arrays and is presented in this section [48–51].

Similar to a three-electrode cell, multiple electrode array systems can be used with various electrochemical detection techniques. Linear sweep voltammetry (LSV), differential pulse voltammetry (DPV), square wave voltammetry (SWV) as well as electrochemical impedance spectroscopy (EIS) have been used most extensively in various labeled or label-free set-ups [50–52].

Eissa et al. demonstrated multiplex detection of survival motor neutron 1 (SMN1), cystic fibrosis transmembrane conductance regulator (CFTR), and Duchenne muscular dystrophy protein (DMD) on a carbon screen-printed electrode array [53]. The electrode system was composed of 8 working electrodes, a round-shaped central Ag/AgCl reference electrode, and a ring-shaped carbon counter electrode. Working electrodes were functionalized with carbon nanofibers, to increase electron transfer efficiency, and with respective antibodies for assay construction. Label-free detection of protein biomarkers was achieved using SWV measurements in ferro/ferricyanide solution with impressive detection limits (SMN1: 0.74 pg/mL, CFTR: 0.9, DMD: 0.7 pg/mL). In another work, the authors report detection of dedicator of cytokinesis 8, phosphoglucomutase 3, and signal transducer and activator of transcription 3 proteins [54]. This time, AuNPs were used to improve surface properties, achieving both a higher surface area, and enhancing the electron transfer. Cysteamine self-assembled monolayers (SAM) was employed for a label-free detection using SWV with detection limits of 3.1 pg/mL, 2.2 pg/mL, and 3.5 pg/mL respectively. Label-free detection has multiple advantages over labeled techniques such as simplicity of the measurement steps needed to obtain the result. However, label-free techniques rely on the intrinsic properties of the analytes. In case of analytes with small molecular weight or zero net charge, these techniques are hard to implement. The same group presented a competitive multiplex immunosensor for the detection of three small metabolites, morphine, tetrahydrocannabinol, and benzoylecgonine, depicted in Fig. 10. After AuNP functionalization, cysteamine and glutaraldehyde linkers were used to link monoclonal antibodies (mAbs) to the surface. For competitive detection to occur, drug-BSA conjugates competed with unbound drugs to attach to the binding sites available on respective mAbs. Ultrasensitive detection limits were

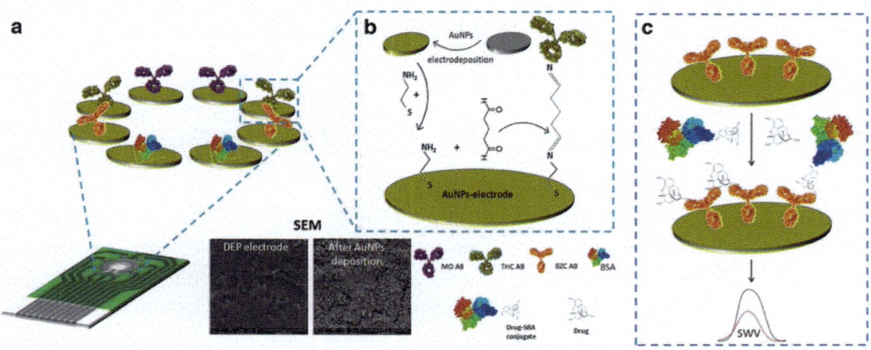

Fig. 10 **a** Competitive label-free multiplexed antibody-based immunosensor for the detection of morphine (MO), tetrahydrocannabinol (THC), and benzoylecgonine (BZC). **b** The working electrodes are functionalized with AuNPs by using electrodeposition. **c** During SWV detection, drug-BSA conjugates compete with unbound drugs for antibody binding sites within the competitive assay. Reprinted from [55], with permission from Springer Nature

achieved at 1.2 pg/mL, 7.0 pg/mL and 8.0 pg/mL, respectively. When using spiked urine sample to test for the three small molecules, recovery percentages averaged between 88 and 115%, demonstrating the potential to be used for drug detection in biological samples [55].

Another miniaturized electrode array for protein detection was demonstrated by Gupta et al. with a nanoelectrode 3×3 array device that could detect C-reactive protein, cardiac troponin-I, and myoglobin in a label-free set-up [56]. To ease the simultaneous detection of the cardiac biomarkers and to avoid the use of a complex microfluidic channel, a hydrophobic resist layer was coated and etched onto the electrodes, eliminating cross talk. Antibodies were attached to the carboxylated vertically aligned carbon nanofibers immobilized on the working electrode, using EDC/NHS chemistry and DPV was used for detection analysis. Tang et al., reported a sensitive electrochemical immunoassay array, integrated with a microfluidic system, for simultaneous detection of four prostate cancer biomarkers [57]. They described a 32-individually addressable microarray that could be multiplexed into a system with 256 working electrodes. To achieve high sensitivity, the detection antibodies were modified with magnetic nanoparticles and linked to HRP. DPV was utilized for detection in diluted calf serum solutions. The limit of detections reported for prostate-specific antigen, prostate-specific membrane antigen, interleukin-6 and platelet factor-4, diluted in calf serum, were 2, 0.15, 0.05 and 0.1 pg/mL, respectively. Overall, this is a high-throughput and sensitive approach that can provide results within one hour, at a low-cost.

In general, the use of signal amplification labels increases the possibility of cross talk, and electrodes need to be placed further apart or electrode antifouling surface chemistry needs to be carefully controlled. The beforementioned CustomArray 12 K™ has also been employed for electrochemical detection of pathogens and endotoxins [58]. *Yersinia pestis*, *Bacillus anthracis*, and the bacterial enterotoxin B were detected in this platform that was adopted for electrochemical EILSA with the total assay time of 3.5 h. The authors reported an avidin-biotin system utilizing tertiary labeling step creating scaffolds of HRP, further enhancing signal amplification.

An example of a commercially available multiplexed device is Abbott's i-STAT. The technology is based on microfabricated silicon chip incorporated in a plastic cartridge, needed for microfluidic sample delivery. Mass manufacturing of the silicon sensors enables reasonable cost of the final cartridge, which is prefilled with calibration solution to perform self-calibration before every individual test. Upon sample addition, the calibration solution is released, and the sensor chip is wetted and calibrated. Next, the sample is flown through the sensors array, which consists of multiple electrodes, functionalized with enzymes or ion-selective membranes. Potentiometric detection is performed for analytes such as sodium, potassium, chloride, and pH, where the concentration of the analyte changes the potential developed on the electrode. Since potentiometric sensors are temperature dependent, calibration is especially important for such analysis. In parallel to potentiometric detection, amperometric reading of glucose is based on glucose oxidase while Clark type electrode with gas permeable membrane is utilized for amperometric detection of oxygen. Hematocrit levels are determined using conductivity measurement between two electrodes, using alternating current passing the two-electrode cell [59]. The combination of multiple electrochemical detection techniques integrated into a single platform demonstrates the potential of electrochemical sensors for multiplexed detection of variable analytes. Although substantial multiplexing capability has been demonstrated by i-STAT, incorporation of immunoassays into the platform is not straightforward. Cardiac protein biomarkers assays for creatine kinase MB, cardiac troponin I, and B-type natriuretic peptide have been incorporated into i-STAT single use cartridges, but are currently only available as single biomarker tests, demonstrating the challenges in multiplexed analysis [60].

Another approach to further increase multiplexing capability is to capture multiple analytes on a single electrode and achieve multiplexing through labels with different properties. Such approaches will be reviewed next.

4 Label-Assisted Multiplexing Techniques

Many recent advances in the development of multiplexed system have been demonstrated with the use of different labels including enzymes, redox markers, magnetic beads, nanoparticles, or quantum dots [61–63]. These strategies are mostly used in immunoassays, where labels are conjugated to secondary antibodies, completing the sandwich assay [64]. This allows ultrasensitive detection, minimizing the effect of non-specific binding and bio-fouling [65].

This section discusses microbeads, enzyme labels, redox-active labeling molecules, and nanoparticles in more detail.

4.1 Use of Microbeads in Multiplexed Biosensing

Magnetic beads are regularly used for sample pre-processing, where they act as capturing agent which can be easily extracted from the sample by a magnet allowing analyte separation or enrichment. Immobilization of capturing probes on microbeads can reduce assay time and increase the dynamic range of the sensor, due to increased immobilization surface presented on the beads. Additionally, biological fouling on the sensing electrodes can be eliminated, as the sample does not have to get in contact with the sensing element [66].

One strategy for bead-based multiplexed electrochemical detection is separating microbeads-capturing probe conjugate into separate sections where detection is performed. This was demonstrated by Ko et al. who presented a biosensor for alpha-fetoprotein (AFP), carcinoembryonic antigen (CEA), and prostate-specific antigen (PSA) based on microbeads in a PDMS based microsystem on a glass surface

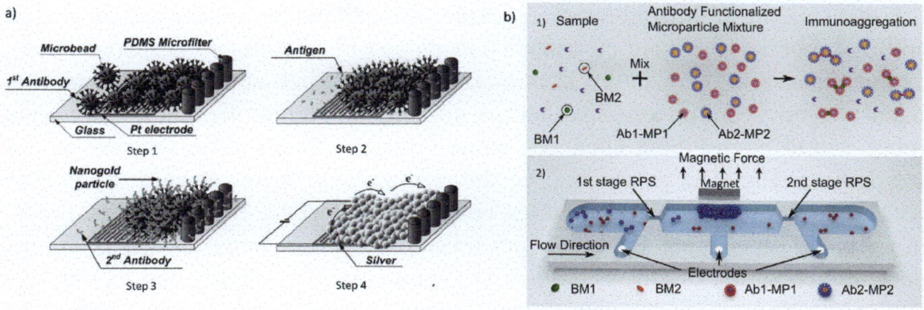

Fig. 11 **a** Microbead based detection strategy. Microbeads conjugated to mAbs are first flown through the channel and captured by micropillar filter, then sample is introduced followed by a AuNP-conjugated labeling mAb and silver enhancer. Reprinted from [67], with permission from John Wiley and Sons. **b** Microbead based resistive pulse sensor. In the first step sample is mixed with mAb-conjugated microbead mixture and the solution is flown through RPS. In the middle chamber, magnetic beads are captured while non-magnetic beads are counted again with a second RPS. Reprinted from [14], with permission from AIP Publishing

[67]. Microbeads were conjugated to specific antibodies and flown to the measurement chamber, which consisted of microfabricated interdigitated electrodes upstream of an array of PDMS micropillars, acting as a filter for capturing of microbeads (see Fig. 11a). Once the microbeads were in place, the sample was flown through the channel and antigen-antibody binding occurred, followed by gold-conjugated labeling antibody and a silver enhancer, which reduced to metallic silver in the presence of gold nanoparticles, catalyzing the reaction. Precipitated metallic silver then causes changes in resistivity between the interdigitated electrodes. Multiplexing was achieved by constructing multiple parallel sensing chambers and channels.

A second approach for multiplexed detection is based on the use of multiple beads with varied properties like size or shape. Han et al. demonstrated multiplexed detection of biomarkers exploiting immunoaggregation with magnetic microbeads [14]. The sensing mechanism is based on sub-micrometre pore and a resistive pulse sensor (RPS), measuring resistivity through the pore, which changes when an immunoaggregate passes through. Microbead size can be distinguished by the sensor, hence different sized microbeads can be used simultaneously for parallel multiplexed detection. An alternative approach is based on the versatile properties of the microbeads e.g. the use of magnetic and non-magnetic microbeads. In the first step, a sample is mixed with antibody functionalized microbead mixture, which induces immunoaggregation (see Fig. 11b). In a two-stage sensor, the authors demonstrated RPS sensor can quantify two biomarkers by obtaining a total concentration of both biomarkers, then the magnetic beads are removed with a magnet and non-magnetic beads are quantified again in a second RPS sensor. The concentration of biomarker targeted by magnetic beads can then be deduced. Detection of anti-rabbit IgG and human ferritin was demonstrated in this work but up to four biomarkers can be quantified by this device when utilizing microbeads versatile in size and magnetic properties.

A third and most utilized approach explores magnetic microbeads as carriers of capturing probes, which are then labeled using varied labelling strategies including enzymes, redox tags, quantum dots or metal ions. When such approach is integrated into microfluidic systems there is no need for immobilization of capturing probes within microfluidic channels, simplifying the final production process for such devices. There are two main procedures of transferring the signal from analyte recognition site (on the magnetic microbead) to the working electrode. First procedure is based on enzymatic 12 reactions taken place on the magnetic beads and the concentration of the produced product is then analyzed separately, by moving the solution from a reaction chamber to measurement chamber [68] or, as an alternative procedure, a magnet is used to guide magnetic beads to the electrode surface, where the signal is read [69].

In both procedures, there is clear need for labeling strategy, which can also determine the level of multiplexed detection. Various labeling strategies will be covered in the next section.

4.2 *Barcoding with Redox Probes*

A method of labeling that has been widely employed for multiplexing is barcoding. Barcoding is the concept of utilizing multiple labeling molecules on a single sensor electrode platform with a variety of probes for the purpose of multiplexing. This is done by labeling with various ligands that contain different electroactive signal properties. These properties may be redox-based, e.g., methylene blue and ferri/ferrocyanide, or they may give varying potentials upon anodic stripping of the electrode using stripping techniques such as in the case of dissolved metal-based probes [70]. By doing so, a simple setup can be fabricated in which multiple unique labels are attached to targets that bind to separate probes simultaneously giving signals that are easily distinguishable [71]. A distinct advantage of the barcoding technique is that signals may can be easily obtained during a single voltammetry/amperometric scan. However, it has been shown in some cases that if the electroactive potentials are similar 'cross-talk' may be observed [65].

The labeling of target molecules with redox-active species is a frequently used technique for detection using electrochemical biosensors, enabling ultrasensitive detection at varying redox potentials. Redox labels have the capability of providing increased sensitivity for detection due to excellent electron transfer ability that non-labeled alternatives may not provide [72]. For example, one study used labeled gold nanoparticles (AuNPs) with either ferrocene (Fc)-based or mercaptohexanol (MCH) spacing molecules co-immobilized with aptamers for prostate cancer detection to enable both impedimetric and amperometric signal acquisition [73]. Other examples of redox-active species used for labeling include Methylene Blue (MB) [74–76], erythrosine, anthraquinones/hydroquinones [77], hemin, thionene, and ruthenium [78, 79]. Redox-active species are those molecules that are capable of both donating and accepting electrons at specific potentials to provide reduction/oxidation (redox) signals. Different species undergo reduction and oxidation at significantly different potentials. This enables multiplex measurements of various targets in samples by labeling each target with different redox-active species.

A study in 2015 utilized two redox-active labels, hemin, and ruthenium (Ru), for multiplex detection of both alpha-fetoprotein (AFP) and carcinoembryonic antigen (CEA) tumor markers within serum samples [78]. Antigen-specific antibodies were first immobilized onto glassy carbon electrodes. The functionalized sensor was then exposed to antigen samples and labels to form a sandwich-type immunoassay with either hemin or Ru labeled antibodies. Both redox-active species were also attached to nanotags for CV and electrogenerated chemiluminescence (ECL) detection techniques (see Fig. 12). Utilizing this assay, the researchers were able to reach ultra-sensitive LODs of 1 and 0.5 pg mL^{-1} within serum samples for AFP and CEA respectively.

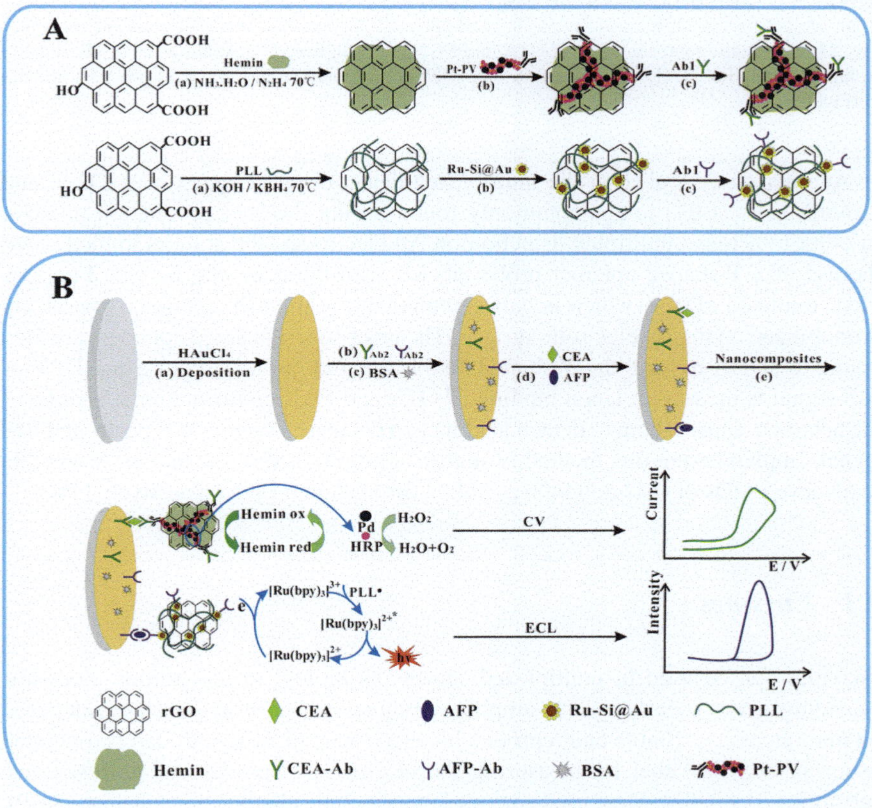

Fig. 12 Schematic representation of; **a** the preparation protocol for CV and ECL nanotags. **b** Immunosensor fabrication process and the signal generation mechanism. Reprinted from [78], with permission from Elsevier

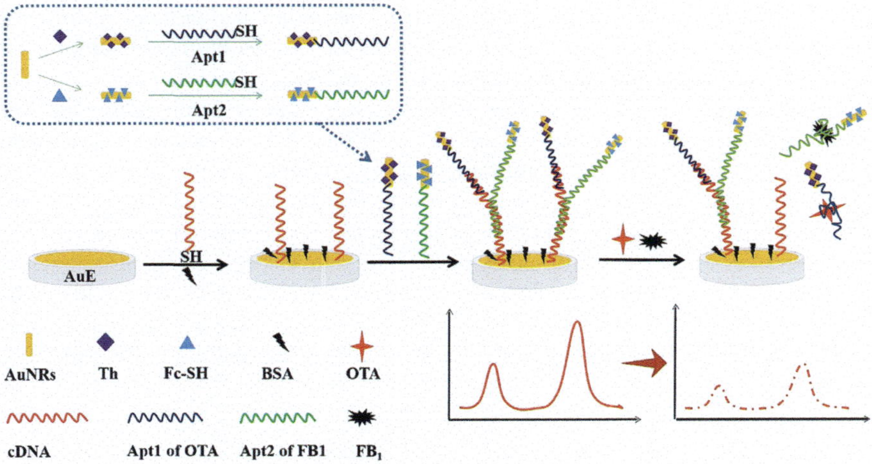

Fig. 13 Schematic representation of OTA and FB1 detection based on Aptamer 2-AuNRs-Fc and Aptamer 1-AuNRs-Th/cDNA/AuE assay using DPV. Reprinted from [79], with permission from Springer Nature

Another study looked at the multiplexed detection of ochratoxin A (OTA) and fumonisin B_1 (FB1) toxins commonly found during the beer production process. In order to achieve multiplex detection on Au electrodes, the authors implemented the use of a Y-shaped aptamer probe labeled with thionene and Fc (see Fig. 13). First, thiolated cDNA, which is half complementary to each aptamer, is passively immobilized on the Au electrode surface. The label-functionalized aptamers are then bound to cDNA to form the Y-shaped double aptamer probe structure and DPV or EIS signal is measured. Upon binding of OTA and FB1, conformational change of the aptamers causes them to detach leading to a decrease in both DPV peaks and EIS signal. Implementing this method of multilabel multiplexing the authors were able to achieve LODs of 0.47 and 0.26 pg mL^{-1} for OTA and FB1 respectively [79].

4.3 *Enzymes*

Enzymes can be used in multiplexed sensors as biological recognition elements, providing sensor specificity or to amplify the signal in an affinity-based sensor based on immunoassays. Stable and reproducible electrochemical signals can be obtained using an enzyme label by measuring the electroactive product of its respective substrate [70, 80–82]. Most used enzyme labels include alkaline phosphatase (ALP), glucose oxidase (GO_x), and horseradish peroxidase (HRP). The use of enzyme-labels can require the presence of a mediator to aid with the electron transfer during the enzymatic reaction [63]. For instance, the oxidation of hydroquinone is catalyzed by HRP in the presence of hydrogen peroxide (H_2O_2), a substrate [63, 83].

When an analyte of interest can be catalyzed with an enzyme, multiplexed detection can be achieved by immobilization of enzyme to an individual electrode in a multiple electrode array set-up. This was demonstrated by Kucherenko et al. who reported the detection of six analytes with detection limits of 1 μM for glutamate, 1 μM for glucose, 2 μM for choline, 3 μM for acetylcholine, 2 μM for lactate, and 5 μM for pyruvate [84]. Such sensors have been successfully used as point-of-care (POC) diagnostics. An example is a commercially available device from Siemens named ePOC system. Amperometric detection of glucose, lactate, and creatinine is demonstrated using multi-level sensor construction. In a lactate sensor, gold electrode is covered with a layer including lactate oxidase, HRP, and ABTS substrate (2, 2′-azino-bis3-ethylbenzothiazoline-6-sulfonic acid diammonium salt). A diffusion barrier layer encapsulates the reagents and when lactate is present, its oxidation produces hydrogen peroxide, which enables the HRP modulated reaction. The creatinine sensor is based on three-layer multi-enzymatic reaction including creatinine amidohydrolase catalysis of creatinine to creatine, creatine amidinohydrolase mediated hydrolysis of creatine to sarcosine and urea and finally sarcosine oxidase needed to produce hydrogen peroxide which enables HRP mediated reaction [85].

The use of enzymatic labels in electrochemical immunoassays has also been demonstrated commercially (e.g. i-STAT), however, these usually rely on individually addressable electrodes. Jia et al. presented a strategy where two capture antibodies can be immobilized on a single electrode and multiplexing is achieved with a label [86]. Secondary antibodies for carcinoembryonic antigen and alpha-fetoprotein were labeled either with thionine or ferocenne and conjugated to a graphene oxide nanosheet which also carried platinum nanoparticles, GO_x and HRP. Thionine or ferocenne acted as electron mediators yielding a separate voltammetric peak, enabling multiplexed detection of the sandwich immunoassay. Using square wave voltammetry (SWV), two distinguishable redox peaks were observed at −0.15 V for thionine and +0.35 V for ferocenne. The detection limit was 1.33 pg/mL and 1.64 pg/mL, respectively.

Combining different assay configurations such as enzymatic sensors and affinity sensors in a single platform for simultaneous detection was demonstrated by Vargas et al., who devised a dual-marker biosensor for diabetic markers glucose and insulin [87]. Glucose sensor was based on glucose oxidase and tetrathiafulvalene mediator incorporated into a chitosan biopolymer layer. On a separate electrode, a sandwich immunoassay was constructed using specific anti-insulin capture antibodies and HRP labeled detection antibodies. Both sensors employed amperometric detection, and due to different modes of operation, glucose was quantified while the insulin sensor was exposed to the sample, hence the labeling step didn't interfere with the glucose measurements. Total analysis time was under 30 min, with a sample volume of 10 μL.

Overall advantages of operating multiplexed electrochemical systems with enzyme labels include higher biocatalytic activity with increased specificity and sensitivity [88, 89]. However, enzymes are known to lose their activity over time and are usually required to be kept at low temperature to maintain functionality. Thus, other labeling strategies will be discussed further in this section.

4.4 Nanoparticles

Nanoparticles are any type of particulate matter that has an individual diameter in the nanometer range (typically 1–100 nm). Given their extremely small size, diverse compositions, and ability to be functionalized with a wide range of molecules, nanoparticles have led to significant advances in CO_2 capture, drug delivery, and disease diagnostics [90]. Nanoparticle-based labeling has also provided several unique methods of achieving multiplexed molecular detection—some of which show great potential for developing comprehensive diagnostic tests in the future. The two main types of nanoparticles commonly employed as multiplex labels include quantum dots and metal nanoparticles [91]. Quantum dots are among the smallest of nanoparticles, with diameters as low as 2 nm, and are composed of semiconducting materials (e.g. ZnO). Because of their size, quantum dots display unique optical and electrochemical properties that make them well-suited for distinctive labeling. Metal nanoparticles are typically larger than quantum dots and composed of a pure metal or compound. Some commonly used metal nanoparticles include silver and gold nanoparticles, which have been employed in a variety of applications (Figs. 14 and 15).

While quantum dots are often exploited for their unique optical properties, they can also be used for their electrochemical characteristics. Since quantum dots are composed of a wide range of compounds that have significantly different redox potentials, different quantum dot labels can enable multiplexed detection using a single working electrode [91]. Vijian et al. developed a genosensor for the multiplexed detection of pathogen RNA that utilized metal sulfide quantum dots as unique electrochemical labels for SWV (Fig. 16) [92]. Using PbS, CdS, and ZnS particles conjugated with DNA reporter probes, the device detected RNA for *V. cholerae*, *Salmonella* sp., and *Shigella* sp. with LODs of 51 aM, 53 aM, and 38 aM, respectively. Similarly, Kong et al. employed CdS and PbS quantum dots coupled to secondary antibodies to simultaneously detect both CEA and AFP biomarkers with LODs of 3.3 pg/mL and 7.8 pg/mL, respectively [93].

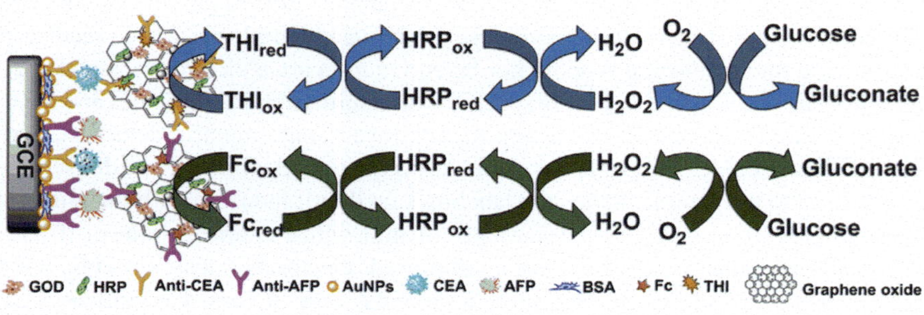

Fig. 14 AuNPs modified working electrode is immobilized with the capture antibodies. Graphene oxide nanosheet equipped with labeling antibodies, thionine or ferocenne electron mediators, platinum nanoparticles and HRP with GO_x. Reprinted from [86], with permission from Elsevier

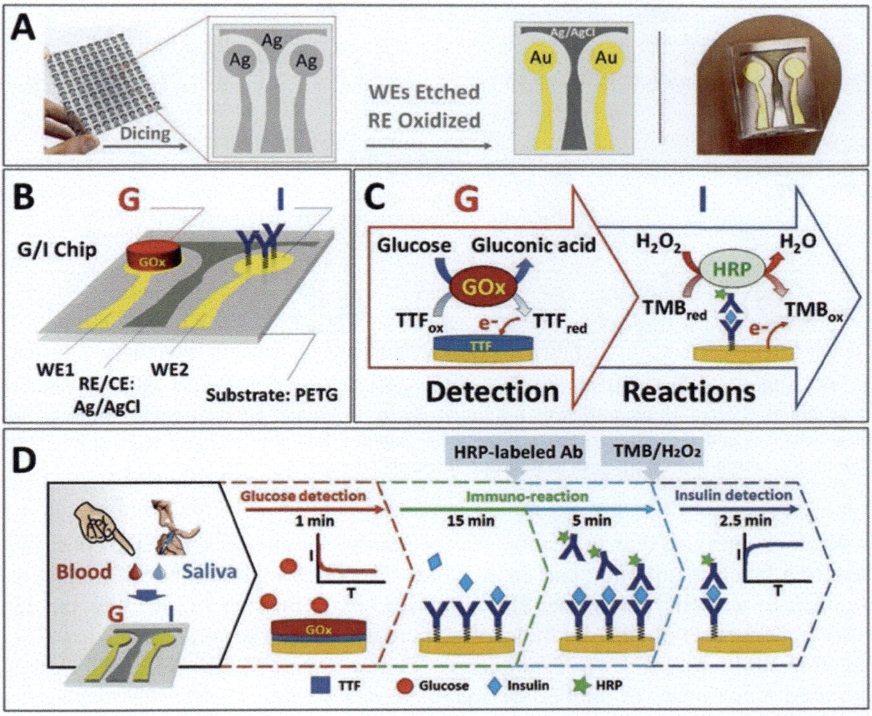

Fig. 15 **a** Array fabrication on the plastic PETG substrate in the form of a two-electrode cell with two working electrodes. **b** Illustration of a single biosensor chip in which the glucose biosensor involves the formation of chitosan film immobilized with GO_x, whereas the insulin sensor involves the use of an insulin capture antibody. **c** Schematic of the amperometric detection principle of glucose and insulin sensors. **d** Detection steps required for the measurement of glucose and insulin via whole blood and human saliva samples. Reprinted from [87], with permission from John Wiley and Sons

Metal nanoparticles have been used to achieve multiplexed detection in several different manners. Wan et al. employed Cu, Ag, and Pd nanoparticles conjugated with antibodies and aptamers to simultaneously detect three different cancer cell lines (MDA-MB-231, SK-BR-3, VCaP) with an LOD of 2 cells per sensor [94]. Since the metals had adequately separated redox peaks, linear sweep voltammetry was able to resolve all three nanoparticles on a single electrode. In addition to redox labeling, metal nanoparticles may also function as catalytic labels, enabling the production of redox-active material. The sensor published by Lai et al. utilized secondary antibody-conjugated gold nanoparticles as labels to detect CEA and AFP biomarkers [51]. After labeling, a silver enhancer solution was added to the electrodes, which enabled the gold-catalyzed deposition of silver nanoparticles onto the sensor. Subsequent anodic stripping analysis quantified the amount of deposited silver with LODs of 3.5 pg/mL and 3.9 pg/mL for CEA and AFP, respectively.

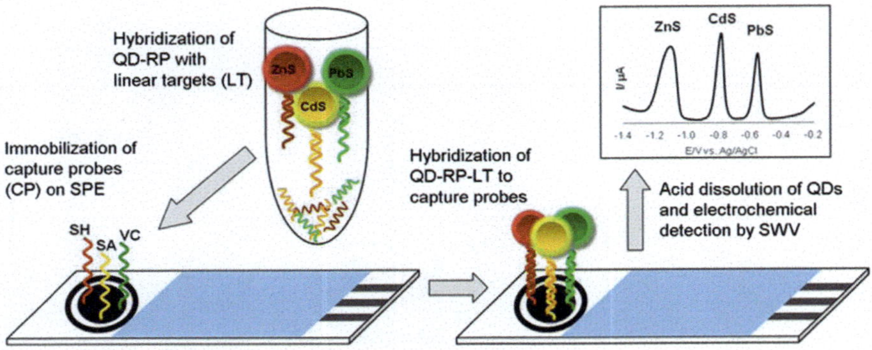

Fig. 16 Addition of quantum dot (QD)-labeled reporter probes (RP) and subsequent binding to linear targets (LT) of *V. cholerae* (VC), *Salmonella* sp. (SA), and *Shigella* sp. (SH) pathogens. Nitric acid was added to encourage dissolution of quantum dots prior to SWV to increase signal readout. Reprinted from [92], with permission from Elsevier

Another approach of using nanoparticles mainly in immunoassays is to conjugate them to a labeling antibody and an enzyme, allowing a high number of molecules to attach to a single conjugate. This allows a single binding event to be labeled with a high number of labeling molecules, increasing signal generation and lowering detection limits [82, 83]. An example was demonstrated by Krause et al., as they also used a conjugate of antibodies, magnetic nanoparticles, and HRP to obtain a label with magnetic properties allowing ultrasensitive detection of four oral cancer biomarkers in a microfluidic immunoarray setup [95]. The limits of detection were 10 fg/mL, 18 fg/mL, 40 fg/mL, and 15 fg/mL for tumor necrosis factor (TNF-α), interleukin-6 (IL-6), interleukin-1β (IL-1β), and C-reactive protein (CRP) in diluted calf serum, respectively. Only 5 μL was required to achieve ultrasensitive detection in 30 min. In comparison, the authors used the conventional ELISA method, which required 400 μL of the biological sample with a total analysis time of about 12 h.

5 Conclusion

Electrochemical biosensors with multiplexing capability offer several advantages over single test devices such as decreased cost per single test and high assay throughput. We expect multiplexed sensors will become frequently used in applications such as cancer biomarker screening, where new biomarkers are still being proposed and there is no ideal single biomarker. Multiple strategies for construction of such sensors for multiplexed analysis have been presented with the focus on the efforts towards ultrasensitive sensors. Detection of multiple analytes can be achieved mainly by spatial separation of detection chambers with the use of microfluidics, where we examined microfabricated devices offering high precision and reliability and are therefore best suited for applications with such needs. Low-cost devices can be fabricated using paper technologies or printed circuit boards, which could enable greater accessibility of such sensors and improve the social impact. To date, commercially most successful platforms have been based on electrode arrays, which offer great scalability due to miniaturization and simple electrode fabrication and functionalization. In the effort to further miniatures such devices, strategies of combining multiple probes on a single electrode and achieve multiplexing using labels with different electrochemical properties have emerged. Novel approaches are continuously being developed with primary focus on the use of nanomaterials for increasing the rate of electron transfer at the measuring electrode or differential redox activity of electroactive labels. Such developments hold great promise but need more work. A limited number of electroactive species with minimal cross-reactivity is currently available, limiting multiplexing capability. Sensors with capability to detect more than two analytes in this format are rare. Despite all recent advances, incorporating high number of immunoassays into a single platform remains challenging, with rare examples detecting more than four analytes. However, there is a lot of potential for electrochemical sensors to be in the forward seat in the innovation in this field and someday achieving reliable, low cost and portable devices that can have a major impact on clinical decision making and consequently public health.

References

1. Betancourt, T., Brannon-Peppas, L.: Micro- and nanofabrication methods in nanotechnological medical and pharmaceutical devices. Int. J. Nanomed. **1**(4), 483–495 (2006)
2. Terry, S.C., Jerman, J.H., Angell, J.B.: A gas chromatographic air analyzer fabricated on a silicon wafer. IEEE Trans. Electron Devices **26**(12), 1880–1886 (1979)
3. Castillo-Léon, J., Svendsen, W.E.: Lab-on-a-Chip Devices and Micro-Total Analysis Systems: A Practical Guide, vol. 1. Springer, Cham, Switzerland (2015)
4. Becker, H., Gärtner, C.: Polymer microfabrication technologies for microfluidic systems. Anal. Bioanal. Chem. **390**(1), 89–111 (2008)
5. Liao, Z., et al.: Recent advances in microfluidic chip integrated electronic biosensors for multiplexed detection. Biosens. Bioelectron. **121**, 272–280 (2018)
6. Hwang, S.Y., et al.: Microfluidic multiplex biochip based on a point-of-care electrochemical detection system for matrix metalloproteinases. J. Electroanal. Chem. **756**, 118–123 (2015)
7. Lee, G., et al.: Single microfluidic electrochemical sensor system for simultaneous multi-pulmonary hypertension biomarker analyses. Sci. Rep. **7**(1), 7545 (2017)
8. Pursey, J.P., et al.: Microfluidic electrochemical multiplex detection of bladder cancer DNA markers. Sens. Actuators B: Chem. **251**, 34–39 (2017)
9. Gao, J., et al.: A multiplex electrochemical biosensor for bloodstream infection diagnosis. SLAS Technol **22**(4), 466–474 (2017)
10. GeneFluidics. 2015 [Accessed on April 20th, 2020]; Available from: http://www.genefluidics.com/
11. Altobelli, E., et al.: Integrated biosensor assay for rapid uropathogen identification and phenotypic antimicrobial susceptibility testing. Eur Urol Focus **3**(2–3), 293–299 (2017)
12. Du, Y., et al.: Microfluidic electrochemical aptameric assay integrated on-chip: a potentially convenient sensing platform for the amplified and multiplex analysis of small molecules. Anal. Chem. **83**(5), 1523–1529 (2011)

13. Gao, J., et al.: Simultaneous detection of glucose, uric acid and cholesterol using flexible microneedle electrode array-based biosensor and multi-channel portable electrochemical analyzer. Sens. Actuators B: Chem. **287**, 102–110 (2019)
14. Han, Y., et al.: A multiplexed immunoaggregation biomarker assay using a two-stage micro resistive pulse sensor. Biomicrofluidics **10**(2), 024109 (2016)
15. Akyazi, T., Basabe-Desmonts, L., Benito-Lopez, F.: Review on microfluidic paper-based analytical devices towards commercialisation. Anal. Chim. Acta **1001**, 1–17 (2018)
16. Ataide, V.N., et al.: Electrochemical paper-based analytical devices: ten years of development. Anal. Meth. **12**, 1030–1054 (2020)
17. Solhi, E., Hasanzadeh, M., Babaie, P.: Electrochemical paper-based analytical devices (ePADs) toward biosensing: recent advances and challenges in bioanalysis. Anal. Meth. **12**, 1398–1414 (2020)
18. Noviana, E., Henry, C.S.: Simultaneous electrochemical detection in paper-based analytical devices. Current Opin. Electrochem. **23**, 1–6 (2020)
19. Wang, Y., et al.: Label-free microfluidic paper-based electrochemical aptasensor for ultrasensitive and simultaneous multiplexed detection of cancer biomarkers. Biosens. Bioelectron. **136**, 84–90 (2019)
20. Li, W., et al.: Development of a 3D origami multiplex electrochemical immunodevice using a nanoporous silver-paper electrode and metal ion functionalized nanoporous gold-chitosan. Chem. Commun. **49**, 9540–9542 (2013)
21. Nantaphol, S., et al.: Janus electrochemistry: Simultaneous electrochemical detection at multiple working conditions in a paper-based analytical device. Anal. Chim. Acta **1056**, 88–95 (2019)
22. Jobst, G., et al.: Mass producible miniaturized flow through a device with a biosensor array. Sens. Actuators B: Chem. **43**(1), 121–125 (1997)
23. Merkel, T., Graeber, M., Pagel, L.: A new technology for fluidic microsystems based on PCB technology. Sens. Actuators, A **77**(2), 98–105 (1999)
24. Kling, A., et al.: Multianalyte antibiotic detection on an electrochemical microfluidic platform. Anal. Chem. **88**(20), 10036–10043 (2016)
25. Panneer Selvam, A., Prasad, S.: Companion and point-of-care sensor system for rapid multiplexed detection of a panel of infectious disease markers. SLAS Technol.: Trans. Life Sci. Innovat. **22**(3), 338–347 (2017)
26. Sanchez, J.L., et al.: Multiplex PCB-based electrochemical detection of cancer biomarkers using MLPA-barcode approach. Biosens. Bioelectron. **82**, 224–232 (2016)
27. Aracil, C., et al.: Portable Lab-on-PCB platform for autonomous micromixing. Microelectron. Eng. **131**, 13–18 (2015)
28. Moschou, D., Tserepi, A.: The lab-on-PCB approach: tackling the μTAS commercial upscaling bottleneck. Lab Chip **17**(8), 1388–1405 (2017)
29. Jolly, P., et al.: A PNA-based Lab-on-PCB diagnostic platform for rapid and high sensitivity DNA quantification. Biosens. Bioelectron. **123**, 244–250 (2019)
30. Moschou, D., et al.: Amperometric IFN-gamma immunosensors with commercially fabricated PCB sensing electrodes. Biosens. Bioelectron. **86**, 805–810 (2016)
31. Dutta, G., et al.: Label-free electrochemical detection of S. mutans exploiting commercially fabricated printed circuit board sensing electrodes, in Micromachines (Basel) (2019)
32. Schmitz, J.E., Tang, Y.-W.: The GenMark ePlex®: another weapon in the syndromic arsenal for infection diagnosis. Future Microbiology **13**(16), 1697–1708 (2018)
33. Pierce, V.M., Hodinka, R.L.: Comparison of the GenMark diagnostics eSensor respiratory viral panel to real-time PCR for detection of respiratory viruses in children. J. Clin. Microbiol. **50**(11), 3458–3465 (2012)
34. Reed, M.R., Coty, W.A.: eSensor®a microarray technology based on electrochemical detection of nucleic acids and Its application to cystic fibrosis carrier screening. In Dill, K., Liu, R.H., Grodzinski, P. (Eds.), Microarrays: Preparation, Microfluidics, Detection Methods, and Biological Applications. Springer New York, New York, NY, pp 247–260 (2009)

35. Díaz-González, M., et al.: Diagnostics using multiplexed electrochemical readout devices. Electroanalysis **26**(6), 1154–1170 (2014)
36. Ling, M.M., Ricks, C., Lea, P.: Multiplexing molecular diagnostics and immunoassays using emerging microarray technologies. Expert Rev. Mol. Diagn. **7**(1), 87–98 (2007)
37. Bumgarner, R.: Overview of DNA microarrays: types, applications, and their future. Current Prot. Molecular Biol. **101**, 1–11 (2013)
38. McCombie, W.R., J.D. McPherson, E.R. Mardis, Next-generation sequencing technologies. Cold Spring Harbor Perspectives in Medicine, vol. 9 (2019)
39. Niedringhaus, T.P., et al.: Landscape of next-generation sequencing technologies. Anal. Chem. **83**, 4327–4341 (2011)
40. Merriman, B., R&D Team, I.T., Rothberg, J.M.: Progress in Ion torrent semiconductor chip based sequencing. Electrophoresis **33**, 3397–3417
41. Ion GeneStudio S5 system - targeted gene sequencing using NGS. [Accessed on May 5th, 2020]; Available from: https://www.thermofisher.com/uk/en/home/life-science/sequencing/next-generation-sequencing/ion-torrent-next-generation-sequencing-workflow/ion-torrent-next-generation-sequencing-run-sequence/ion-s5-ngs-targeted-sequencing.html
42. Deamer, D., Akeson, M., Branton, D.: Three decades of nanopore sequencing. Nat. Biotechnol. **34**, 518–524 (2016)
43. Dekker, C.: Solid-state nanopores. Nat. Nanotechnol. **2**, 209–215 (2007)
44. Nanopore. Oxford nanopore technologies: product comparison. 2020 [Accessed on May 19th, 2020]; Available from: https://nanoporetech.com/products/comparison
45. GenScript. Custom microarrays and oligo pools. 2020 [Accessed on April 20th, 2020]; Available from: http://www.customarrayinc.com/products_main.htm
46. Bolotin, S., et al.: Verification of the Combimatrix influenza detection assay for the detection of influenza A subtype during the 2007–2008 influenza season in Toronto, Canada. Virology Journal, vol. 6 (2009)
47. Leng, C., et al.: Gold nanoparticle as an electrochemical label for inherently crosstalk-free multiplexed immunoassay on a disposable chip. Anal. Chim. Acta **666**(1), 97–101 (2010)
48. Du, D., et al.: Multiplexed electrochemical immunoassay of phosphorylated proteins based on enzyme-functionalized gold nanorod labels and electric field-driven acceleration. Anal. Chem. **83**(17), 6580–6585 (2011)
49. Dincer, C., et al.: Multiplexed point-of-care testing—xPOCT. Trends Biotechnol. **35**(8), 728–742 (2017)
50. Zhao, C., et al.: Multiplexed electrochemical immunoassay using streptavidin/nanogold/carbon nanohorn as a signal tag to induce silver deposition. Anal. Chim. Acta **847**, 37–43 (2014)
51. Lai, G., et al.: Electrochemical stripping analysis of nanogold label-induced silver deposition for ultrasensitive multiplexed detection of tumor markers. Anal. Chim. Acta **721**, 1–6 (2012)
52. Tian, F., et al.: A polymeric microfluidic device integrated with nanoporous alumina membranes for simultaneous detection of multiple foodborne pathogens. Sens. Actuators B: Chem. **225**, 312–318 (2016)
53. Eissa, S., et al.: Carbon nanofiber-based multiplexed immunosensor for the detection of survival motor neuron 1, cystic fibrosis transmembrane conductance regulator and Duchenne Muscular Dystrophy proteins. Biosens. Bioelectron. **117**, 84–90 (2018)
54. Eissa, S., et al.: Multiplexed detection of DOCK8, PGM3 and STAT3 proteins for the diagnosis of Hyper-Immunoglobulin E syndrome using gold nanoparticles-based immunosensor array platform. Biosens. Bioelectron. **117**, 613–619 (2018)
55. Eissa, S., Almthen, R.A., Zourob, M.: Disposable electrochemical immunosensor array for the multiplexed detection of the drug metabolites morphine, tetrahydrocannabinol and benzoylecgonine. Microchim. Acta **186**(8), 523 (2019)
56. Gupta, R.K., et al.: Multiplexed electrochemical immunosensor for label-free detection of cardiac markers using a carbon nanofiber array chip. J. Electroanal. Chem. **773**, 53–62 (2016)
57. Tang, C.K., et al.: High-throughput electrochemical microfluidic immunoarray for multiplexed detection of cancer biomarker proteins. ACS Sens. **1**(8), 1036–1043 (2016)

58. Wojciechowski, J., et al.: Multiplexed electrochemical detection of Yersinia pestis and staphylococcal enterotoxin B using an antibody microarray. Sensors **10**(4), 3351–3362 (2010)
59. Peled, N.: Design and implementation of a microchemistry analyzer. Pure Appl. Chem. **68**(10), 1837–1841 (1996)
60. Abbott. CHEM8 + Cartridge. 2020 [Accessed on April 4th, 2020]; Available from: https://www.pointofcare.abbott/int/en/offerings/istat/istat-test-cartridges/Chem8+
61. Arduini, F., et al.: Electrochemical biosensors based on nanomodified screen-printed electrodes: aecent applications in clinical analysis. TrAC Trends Anal. Chem. **79**, 114–126 (2016)
62. Otieno, B.A., Krause, C.E., Rusling, J.F.: Bioconjugation of antibodies and enzyme labels onto magnetic beads. In: Kumar, C.V. (Ed.) Methods in Enzymology. Academic Press, pp. 135–150 (2016)
63. Han, K.N., Li, C.A., Seong, G.H.: Microfluidic chips for immunoassays. Ann. Rev. Analyt. Chem. **6**(1), 119–141 (2013)
64. Pakchin, P.S., et al.: Recent advances in simultaneous electrochemical multi-analyte sensing platforms. Trends Anal. Chem. **92**, 32–41 (2017)
65. Liu, F., Ni, L., Zhe, J.: Lab-on-a-chip electrical multiplexing techniques for cellular and molecular biomarker detection. Biomicrofluidics **12**(2), 021501–021501 (2018)
66. Wang, D., et al.: Signal amplification for multianalyte electrochemical immunoassay with bidirectional stripping voltammetry using metal-enriched polymer nanolabels. Sens. Actuators B: Chem. **197**, 244–253 (2014)
67. Ko, Y.-J., et al.: Microchip-based multiplex electro-immunosensing system for the detection of cancer biomarkers. Electrophoresis **29**(16), 3466–3476 (2008)
68. Kong, F.-Y., et al.: Simultaneous electrochemical immunoassay using CdS/DNA and PbS/DNA nanochains as labels. Biosens. Bioelectron. **39**(1), 177–182 (2013)
69. Conzuelo, F., et al.: Disposable amperometric magneto-immunosensor for direct detection of tetracyclines antibiotics residues in milk. Anal. Chim. Acta **737**, 29–36 (2012)
70. Munge, B.S., et al.: Multiplex immunosensor arrays for electrochemical detection of cancer biomarker proteins. Electroanalysis **28**(11), 2644–2658 (2016)
71. Yáñez-Sedeño, P., Campuzano, Actuators, Pingarrón, M.J.: Multiplexed electrochemical immunosensors for clinical biomarkers. Sensors **17**(5) (2017)
72. Wang, L., Rong, Q., Ma, Z.: Construction of electrochemical immunosensing interface for multiple cancer biomarkers detection. Electroanalysis **28**(8), 1692–1699 (2016)
73. Jolly, P., et al.: Self-assembled gold nanoparticles for impedimetric and amperometric detection of a prostate cancer biomarker. Sens. Actuators B: Chem. **251**, 637–643 (2017)
74. Zhang, H., et al.: A surface-confined DNA assembly enabled target recycling amplification for multiplexed electrochemical DNA detection. J. Electroanal. Chem. **833**, 290–296 (2019)
75. Fu, P., et al.: Peptide nucleic acid-based electrochemical biosensor for simultaneous detection of multiple microRNAs from cancer cells with catalytic hairpin assembly amplification. Sens. Actuators B: Chem. **305**, 127545 (2020)
76. Xu, S., et al.: One DNA circle capture probe with multiple target recognition domains for simultaneous electrochemical detection of miRNA-21 and miRNA-155. Biosens. Bioelectron. **149**, 111848 (2020)
77. Fang, C.S., et al.: Ultrasensitive electrochemical detection of miRNA-21 using a zinc finger protein specific to DNA–RNA hybrids. Anal. Chem. **89**(3), 2024–2031 (2017)
78. Feng, X., et al.: A novel strategy for multiplexed immunoassay of tumor markers based on electrochemiluminescence coupled with cyclic voltammetry using graphene-polymer nanotags. Electrochim. Acta **170**, 292–299 (2015)
79. Wei, M., et al.: Simultaneous electrochemical determination of ochratoxin A and fumonisin B1 with an aptasensor based on the use of a Y-shaped DNA structure on gold nanorods. Microchim. Acta **187**(2), 102 (2020)
80. Yang, H.: Enzyme-based ultrasensitive electrochemical biosensors. Curr. Opin. Chem. Biol. **16**(3), 422–428 (2012)
81. Rusling, J.F.: Multiplexed electrochemical protein detection and translation to personalized cancer diagnostics. Anal. Chem. **85**(11), 5304–5310 (2013)

82. Pakchin, P.S., et al.: Recent advances in simultaneous electrochemical multi-analyte sensing platforms. TrAC Trends Anal. Chem. **92**, 32–41 (2017)
83. Florea, A., et al.: Electrochemical biosensors as potential diagnostic devices for autoimmune diseases. Biosensors **9**(1), 38 (2019)
84. Kucherenko, D.Y., et al.: A highly selective amperometric biosensor array for the simultaneous determination of glutamate, glucose, choline, acetylcholine, lactate and pyruvate. Bioelectrochemistry **128**, 100–108 (2019)
85. SIEMENS, Epoc blood analysis system: Summary of analytical methods and performance. 2017, Siemens Healthcare Diagnostics Inc.
86. Jia, X., et al.: Triple signal amplification using gold nanoparticles, bienzyme and platinum nanoparticles functionalized graphene as enhancers for simultaneous multiple electrochemical immunoassay. Biosens. Bioelectron. **53**, 65–70 (2014)
87. Vargas, E., et al.: Enzymatic/Immunoassay dual-biomarker sensing chip: towards decentralized insulin/glucose detection. Angew. Chem. Int. Ed. Engl. **58**(19), 6376–6379 (2019)
88. Grieshaber, D., et al.: Electrochemical biosensors—sensor principles and architectures. Sens. (Basel, Switzerland) **8**(3), 1400–1458 (2008)
89. Cho, I.H., et al.: Current technologies of electrochemical Immunosensors: perspective on signal amplification. Sensors **18**(1) (2018)
90. Khan, I., Saeed, K., Khan, I.: Nanoparticles: Properties, applications and toxicities. Arab. J. Chem. **12**, 908–931 (2019)
91. Kokkinos, C.: Electrochemical DNA biosensors based on labeling with nanoparticles. Nanomater. (Basel, Switzerland) **9**(10), 1361 (2019)
92. Vijian, D., et al.: Non-protein coding RNA-based genosensor with quantum dots as electrochemical labels for attomolar detection of multiple pathogens. Biosens. Bioelectron. **77**, 805–811 (2016)
93. Kong, F.-Y., et al.: Simultaneous electrochemical immunoassay using CdS/DNA and PbS/DNA nanochains as labels. Biosens. Bioelectron. **39**, 177–182 (2013)
94. Wan, Y., et al.: Highly specific electrochemical analysis of cancer cells using multi-nanoparticle labeling. Angew. Chem. **53**, 13145–13149 (2014)
95. Krause, C.E., et al.: Ultrasensitive microfluidic array for serum pro-inflammatory cytokines and C-reactive protein to assess oral mucositis risk in cancer patients. Anal. Bioanal. Chem. **407**(23), 7239–7243 (2015)

An IoT Enabled Enzyme Embossed Biosensor for Determination of Vitamin D Level in Human Blood Sample

Monidipa Ghosh and Chiranjib Koley

Abstract Biosensors are the state-of-the-art devices which enact in a perfect synchronization with biorecognition material, target analyte, and a close contact transducer for the detection of biomolecules. Electrochemical enzyme-based biosensors are one of the most successful group of sensors where nanomaterials are integrated to increase sensitivity, stability, response rate, limit of detection, and other analytical parameters. This has led to the development of precise and sensitive analytical device, using biological sensing element as biosensor. Not only that, it's imperative utilization in the field of drug discovery, biomedicine, food safety standards, defense, security, and environmental monitoring, the electrochemical sensor has gained paramount importance. The present study provides a highlight on a point of care system for measuring vitamin D3, the biomarker for cardiovascular disease, obesity, and diabetes, from serum samples, comprising a three-electrode system biosensor. An enzyme responsive to the select biomarker is immobilized on the working electrode surface, enabling a change in current flow between the working electrode and the auxiliary electrode. This change helps the device, in detecting the physiological range of the said biomarker, present in the medium. The process for the estimation of 25-Hydroxyvitamin D3 in a serum sample involves; the use of anti-VD3 1 A hydroxylase enzyme antibody, for estimation of the sample analyte, vitamin D3. The electrode system is coated with a redox mediator along with a linker molecule, to immobilize the anti-Vit D3 antibody on the electrode surface. Advantageously, the said system provides a cost-effective and high fidelity point of care detection device, finding it's application in health care.

Keywords 25-hydroxyvitamin D3 · Internet-of-Things · Biosensor · Disposable strip sensor · Blood test

---

M. Ghosh (✉)
Department of Biotechnology, National Institute of Technology Durgapur, Durgapur, West Bengal 713209, India
e-mail: monidipa.ghosh@bt.nitdgp.ac.in

C. Koley
Department of Electrical Engineering, National Institute of Technology Durgapur, Durgapur, West Bengal 713209, India

G. Dutta et al. (eds.), *Modern Techniques in Biosensors*, Studies in Systems, Decision and Control 327, https://doi.org/10.1007/978-981-15-9612-4_4

1 Introduction

In recent decades, biosensors have integrated their results in respective fields, inclusive of biomedical applications. Superiority of high selectivity and sensitivity of modern biosensor can also estimate the analytes in real-time and that is very convenient for the monitoring of rapid changes in biological samples. A biosensor can be designated as "a self-contained analytical device that combines a biological component with a physicochemical device for the detection of an analyte of biological importance" [1]. It is composed of a bio analyte recognizing component and a transducer which transforms the chemical interaction into a quantifiable signal. In case of chemical biosensor, biological specimen is recognized by the specific analyte, which is stable under normal circumstances [2], The conceding elements such as receptors, nucleic acids, whole cells, antibodies, and different classes of enzymes have been used in biosensors. Classification of biosensor depends on the transduction method. Hence, the transducer of a biosensor converts the array of biochemical reactions into an electrical signal, on which optical, calorimetric, or acoustic biosensors are developed and analyzed. But the electrochemical property led biosensors are the most popular one. The first electrochemical biosensor was developed for the detection of glucose oxidase [3]. Through electrochemical method, biochemical signal is transformed into a measurable amperometric signal. But analysis can also be done by impedimetry or potentiometry [4]. In case of Enzyme-based amperometric biosensors, a fixed potential current of two electrodes is monitored, and widely studied over last few decades, as the technique is very robust and able to operate a minute volume [5]. Designing of a biosensor requires the proper understanding of target analyte and the matrix chemistry. Electrochemical measurement is dependent on the chemical property of the working electrode material [6]. Nanomaterials (NM) possess some unique characteristics like, a high surface-to-volume ratio, high electrical conductivity, magnetic properties, catalytic activity, etc. which enables significant increase in the transducer surface sensitivity and more precise enzyme immobilization. Moreover, chemical functionalization of NMs, essential for the interaction with biomaterial in biosensors, can be easily attained. That is why the basic analytical properties of biosensors, such as sensitivity, limit of detection (LOD), linear detection range, selectivity, reproducibility, are improved by the integration of different nanomaterials [7].

Since the end of 1980s, carbon paste electrodes are being used to develop the amperometric biosensors [8]. In electrochemical biosensor, carbon is the most widely used component for biosensing, as it possesses broad surface area, ample biostability and advantageous physicochemical as well as catalytic properties [9, 10]. In case of electrochemical biosensors, transducers are also made by gold, platinum, or palladium. They facilitates easy electron transfer, produced from hydrogen peroxide by oxidase based biosensors and successful signal generation from the electro-oxidation [11, 12]. The above characteristics make the biosensors a convenient tool for biomedical research and diagnostic application. These low-cost biosensors are user friendly, reliable, wireless, and rapid diagnosis technique [13]. The enzyme-based amperometric biosensors could be used to measure target analytes from collected biological samples or implanted device for the detection of extracellular analyte concentration change or sampling device, implanted in the body, like microdialysis probe. Amperometric biosensor is based on enzyme-linked technique, which analyses the interferants of the sample matrix-like small molecule metabolites, proteins, macromolecules, and cells, producing analysis error [14]. In this study, the working principles, optimization modules, and analytical performance of enzymatic biosensors, in detection of serum VitaminD3 will be discussed.

For the activation of vitamin D, a steroidal compound needs collective metabolic steps in human body. UVB component of sunlight is the vital source of vitamin D in our body. Vitamin D or cholecalciferol synthesis is stimulated by the incorporation of UVB radiation in cutaneous layer of human skin, after that in liver it is converted into the form of 25-hydroxyvitamin D [25(OH)D] through hydroxylation process [15]. Vitamin D is crucial for the human body metabolism and recent research has linked it's malfunctioning with diabetes and cardiovascular dysfunction. Maintenance of desired levels (30–50 ng/ml) of vitamin D controls diabetic conditions and also improves the cardiovascular functioning. Therefore early detection of low (less than 12 ng/ml) Vitamin D3 level is an important biomarker for the prediction and prevention of a wide range of human diseases. The ultraviolet B ray makes active forms of vitamin D. During the cholesterol metabolism in the body, 7-dihydrocholesterol produced after eating food, is broken by Ultraviolet B in the skin, synthesizing vitamin D3. Vitamin D3 goes to liver through the blood stream, then becomes converted into 25-vitamin D which circulates to the kidneys and turns into active vitamin D3. In case of vitamin D level monitoring, 25-hydroxyvitamin D is evaluated, as it shows the contribution from both diet and synthesis in the skin [16, 17]. The Institute of Medicine, DRI committee on calcium and Vit D, uses serum or plasma to establish the dietary reference value of it. It is shown by several reports that, 25(OH) VitD assays can yield variable results. This has led to international efforts to begin a evidence-based guidelines. The measurement of vitamin D present in serum is internationally standardized by the development of VSDP or vitamin D standardized program. This is a collaborative initiative between the office of the diatory supplement of NIH, the CDC, the National Institute of Standards and Technology, and a number of the national health surveys around the world. Dr. Rehinhard Voll from Germany developed the principle of "EAV (biocurrent measurement) test", which is able to estimate the current of each apex in the human body which helps to dictate it's status, applied to the products. In case of EVA (electro acupuncture according to Voll) system, the cell act as a single ion battery, that uses for the characteristic analysis of meridian reaction by the stimulation of weak current externally. When the meridian is touched with the pen-like vitastiq that corresponds to the elements to be measured, the nutritional status of body is measured without a blood test [18].

Current methods for the detection of Vitamin D, like Radioimmunoassays (RIA), high-pressure liquid chromatography (HPLC) and liquid chromatography mass spectrometry (LC-MS/MS) are time-consuming, expensive, require long analytical periods and trained personnel etc. Although various measuring devices are presently available, most of them require multiple sample pretreatment steps. Hence there is still a need for a simple, cost-effective, rapid, efficient system for the detection and quantification of Vitamin D3 in biological samples overcoming the above limitations.

2 Methodology

2.1 *Working Principle of Enzyme Based Electrochemical Sensor*

The proposed strip based sensor is of single-use type but the detection system and the excitation system are of multi-use type. The disposable sensor strip is to be attached with the multi-use detection system before starting the complete measurement process, i.e., excitation and then the measurement of current, etc. A system for the measurement of the Vitamin D level in biological samples is comprised with a three-electrode electrochemical system, including a working electrode, auxiliary electrode, and a reference electrode. An electrode excitation and sensing means; current sensing and impedance matching unit comprising of; ramp waveform generator, current sensing, impedance matching; and microprocessor means for extraction of relevant parameters including estimation of 25-Hydroxyvitamin D3, indicative of the Vitamin D level in serum samples.

Screen-printed electrodes (SPE) are immobilized with the anti-human monoclonal antibody (ANTI CYP27B1) to detect the antigen, 25-(OH)VitD3 in the sample. Oxidized Graphene of the Screen Printed Electrode (SPE) is functionalized with a 4 aminophenylacetic acid (4-APA) linker, upon which redox, cobalt chloride dihydrate has been applied before anti-CYP27B1 antibody application. The activated carboxyl group of 4(APA) would conjugate top the terminal amine group of the (CYP27B1 or 25 hydroxylic D1 X hydroxylase or isotype control) antibodies through amide bond. The linker eliminates the possibility of any cross-reaction between CYP27B1 hydroxylase enzyme and the other Vit D metabolites in serum as substrate and chances of false positivity in the in vivo system.

The electrolyte behaves as the conducting medium and the conductivity dependents on the concentration of ions. As the working electrode is excited and the auxiliary (counter) electrode is grounded through a current sensor, therefore the current passes due to potential differences between the working and auxiliary electrodes, which is carried by the ions in the electrolyte. It is to be noticed here that due to the combination of the antigen (sample i.e. 25-hydroxy Vitamin D3) with the antibody (Enzyme 25-Hydroxyvitamin D-1 hydroxylase), hydrogen peroxide is generated, which is unstable in solution and therefore breaks into H^+ and OH^- ions. This, in turn, increases the ions in the electrolyte and thus the current passes through the working and auxiliary electrodes increases.

A small volume serum layer at the electrode surface leads to the exchange of ions between the working and the auxiliary electrode. This current was measured

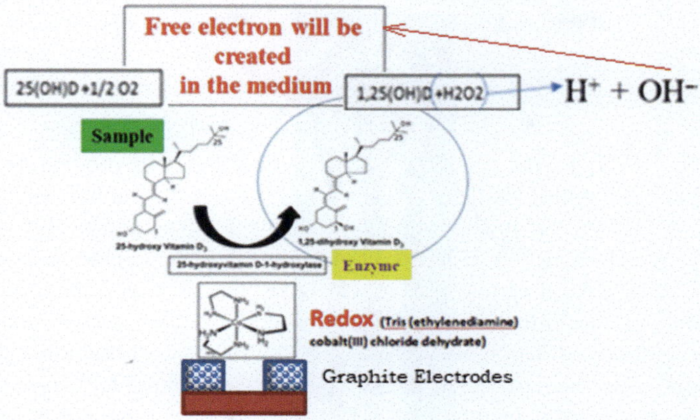

Fig. 1 Illustrates schematic diagram of the system for measurement of Vitamin D level from serum

with the help of specially designed current to voltage converters in combination with high input impedance instrumentation amplifier, having high CMRR. The specimen physically adsorbed by the carbon surface of the electrode, change the dielectric of the medium. As a result, the device conduction is changed, enabling the measurement of particular amount of biospecimen present on the medium (Fig. 1).

2.2 Electrode System and the IoT Enabled Biosensor for Monitoring the Reaction Rate

The electrode system is coated with above-mentioned redox mediator along with a linker molecule, and the IoT enabled biosensor that has been developed for the measurement of Vitamin D3 is presented here.

Electrode System: The electrode system consists of a circular carbon working electrode [WE] (area: 3 mm^2) inscribed in hemispherical auxiliary/counter [AE/CE] carbon electrode (area: 10 mm^2) and hemispherical Ag/AgCl reference electrodes [RE] (area: 2 mm^2). Three electrodes working, reference, and counter, were printed on the non-conducting ceramic substrate. A schematic diagram of the electrode system is shown in Fig. 2. Specific shapes of electrodes are helpful to maintain a uniform distance between the working-auxiliary and working-reference electrodes, and it uses less metal during the manufacturing of the electrodes.

IoT Enabled Biosensor Module: An electronic device for the sensing of the Vitamin D3 level with the help of the electrochemical system is being presented in this subsection. When a serum layer of the specific volume is formed at the above-mentioned electrode surface, the exchange of ions between the working electrode and the auxiliary electrode occurs. This leads to a flow of current, and this current

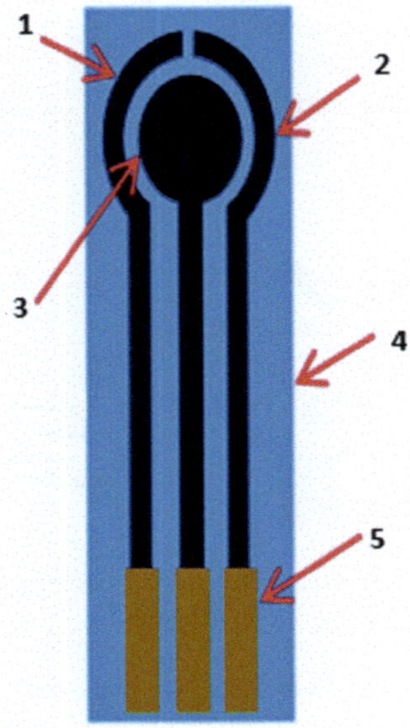

Fig. 2 Schematic diagram of the electrode system, 1—CE, 2—RE, 3—WE, 4—non conducting base, and 5—copper connecting terminals

can be measured with the help of specially designed current to voltage converters with the help of high input impedance instrumentation amplifier having high CMRR. A schematic diagram of the device that can be used for measurement of vitamin D3 is presented in Fig. 3.

The device consists of a current driving unit, an ARM-based Cortex M3 series microcontroller, fast A/D, D/A converters, and a switchable potentiostat. The system may also facilitate an automatic calibration by making use of the standard addition technique.

For cyclic voltammetry based measurement, a suitable ramp type excitation of optimized frequency is to be given across the working and reference electrode with the help of waveform generated in the microcontroller, and then sending it across the onboard Digital to Analog (D/A) converter and finally through excitation driving unit as shown in Fig. 3. The current that flows from the auxiliary electrode (AE) to the working electrode (WE) will be converted into suitable voltage with the help of high impedance Instrumentation Amplifier. The amplified signal will then be converted into digital with the help of onboard Analog-to-Digital (A/D) converter. The ARM-based processor will then pre-process the digitized signal for reduction of noise using the suitable digital filter. The pre-processed signal will then be utilized for further processing and extraction of relevant parameters that indicate the level of Vitamin D3 present in the sample.

Detection Mechanism: The voltammeter analysis of Vitamin D is performed at room temperature. For cyclic voltammetry based measurement, a suitable ramp type

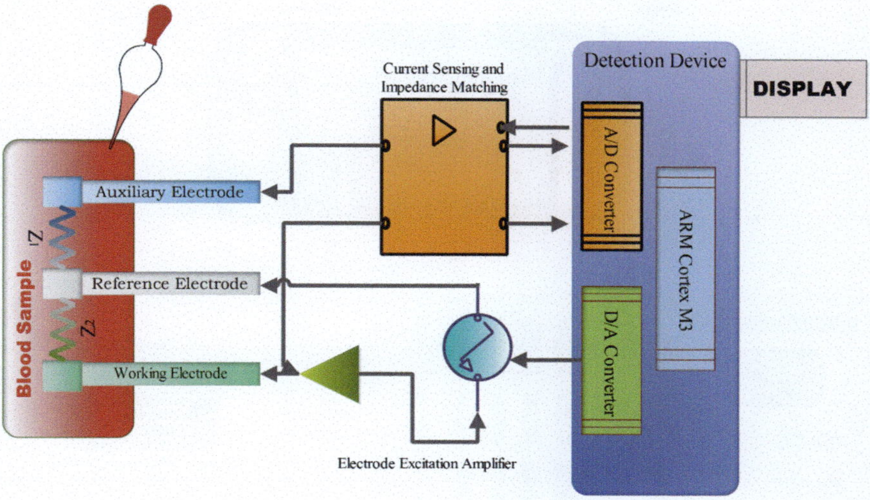

Fig. 3 Schematic diagram of the electronic device for measurement of Vitamin D3 level

excitation of 0.3 Hz with an amplitude of ± 0.5 V is given across the working and reference electrode with the help of waveform generated in the microcontroller, and then sending it across the onboard Digital to Analog (D/A) converter and finally through excitation driving unit as shown in Fig. 3. The current that flows from the auxiliary electrode (AE) to the working electrode (WE) is converted into suitable voltage with the help of high impedance Instrumentation Amplifier. The amplified signal is then converted in digital with the help of onboard Analog-to-Digital (A/D) converter. The ARM-based processor then pre-processes the digitized signal for reduction of noise using the suitable digital filter. The pre-processed signal is then utilized for further processing and extraction of relevant parameters that indicate the level of Vitamin D present in the sample, which is determined from the average value of the current during the transition of excitation voltage from negative to positive.

After the validation test, software for the microcontroller can be design for the identification of the disease. Finally, the prototype may be designed and packaged for field deployment. The WiFi communication protocol can be used for ease of access through smartphone and necessary IoT based implementation. A schematic diagram of the proposed device is shown in Fig. 4.

3 Results and Discussions

This data have been obtained under different conditions, that validates the proposed system for the determination of Vitamin D3 level. Various tests that have been conducted are of three categories. These are a performance characterization system

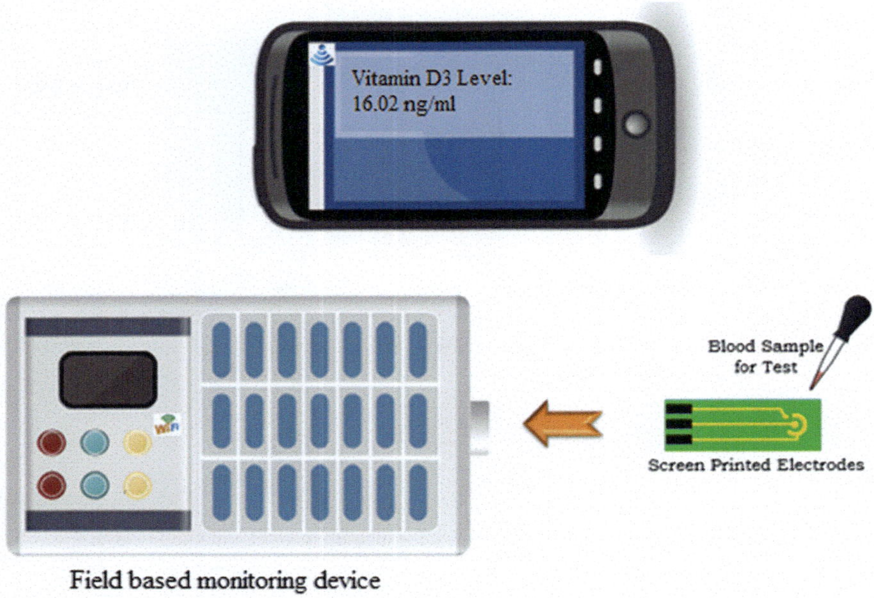

Fig. 4 Schematic of proposed sensing system

with commercial vitamin D3 sample to find a suitable indicator of Vitamin D3 level: calibration of the system with commercial vitamin D3 sample, and detection of vitamin D3 in the serum samples, collected from different individuals.

3.1 Validation with Commercial Vitamin D3 Sample

In this test, the surface in between the three electrodes (reference, working, and auxiliary) is coated with the redox mediator, an antibody capturing linker molecule and a thin layer of the enzyme is immobilized on the surface. Different concentrations (10–200 ng/ml) of commercial sample of vitamin D3 (25, hydroxyvitamin D) are dispersed on the surface of the said device and reaction efficacy are checked through the measurement of excitation voltage and the current flowing through the auxiliary electrode to the working electrode. The excitation that is applied, is of triangular wave shape, with an amplitude of −500 to +500 mV. The recorded current in between working and auxiliary during application of such triangular excitation is shown in Fig. 5, for a different level of Vitamin D3 concentration. From Fig. 5, it can be observed that as the concentration of Vitamin D3 varies, the sigmoid curve changes, indicating different rates of ion generation for different Vitamin D3 concentrations. This also proves the effectiveness of the proposed Enzyme. However, it can be noted that for automated detection, a suitable feature that can be computed from the sigmoid

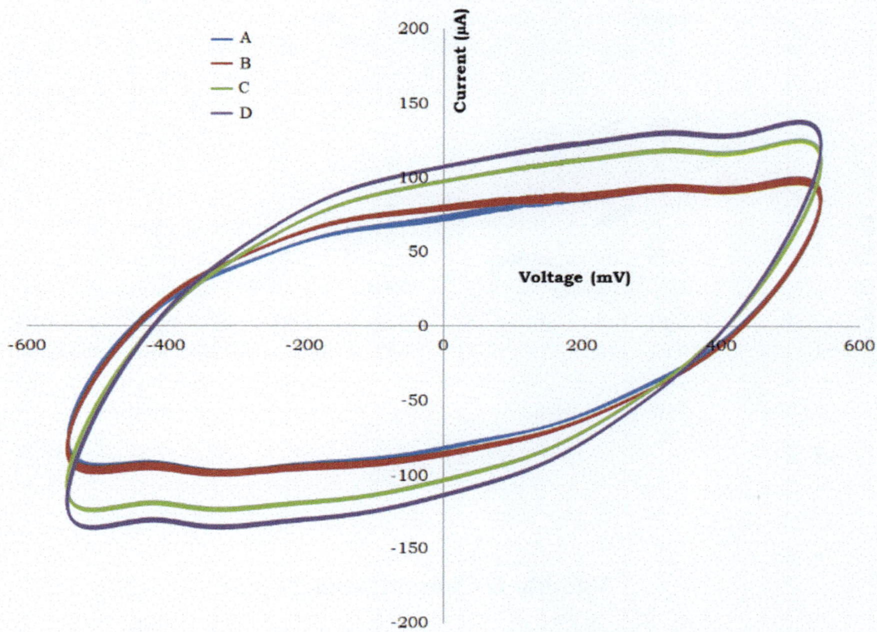

Fig. 5 Cyclic voltammetry plot for different Vitamin D3 concentrations, A. 10 ng/ml, B. 12 ng/ml, C. 20 ng/ml, and D. 40 ng/ml from four different subjects

curve is required, that should not only be sensitive for the variation of Vitamin D3 concentration but also need to be linear.

In order to find a suitable indicator of Vitamin D3 level, various kinds of studies have been performed, which are presented below.

Average Value of the Current: In this case, the average value of the current, measured from at least three complete cycles of excitation has been used as an indicator of Vitamin D3 concentration. The obtained result is shown in Fig. 6, with a dotted best-fitting line. From Fig. 6, it can be observed that the average value of the current initially decreasing with the increasing concentration of Vitamin D3. However, this cannot be a good indicator as there is a significant amount of error and saturation occuring at higher concentration.

The Peak of the Current: In this case, the peak value of the current measured from at least three complete cycles of excitation has been used as an indicator of Vitamin D3 concentration. The obtained result is shown in Fig. 7, with a dotted best-fitting line. From Fig. 7, it can be observed that the indicator though initially increasing with the increasing level of Vitamin D3 concentration, but later saturates with the higher concentration of Vitamin D3.

Region-Wise Average Current: Another category of features that have been investigated is the region-wise average values of the current. That has been calculated with

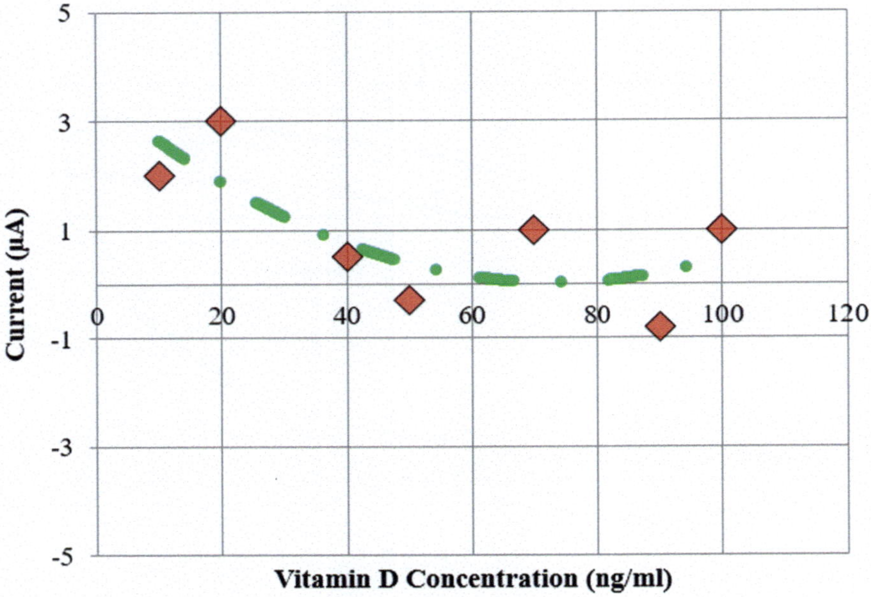

Fig. 6 Illustrates average value of current as an indicator of Vitamin D3 concentration

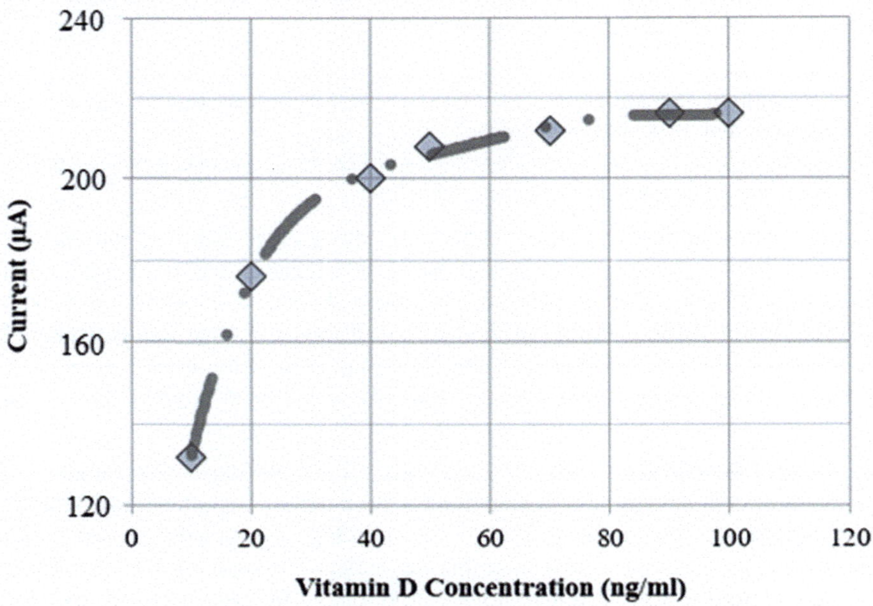

Fig. 7 Illustrates peak current values of the current as an indicator of Vitamin D3 concentration

the help of Eq. (1), where P indicates the region-wise average value, is the response current that passes from the working electrode and θ is the phase angle.

$$P = \int_{\theta=1.93\pi}^{\theta=2\pi} i(\theta)\mathrm{d}\theta \tag{1}$$

Figure 8 shows the variation of the average current (P) while θ, varies from 0.5π to 0.7π, i.e. when the triangular excitation transit from positive polarity to negative. From the figure, it can be observed that though the extracted feature is sensitive to the changes in Vitamin D3 concentration, however, a nonlinearity exists in between. In other cases of the investigation by varying the θ, so that average values of the current can be taken when the triangular excitation is in the positive and negative peak have been observed. Figure 9 shows such variation during positive peak, a similar kind of observation has been observed for the negative peak also.

Figure 10 shows the variation of the average current (P) while θ, varies from 0.9π to π, i.e. when the triangular excitation transit from positive polarity to negative. From the figure, it can be observed that this parameter is sensitive to the variation of Vitamin D3 concentration and also almost liner. In further investigation by obtaining the average current by increasing the rage of θ from 0.8π to π, it has been observed that linearity, as observed in previous, is not present, as shown in Fig. 11.

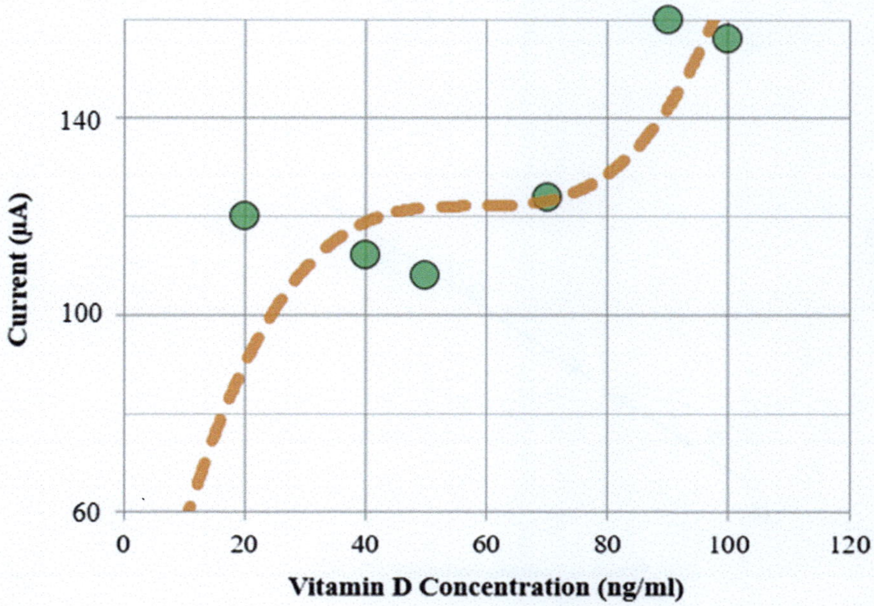

Fig. 8 Illustratesaverage values of the current measured when the triangular excitation transit from positive polarity to negative

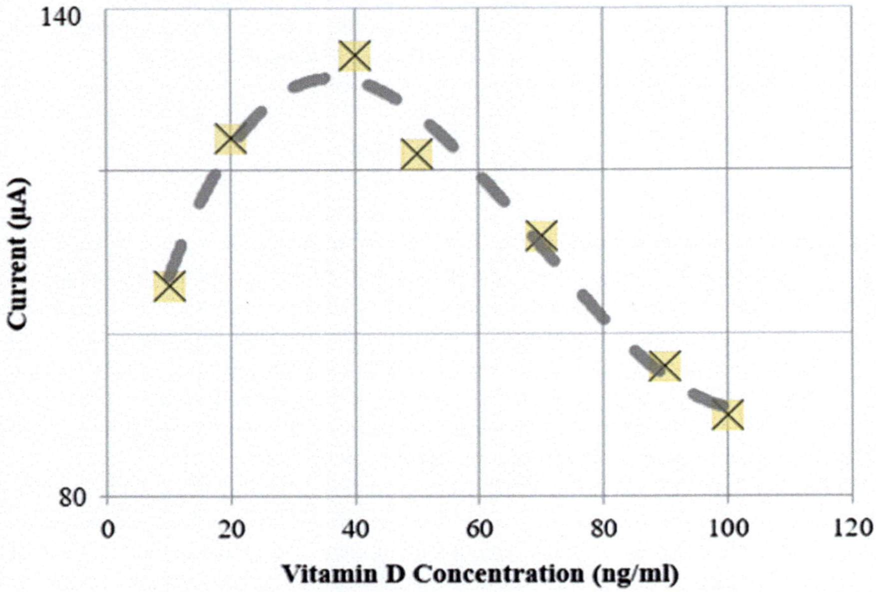

Fig. 9 Illustratesaverage values of the current measured when the triangular excitation is at positive peak value

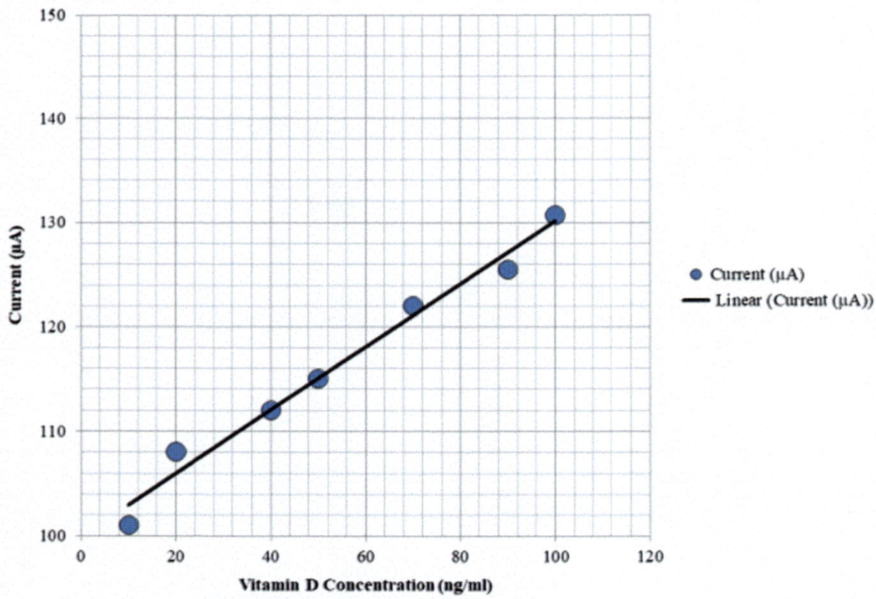

Fig. 10 Illustrates average values of the current measured when the triangular excitation transit from negative polarity to positive for the θ, varies from 0.9π to π

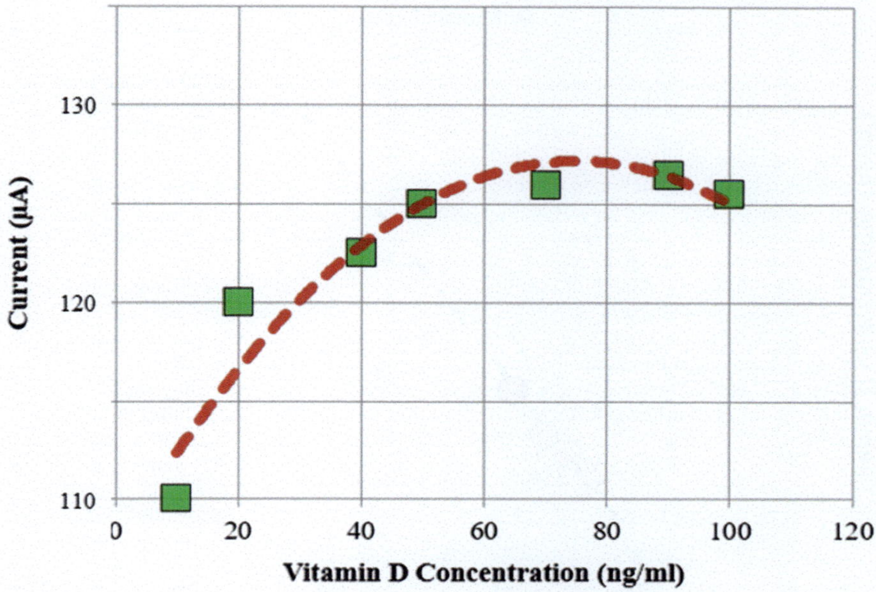

Fig. 11 Illustratesaverage values of the current measured when the triangular excitation transit from negative polarity to positive for the θ, varies from 0.8π to π

3.2 Detection of Vitamin D3 in the Serum Samples, Collected from Different Individual

Serum samples, collected from the normal individuals (with ethical consent) are used for the detection of Vitamin D 3, and the value obtained was used for the estimation of Vitamin D level in the blood circulation. After noting the age and sex, small volume blood samples (approx. 40 μl) is collected by simple sterile needle prick after a normal diet for subsequent use for the test. From Fig. 12, it can be observed that the experimentally obtained results follow the calibration line, which is obtained with the known level of pro-Vitamin D concentration. The indicator used in this case is as given in Eq. (1). This proves the effectiveness of the new indicator as obtained using Eq. (1). The results demonstrate the detection of 25(OH)D3 in a buffer within the physiological range (10–150 ng/ml).

4 Conclusions

Amperometric enzyme-based biosensors are intricate detection systems, combining cross-disciplinary knowledge of electrochemistry, materials science, and biochemistry. This chapter presents a system for measurement of the Vitamin D level in serum

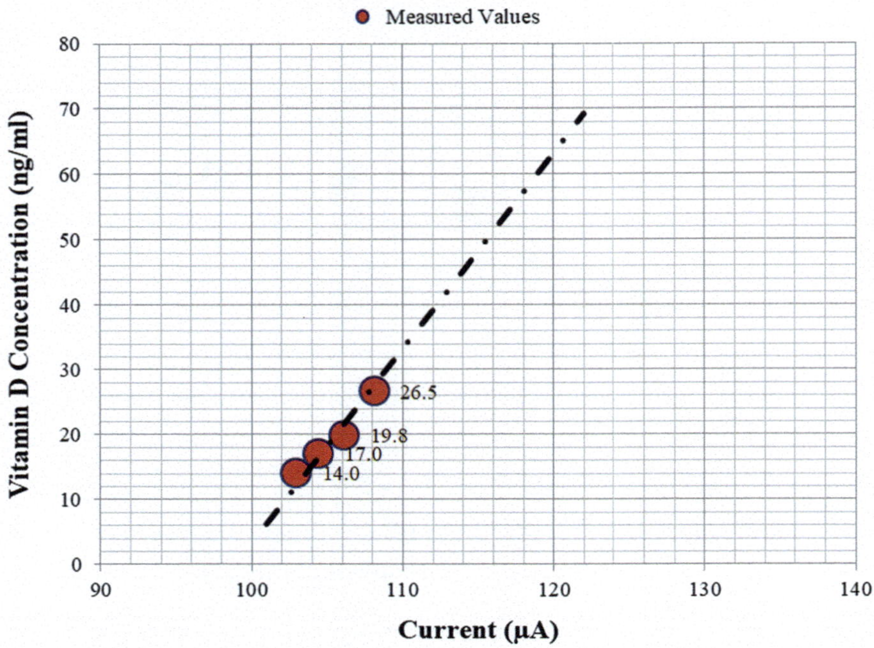

Fig. 12 Vitamin D3 concentration as measured in actual blood sample

samples and the electrode excitation and sensing system which comprises of the ramp waveform generator, current sensing, impedance matching, and microprocessor-based system, for the extraction of relevant parameters, indicative of the Vitamin D level in a serum sample. The process for the estimation of 25-Hydroxyvitamin D3 in a serum sample by the said system involves the use of anti-VD3 1 A hydroxylase (Anti-CYP27B1, C-TERM) antibody for the detection of 25-Hydroxyvitamin D3 antigen. The electrode system, with a coating of Cobalt Chloride Dehydrate, as a redox mediator, along with a linker and analyte antigen is to be used one time for the estimation. The use of average current when the excitation voltage transit from negative to positive is used as an indicator of 25-Hydroxyvitamin D3 presence in biological samples. Advancement in deployment of enzymes and nanoscale electrode materials for the evaluation of biomarker level in the sera has been proved to be a promising advent having huge potential in terms of new functionality, sensitivity, high fidelity, robustness, and miniaturization, in solving a wide range of assays in field of biosensors. The use of the disposable, smart and user-friendly sensing devices are being the new generation preference in disease diagnosis. Undoubtedly, the enzyme-based electrochemical biosensors will show an impending growth to reform the future personalized health care diagnostics with very promising future prospects.

References

1. Vigneshvar, S., Sudhakumari, C.C., Senthilkumaran, B., Prakash, H.: Recent advances in biosensor technology for potential applications—an overview. Front. Bioeng. Biotechnol. **4** (2016)
2. Grieshaber, D., MacKenzie, R., Vörös, J., Reimhult, E.: Electrochemical biosensors—sensor principles and architectures. Sensors **8**, 1400–1458 (2008)
3. Clark, L.C., Lyons, C.: Electrode systems for continuous monitoring in cardiovascular surgery. Ann. N.Y. Acad. Sci. **102**, 29–45 (1962)
4. Pohanka, M., Skládal, P.: Electrochemical biosensors-principles and applications. J. Appl. Biomed. **6**, 57–64 (2008)
5. D'Orazio, P.: Biosensors in clinical chemistry. Clin. Chim. Acta **334**, 41–69 (2003)
6. Kucherenko, I.S., Soldatkin, O.O., Kucherenko, D.Y., Soldatkina, O.V., Dzyadevych, S.V.: Advances in nanomaterial application in enzyme-based electrochemical biosensors: a review. Nanoscale Adv. **1**, 4560–4577 (2019)
7. Wang, J.: Real-time electrochemical monitoring: toward green analytical chemistry. Acc. Chem. Res. **35**, 811–816 (2002)
8. Švorc, J., Miertuš, S., Katrlík, J., Stred'anský, M.: Composite transducers for amperometric biosensors. The glucose sensor. Anal. Chem. **69**, 2086–2090 (1997)
9. Zhu, Z., Garcia-Gancedo, L., Flewitt, A.J., Xie, H., Moussy, F., Milne, W.I.: A critical review of glucose biosensors based on carbon nanomaterials: carbon nanotubes and graphene. Sensors **12**, 5996–6022 (2012)
10. Peña-Bahamonde, J., Nguyen, H.N., Fanourakis, S.K., Rodrigues, D.F.: Recent advances in graphene-based biosensor technology with applications in life sciences J. Nanobiotechnol. **16**, 75 (2018)
11. Pingarròn, J.M., Yáñez-Sedeño, P., González-Cortés, A.: Gold nanoparticle-based electrochemical biosensors. Electrochim. Acta **53**, 5848–5866 (2008)
12. O'Neill, R.D., Chang, S.C., Lowry, J.P., McNeil, C.J.: Comparisons of platinum, gold, palladium and glassy carbon as electrode materials in the design of biosensors for glutamate. Biosens. Bioelectron. **19**, 1521–1528 (2004)
13. Zhou, W., Huang, P.J., Ding, J., Liu, J.: Aptamer-based biosensors for biomedical diagnostics. The Analyst **139**, 2627–2640 (2014)
14. Belluzo, M.S., Ribone, M.E., Lagier, C.M.: Assembling amperometric biosensors for clinical diagnostics. Sensors **8**, 1366–1399 (2008)
15. Nair, R., Maseeh, A.: Vitamin D: the "sunshine" vitamin. J. Pharmacol. Pharmacother. **3**(2), 118–126 (2012)
16. Institute of Medicine, Food and Nutrition Board: Dietary Reference Intakes for Calcium and Vitamin D. National Academy Press, Washington, DC (2010)
17. Jones, G.: Pharmacokinetics of vitamin D toxicity. Am. J. Clin. Nutr. **88**, 582S-S586 (2008)
18. Ma, S.-X., Mayer, E., Lee, P., Li, X., Gao, E.Z.: Transcutaneous electrical stimulation increased nitric oxide-cyclic GMP release biocaptured over skin surface of pericardium meridian and acupuncture points in humans. Acupunct. Electrother. Res. **40**(2), 73–86 (2015)

Computer Aided Diagnosis: Approaches to Automate Hematological Tests

Archita Ghosh, Sricheta Parui, Debasis Samanta, Jayanta Mukhopadhyay, and Nishant Chakravorty

Abstract Hematological disorders are disorders pertaining to blood cell anomalies. Diseases varying from anemia, malaria to leukemia are diagnosed via careful scrutiny of peripheral blood smears (PBS) prepared on glass slides. Usually, the tests are performed in laboratory settings by skilled professionals and clinicians to reach a diagnosis. However, human error due to fatigue and observational oversights might lead to a misdiagnosis. Automated systems are therefore developed to assist the diagnostic procedure of clinicians and provide them with meaningful insights. Computer aided diagnosis is mostly concerned with the evaluation of medical images by employing various image processing techniques. The chapter gives an overview of the stages in developing a computer aided diagnosis system. It concentrates toward the common approaches in designing and developing an assisting system to detect abnormalities in blood samples using image processing tools alongside tools like machine learning and deep learning. The chapter addresses steps like pre-processing, image segmentation, feature extraction, pattern recognition and classification while focusing on specific studies on blood disorders.

Keywords Blood cell images · Computer aided diagnosis · Image processing · Segmentation · Classification · Hematological disorders

A. Ghosh · N. Chakravorty (✉)
School of Medical Science & Technology, Indian Institute of Technology Kharagpur, Kharagpur, West Bengal 721302, India
e-mail: nishant@smst.iitkgp.ac.in

S. Parui
Advanced Technology Development Centre, Indian Institute of Technology Kharagpur , Kharagpur, West Bengal 721302, India

D. Samanta · J. Mukhopadhyay
Department of Computer Science & Engineering, Indian Institute of Technology Kharagpur, Kharagpur, West Bengal 721302, India

G. Dutta et al. (eds.), *Modern Techniques in Biosensors*, Studies in Systems, Decision and Control 327, https://doi.org/10.1007/978-981-15-9612-4_5

1 Introduction

Computer aided diagnosis (CAD) is a discipline that assists clinicians and pathologists in their interpretation of medical images and/or clinical information by utilization of computational methods and algorithms [1, 2]. Significant advancements have been made in the field of computer aided diagnosis since its onset [3]. The evaluation of medical images and information is pivotal to diagnosis. Although conventional diagnostic tests are performed by trained personnel and evaluated by clinicians and experts, the risk of observational oversights and human error may lead to errors and misdiagnosis occasionally [4]. The need to develop intelligent diagnostic algorithms emerged as a potential solution to avoiding such hurdles. These computational methods are used to provide reliable decision support to clinicians and pathologists. Over the years, a plethora of CAD systems have made it into practice. There are widespread applications of computer aided diagnosis in medicine [5–8]. Computer aided diagnosis has been extensively used in radiology, for the detection of cancerous cells and prognosis of cancer. It has also been implemented in several other imaging techniques like CT scans, MRI, ultrasound, microscopic images and X-rays [9].

Diagnostic tests on bodily fluids, like blood, urine, sputum, saliva, sweat; analysis of organs and tissues surgically removed for biopsy, are instrumental components of pathology—the study of disease [10]. The procedure of conducting pathological tests involves collection of samples from respective patients and their thorough examination in the laboratory for the detection of anomalies, if any and their clinical interpretation. In this chapter, we use blood as a representative model to discuss how computer based techniques may support and enhance clinical decision making. Blood tests are one of the major pillars of diagnosing clinical conditions. The quality of test reports and consequently the diagnosis depend on the analysis, interpretation and quantification of parameters obtained from patient samples [11]. There are various tests, equipment and methods to diagnose a pathological condition from hematological parameters. Complete blood count (CBC) and peripheral blood film profiling are common such medical diagnostic tests that require specialization and precision in technique. The blood specimens are visually inspected under a magnifying microscope by a thoroughly trained and skilled technician and blood cells are examined for abnormalities in its morphology, color, size etc for the purpose of diagnostic decision making [12].

Clinicians generally have an intuitive approach towards recognizing underlying patterns in patients befitting any known disease condition which is based on years of experience, learning and literature evidence [13]. The symptoms in the patient are recognized and clustered to match plausible diagnosis and compared with prior probabilities of the disease occurrence to reach a diagnosis [14]. It might be difficult for a clinician to diagnose a condition if the instances encountered of the same kind are very less [15]. At this point, it becomes convenient to draw an analogy between the process adapted by clinicians and that by a computer aided diagnosis. The computer aided diagnostic approach is also quiet similar, if one delves into it. A parallel can be drawn from Fig. 1 where the steps of both approaches are shown diagrammatically.

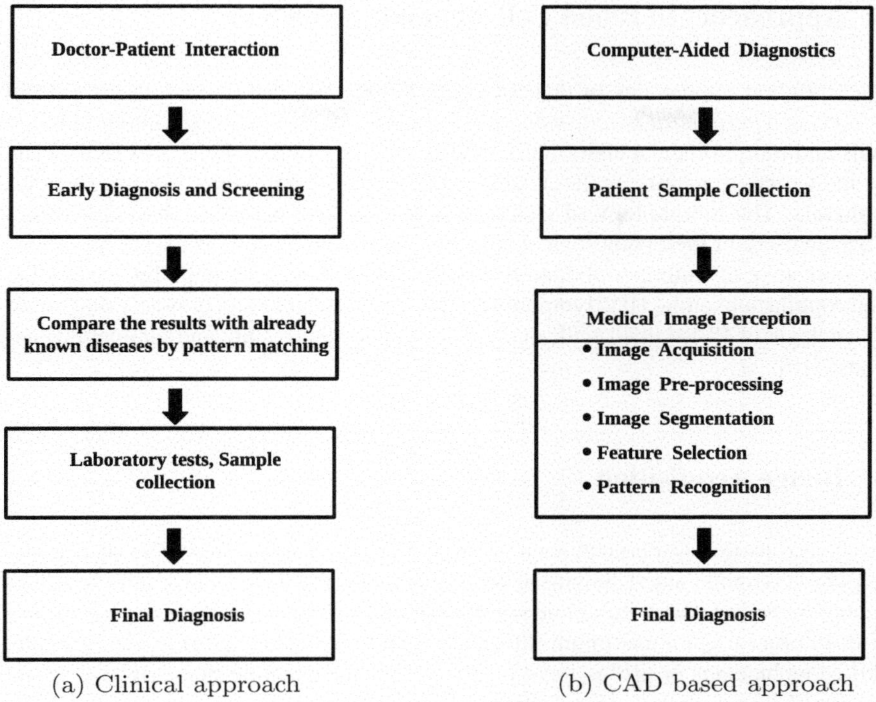

Fig. 1 **a** A typical diagram explaining the approach in clinical diagnosis, **b** A typical diagram explaining the approach in computer aided diagnosis

The specimens or samples collected from the patient are subjected to visual inspection by the image based system. The objective of designing a computer aided diagnostic system is to make use of the feature descriptors of an image and design an algorithm to train the system to learn the underlying patterns. A supervised classification approach is mostly used in medical diagnostics applications. The systems are usually used to provide an assistance and a secondary clinical opinion. However, the medical practitioner can override the results based on individual judgment and expertise.

Computer aided diagnosis would need a step wise computational alternative for all the procedures manually followed by a trained pathologist. There are a few modules that are typically applied on the input data to the computer aided diagnosis system, namely- (a) image acquisition, (b) pre-processing of the image, (c) marking the region of interest, (d) feature selection and extraction (e) pattern recognition and (f) classification. Other than classification in medical images, tasks like parameter estimation and measurement can also be observed in assisting devices and tools.These techniques would be applicable for any other data sample apart from blood as well.

2 Approaches in Blood Cell Anomaly Detection

The common approaches in detecting any blood related disease condition, from observing a *peripheral blood smear* (PBS) sample are evaluating the count of blood cells and morphological abnormalities. Any anomaly in red blood cells (RBC) can be indicative of anemia, malaria, sickle cell anemia, hemolytic syndrome or polycythemia. The morphology of sickle cells and malaria parasites observed from a PBS are used for their pathological detection. White blood cell (WBC) irregularities are indicative of a number of conditions, like, neutropenia, eosinophilia, basophilia, neutrophila, leukemia, HIV, lymphoma, infections, but leukemia is mostly diagnosed by analysis of PBS (albeit with many other supporting results and clinical evaluations) [16].

3 Image Acquisition

Image acquisition is a crucial stage in computer aided diagnostics. The succeeding stages of computer aided diagnosis; viz. segmentation of image and feature selection cannot be undertaken with efficiency if the initial image acquisition stage is imprecise. In order to gain maximum information on a particular hematological sample, skilled technicians or professional pathologists must examine the peripheral blood smear (PBS) closely [17, 18]. Significant advancement in digital microscopy (imaging with digital camera mounted microscope) has facilitated the analysis of PBS for researchers and clinicians. Another common and efficient method for digital microscopy is that of whole slide imaging, which provides high resolution scanned images of the PBS glass slides.

For example, densely clustered leukocytes were segmented from obtained digital images in a study conducted by Nilsson et al. [19]. The samples were stained by May-Grünwald-Giemsa (MGG) and images were captured underneath a 50× microscope mounted with a Sony 9100 color CCD camera. Another study presented by Tomari et al. [20] uses a DinoEye Eyepiece Camera mounted on light microscope working at 40 times (40×) objective which is approximately at 400 magnification. In [21], a 4× optical zoom CCD camera with the microscope at 1000× was used to acquire images of thin blood smear. JPEG images (2048 × 1536 pixels) were obtained at maximum resolution of the camera.

In Fig. 2, the steps of image acquisition have been illustrated diagrammatically. The image shows the steps of preparing a PBS on a glass slide (left), placing the sample under a microscope (middle) and finally obtaining a digital microscopic image of the blood sample (right).

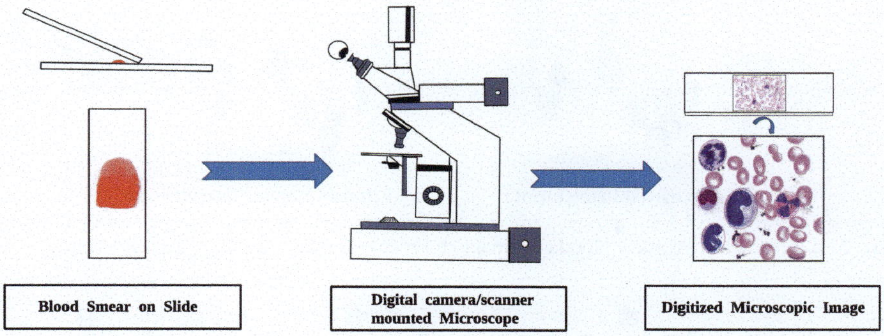

Fig. 2 Digital image acquisition from a peripheral blood smear

4 Pre-processing of Images

The acquired image can bear several defects in it, the presence of which can often lead to misdiagnosis, especially in computer aided systems. For example, malarial parasites and noise due to the camera lens seem to be similar (almost like small patches) in a microscopic blood smear image. If de-noising is not carried out attentively, the noise patches can be misclassified as parasites at a later stage [22]. Pre-processing techniques aid in the improvement of the quality of image data by removal of blur, enhancement of the contrast, sharpening of the image, resizing etc [23, 24]. The process of image sharpening involves enhancement of the edges that have been blurred. Image smoothing is an inverse operation of image enhancement that involves pixel averaging to integrate the pixels and soften the edges. Sharpening and smoothing of images can be achieved through spatial or frequency domain filtering methods. Image enhancement is one of the most commonly used techniques in the image pre-processing stage. Enhancement of an image makes it more lucid and easy to interpret to the naked human eye [25]. The probability of intensity values corresponding to each pixel in an image is graphically represented in the histogram of that image. Histogram equalization is the process where the pixels of an image are mapped to corresponding pixels, giving a more uniformly distributed intensity distribution. This technique is widely utilized in medical diagnosis to enhance an image [26, 27].

In microscopically obtained images, for example, noise removal is an important area to be considered. In Fig. 3a the representative image contains salt and pepper noise. The median filter has been used to remove said noise from the image as shown in the processed output image in Fig. 3b. The median filter here acts as a smoothing filter for the noisy image.

Besides noise removal, enhancing the quality of image is an important factor in microscopic images as precision is pivotal in problems of the kind. Figure 4 shows the steps of enhancement of image quality by contrast stretching method. The contrast

(a) Noisy image (input) (b) Filtered image (output)

Fig. 3 **a** Representative noisy blood cell image, **b** Processed image as output after filtering

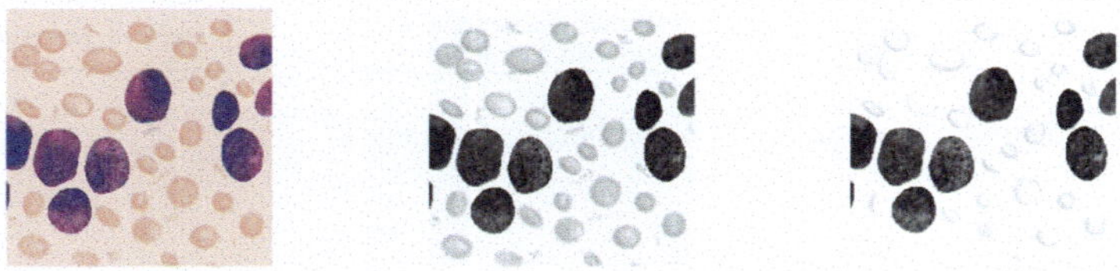

(a) Representative image (b) Gray scale image (c) Enhanced image

Fig. 4 **a** Representative blood cell image, **b** Converted image into gray scale, **c** Enhanced image after contrast stretching

stretched image (Fig. 4b) has a more distinctive foreground and background in comparison to the original gray scale (Fig. 4c). This approach can also be implemented in cases where the edges of cells are blurred or unclear to enhance them.

5 Determination of Region of Interest-Image Segmentation

5.1 An Overview of the Existing Literature

Image segmentation is the method that splits up an image into distinct regions based on discontinuity and similarity for convenience in further analysis. Several segmentation techniques use color, texture or shapes as distinguishing attributes among the regions. The segmentation of an image is vital as it provides a region of interest to work with and thereby help in the subsequent steps.

5.1.1 Threshold Based Segmentation

Image thresholding is a popular technique applied for the segmentation of regions based on variations in their intensity. It is found to be most useful in separation of foreground and background using the most suitable threshold value and binarization of pixels depending on whether the values are higher or lower compared to the threshold. The selection of the threshold value as an intensity in the gray scale image is the most critical part here. It can be achieved by either a heuristic or an automated approach [28]. In blood cell image analysis, studies have shown that automated thresholding can be achieved in gray scale by using color difference values as mathematical combinations of RGB values [29, 30]. An alternative clustering based approach, Otsu's thresholding method is the most widely used optimum global thresholding method. The method maximizes the variances in between the classes, and selects the threshold value that would best tell apart the intensity values of the classes [24, 31]. An adaptive thresholding approach can be used to address the issue of uneven background intensity across an image as the aforementioned method takes up region wise local threshold values [32].

In Fig. 5a, a typical representative microscopic image of blood cells is shown. The image has been processed to show the steps of a simple WBC (nucleus) segmentation problem. Here, segmentation based on Otsu's threshold method has been implemented to identify the area of interest, in this case, the WBC (nucleus), from the background cells. The threshold operation alone has Fig. 5b as the input image and the binary image in Fig. 5c as the output image.

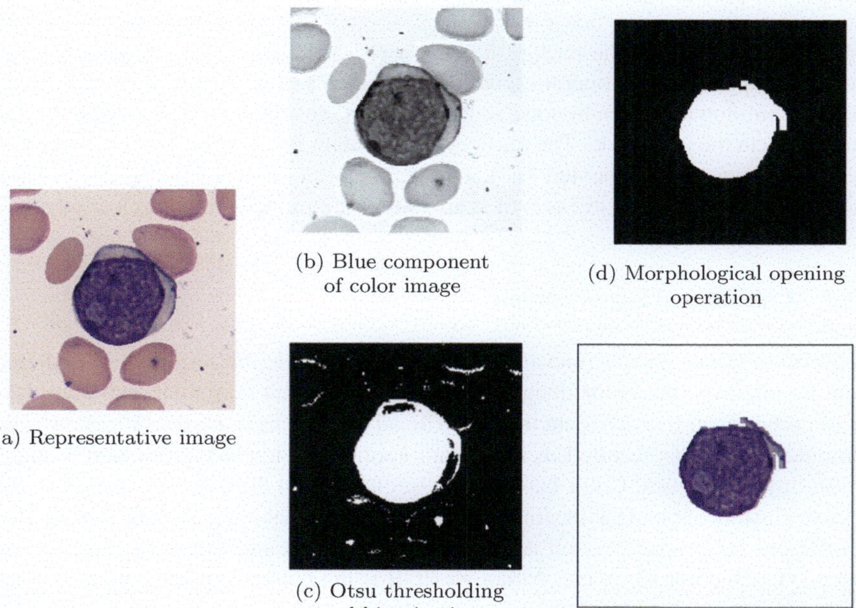

(a) Representative image

(b) Blue component of color image

(c) Otsu thresholding and binarization

(d) Morphological opening operation

(e) Segmentation of (nucleus)

Fig. 5 **a** Representative microscopic image of a blood sample, **b** Blue component extracted from the RGB image, **c** Otsu thresholding applied on blue component and binary image obtained, **d** Morphological operation opening performed on binary image, **e** Segmented WBC (nucleus) obtained from the original image

5.1.2 Morphology Based Detection

Morphology deals with the shapes and structures of objects. There exist a set of mathematical operations involving structures, known as morphological operations, such as opening, closing, dilation, erosion etc. Using these operations structure and shape of objects are analyzed. The morphological operations in image processing can be utilized for the better segregation and in turn better visualization of overlapping or connected cell particles [22, 33, 34]. Another powerful tool in morphological segmentation, often used in image analysis is the watershed transform that is a combination of edge detection and region growing techniques [35, 36]. The region growing algorithm is designed to define regions in an image where an initial seed pixel is identified and an iterative process is undertaken to grow the region based on some criteria. The criteria is meant to identify pixels in the image that could actually belong to the region. In the watershed segmentation approach, the image is considered a topological surface and the homogeneous gray scale areas as basins which are flooded. On the account that two water bodies meet, a dam is constructed (here, a line) [37].

For example, in Fig. 5, the segmentation of a WBC (nucleus) has been shown in subsequent steps. In Fig. 5d, a morphological operation opening has been used to remove the noise from the background and extract the area of the (nucleus) more precisely. Opening is an operation which implements erosion followed by dilation and is a common approach in noise removal, as the erosion not only erodes the noise but objects in the image too. The mask of the resultant image (Fig. 5d) when used on the original one, the segmented WBC (nucleus) can be observed in Fig. 5e. Erosion can also be implemented in cases of segregation of clumped cells for clear edges.

5.1.3 Color Based Segmentation

Color of an object carries relevant information alongside the information gathered from its intensity, but color images also have an inherent computational complexity in comparison to gray scale images. The advancement in image processing and computer vision has resulted in significant approaches towards color based image processing techniques. Color based image segmentation divides an image into distinctive clusters of pixels with similar color properties. Any image can be represented in different color spaces, such as HSI (Hue Saturation and Intensity), LAB color space, YCrCb color space etc. Although RGB is the natural representation of color images owing to the image acquisition system and the color monitors, HSI color space is very similar to how humans perceive color and description of an object's color is more convenient in this color space.

The gray scale image obtained from the colored representative image in Fig. 5, was observed to give poor segmentation results as the cells in the background did not have enough contrast with the WBC in the foreground. However, the color components, like Red, Green and Blue planes contain more information and the Blue plane (Fig. 5b) was observed to provide more precise segmentation results as seen in Fig. 5c.

5.1.4 Edge Detection Based Segmentation

A classical approach in edge detection involves the detection of abrupt gradients in intensity values within the image. However, at times, the edges might not be clearly discernible or the cells might be overlapped [38]. Edge detection algorithms can be used alongside morphological operators, watershed transform or filters can be utilized for better results [39].

In Fig. 6, example of edge detection has been shown with a representative blood cell image. The image has been subjected to gray scale conversion (Fig. 6b) followed by a smoothing filter (Fig. 6c) to smooth the edges of the cells. The different shapes of the cells are made visible in Fig. 6d using the canny edge detection technique. The image shows how the sickle cells have a different shape in comparison to normal RBCs.

5.2 Segmentation Techniques in Blood Cell Image Analysis

Literature reports numerous segmentation methods for different problems in analyzing blood cell images [40–50]. Primarily, the approach towards the detection of a region of interest is a combination of separation of foreground and background and the identification of individual cell components [51]. Threshold based techniques [38, 42, 52], region growing [44, 53, 54] and edge detection [55] are the most common approaches adopted in WBC segmentation problems. However, noise in the form of fragments of other cell particles can be present in the image. Contour based segmentation [56] techniques and morphological operators [57, 58] have been used to address these issues. Color based segmentation is also actively used in the area of WBC detection. The images are usually analyzed in the RGB color space, however, research has also been done in alternative color spaces like HSI [59, 60] and LAB

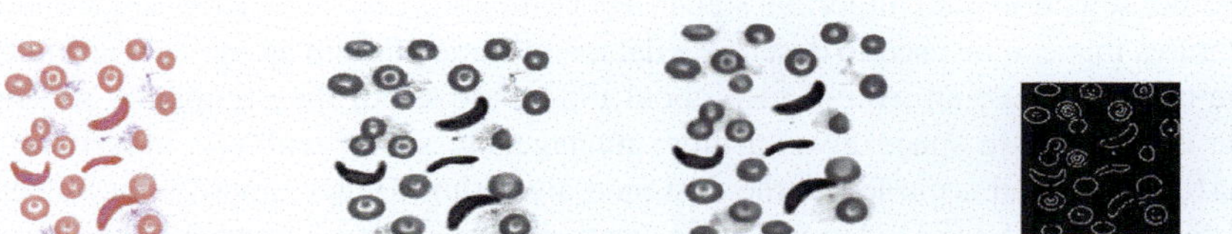

(a) Representative image (b) Gray scale image (c) Filtered image (smoothing) (d) Edge detected output

Fig. 6 **a** Representative blood cell image, **b** Converted image into gray scale, **c** Filtered image using smoothing for smooth edges, **d** Edge detection using Canny edge detector for contour identification of normal RBC and Sickle cells

[61]. Clumped cells are also a common occurrence in images for blood cell analysis, and watershed transform is used for this sort of problems mostly [62, 63].

Studies have shown that morphological operations, filtering, edge detection and thresholding are common approaches in the problems related to RBC analysis. Circular Hough transform has been used quite often [20, 33, 64–66] in counting of red blood cells with accuracy ranging from 80 to 100%. The mentioned approaches can be adopted in the detection of sickle cells [67–70] and malarial parasites [71] too. Shape and texture based approaches have attained high accuracy in malarial parasite detection.

6 Feature Extraction

In problems pertaining to machine learning, we get hold of raw data and there is a necessity to convert this complex data (for example, an image) into a simpler input form (for example, numeric pixel intensity values) as required for the algorithm [72]. Attributes, features of the items of interest when carefully chosen are indicative of the maximum relevant details that a lesion has to be provided by the picture for a complete characterization. Methodologies of selection of features examine artifacts and images to determine the most popular features indicative of the different types of artifacts [73]. For many biomedical applications pictures are significant. Multiple modalities are used to capture images including magnetic resonance imaging, absorption tomography, x-projection, x-computed tomography, ultrasonic, impedance tomography, electrical and magnetic source photography, and advanced photographic techniques. Such photographs are studied, often used as analytical instruments which can provide an glimpse into the inner workings of the mechanism under study quite frequently [73]. In image analysis of blood cells, several quantitative features [74] are used, as shown in Table 1.

7 Feature Selection

Feature selection is an important step in a classification problem. There are numerous features that can be used to describe a data, and some of them are more relevant than the others to give an overall idea about the whole set of data points. A subset of the entire feature space gives a more manageable set of features, improving the performance and accuracy of the designed algorithm. Hence, appropriate selection of features is vital [81]. Among other disadvantages, a high dimensional feature set can also lead to over fitting of data, a common and undesirable issue.

In addition, the features that can serve a method's best interest also need to be identified prior to the classification stage to avoid the inclusion of dependent or redundant features. Principal component analysis is a method capable of addressing both the issues as helps in reduction of the feature space dimension by selecting the

Table 1 Common feature descriptors used in blood cell image analysis

Features	Descriptors	Work done
Color based features (statistical properties)	Mean	In [75], color features are used for quantification of cytoplasmic granulation and cytoplasmic basophilia for the classification of lymphoid cells
	Standard deviation (SD)	
	Skewness	
	Kurtosis	
	Energy (uniformity) entropy (variability)	
Geometry based features	Area	In [76], the geometric features have been used for characterization of lymphocytes, to differentiate between normal and abnormal ones
	Perimeter	
	Circularity	
	Diameter	
	Eccentricity	
	Elongation	
	Roundness	In [77],geometric features have been used for morphological differentiation three groups of abnormal lymphoid cells
	Convexity	
	Nuclear size	
	Cell size	
	(nucleus)-cytoplasmic ratio	
	Nuclear eccentricity	
Texture based features	Gray level co-occurrence matrix (GLCM) granulometry	In [78], GLCM features have been used in bone marrow images to distinguish four erythrocyte cell stages
		In [79], the texture feature GLCM are used in differentiating blast lymphoid cells and normal leukocytes
		In [80], granulometry has been used for differentiating 12 lymphoid cell types

most significant elements from the feature vector. For example, Vogado et al. [82] have used a convolution neural network for feature extraction from the input image to diagnose leukemia using images from blood smear and then to get the most significant features, principal component analysis has been applied on the features (selected first 4096 features of the 12288). They used two image repositories, one containing only one leukocyte per image and the other containing multiple leukocytes. For Feature Selection there are several different approaches that can be implemented (Table 2). Some of the most important are:

Filter Method: filtering the data set and taking only one subset of it that includes all the relevant features.

Wrapper Method: meets the same Filter method goal but uses a machine learning model as the assessment criterion. We feed some features into our Machine Learning model, evaluate their performance, then determine whether to add or delete the feature to improve accuracy. As a consequence, this approach can be more efficient than filtering but is more costly in terms of computation.

Ensemble Method: The Embedded Model also uses a Machine Learning model, including the Filter Model. The distinction between the two approaches is that the embedded approach analyses our ML model's various training iterations and then determines the value of each function based on how much each of the features contributed to the ML model training.

Table 2 Types of feature selection techniques

Type	Feature selection technique
Filter method	Correlation feature selection (CFS) [83]
	Relief method [84]
	Gain information [85]
Wrapper method	Forward selection [86]
	Stepwise regression [87]
	Backward selection/elimination [86]
	Genetic algorithm [88]
	Hill climbing [89]
	Gray relational analysis [90]
	Exhaustive search [91]
	Random search [92]
	Fuzzy logic [93]
	Best first [94]
Ensemble method	Mutual information [95]
	Feature selection using clustering [96]

8 Classification

The next step after feature selection is Classification. This step has a major role in computer aided diagnosis. Many classification algorithms are available for applications in the real world such as Naive-Bayes [97], Support Vector Machine [98], k-Nearest Neighbor algorithms [99], Minimum Mean Distance (MMD) [100], Neural Networks [101], Decision Trees [102], Hidden Markov System [103], K-means clustering algorithms [104] etc. These algorithms are ideal for specific implementations, and classifier selection is dependent on the algorithm's output in a particular application domain [105, 106] The classification part has three steps. First training, validation and then testing. So we need to apply an optimization algorithm during training phase and store the data and process that huge amount of data for further evaluation. At first, the model suits on a training data set. The current model algorithm is applied to training data set and produces a result. The result from the model will be further compared with target value and based on that comparison the model and the learning algorithm used the parameters inside the model are manipulated accordingly. The fitting method is used for variable selection as well as parameter estimation.

Next step of the method is validation. For testing a given pattern, the validation set is used, although this is for frequent evaluation. In practical, data scientists use this knowledge to fine-tune the hyper parameters of the model [107]. The validation set results are used to update higher level hyper parameters. So it can be said that the validation set affects the model indirectly [107].

Finally, testing comes into picture. The Test data set establishes the gold standard used for the sample assessment. This is used only after a model is fully trained (using the train and validation sets). The research collection is usually used for testing competing models [107]. There are so many works where the validation set is used as test data set but that is not authentic process [107]. If in testing the data in the test data set has never been used the test data set is often considered a holdout data set.

Machine-learning methods have been applied most recently in the area of medical imaging.There are a number of algorithms for machine learning classification. Let's take a look at some machine learning classification algorithms.

8.1 Neural Network

Neural network is an interconnected network of neurons much like the biological nervous system in human beings. The neurons in our brain are connected to each other and it helps in the processing and flow of information in our brain. Likewise, the data information fed to a neural network is processed through mathematical operations in this vast system of neurons to give the desired output [108, 108]. Neural network implementations of computer-assisted diagnostics are the primary computational

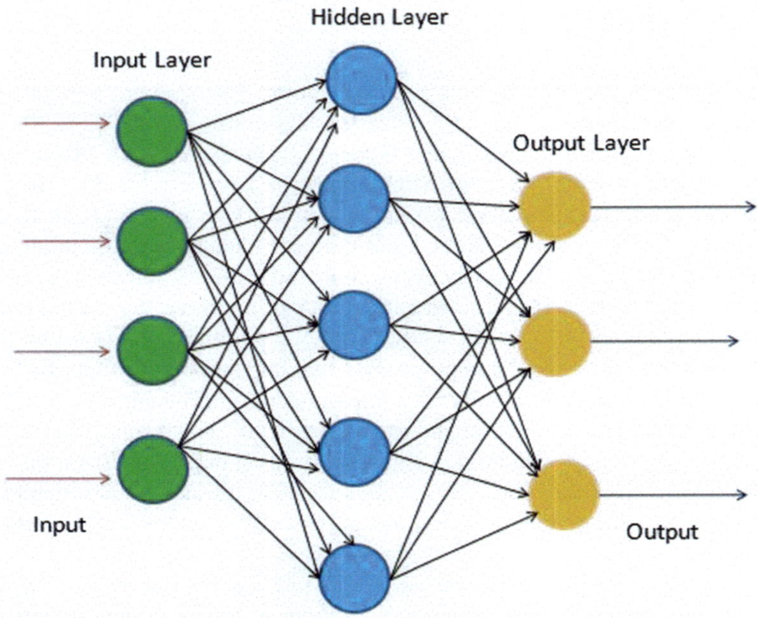

Fig. 7 Basic structure of neural network

knowledge source of medical imaging [109–111]. The Neural Network is developed from 3 types of layers (Fig. 7):

Input layer—This is the first layer in the neural network system. Each layer in the network contains predetermined number of neurons in them. This layer has only as many neurons as the dimension of the feature vector for our data. There might be added an additional neuron in this layer to introduce bias.

Hidden layers—Hidden layers are placed between the input and the output layers of the network. This layer is what makes neural networks so efficient and versatile. The hidden layers receive the input values in each of the neurons and apply an activation function to the weighted sum of said inputs. The activation functions introduce non-linearity to the network increasing the expanse of their applications. If not for the activation function, a neural network would be similar to a simple regression approach. Some of the commonly used activation functions are circular, phase, sigmoid, tanh and rectified circular unit (ReLu). The number of layers and the number of nodes in each of the layers are primary hyper parameters in a neural network architecture.

Output layer—The output layer provides suitable output for the inputs provided. The number of output layer depends on which mode the neural network is working. Depending on a single or multi response, the output layer as a single node or multiple nodes.

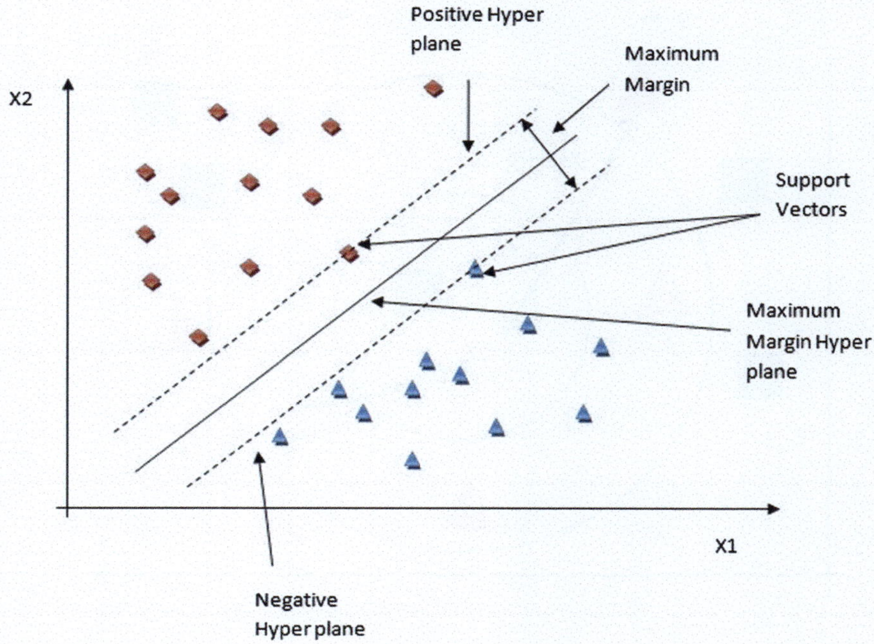

Fig. 8 Classification by SVM algorithm

8.2 *Support Vector Machine*

The support vector machine designs a model such that the data points belonging to different classes are separated by a decision boundary that has the maximum margin[112]. In this context, maximum margin of the hyperplane refers to the maximum distance between data points of separate classes. The algorithm. Support vectors are data points closer to the hyperplane, which affect the hyperplane's direction. Using such help vectors, we optimize the classifier's range (Fig. 8).

Hyper planes are boundaries for decision making which help to distinguish data points. Data points which fall on either side of the hyper plane can be assigned to different groups.

8.3 *K Nearest Neighbor*

The K-nearest neighbor (KNN) [113, 114] algorithm is one of the simplest algorithms conceptually. An object is ranked in this algorithm by a plurality vote of its neighbors [115]. An unknown input to a KNN model is classified to the class its nearest neighbor belongs to, as the nomenclature would suggest [113] (Fig. 9).

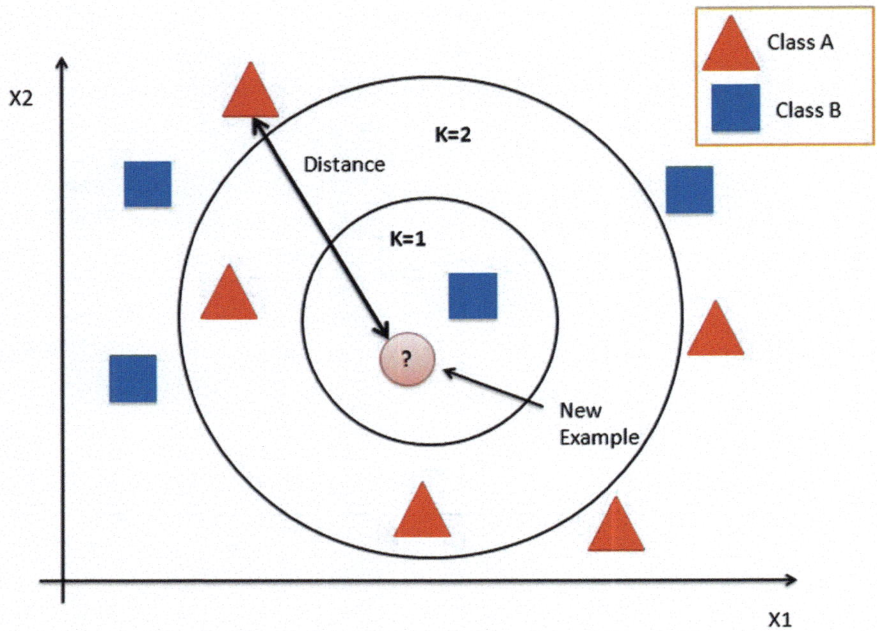

Fig. 9 Classification by KNN algorithm

Now the question is what is K. K is actually the number of neighbors to be considered. If $k = 1$, then test examples in the training set are given the same label as the nearest example. If $k = 2$ is verified for the labels of the two nearest classes. The value of k in the algorithm is a hyper parameter and selecting the appropriate k value affects the accuracy of the problem. But generally, the k value is set to the square root of the number of records in your training set. KNN is popular in medical image processing [115–118] field for its simplicity and performance.

8.4 Decision Tree

All the machine learning approaches listed up to this point have one major disadvantage: Generally, the values used in the weights and activation functions can not be extracted to obtain any sort of knowledge that humans can perceive [119]. A decision tree has a structure similar to that of a tree and the nodes and connectors are compared to the branches and leaves of the tree. In a decision tree, the nodes are a condition or a question and the resultant outcomes are the branches or connectors between the nodes. The terminal nodes, known as the leaf node are a class label. Decision tree is one of the most popular supervised learning algorithms, but has a rather common tendency of over fitting. We map the non-linear structures very well in tree based algorithms, unlike linear models.

8.5 *Naive Bayes Algorithm*

The naive Bayes classifier significantly simplifies learning by suggesting that features are class independent [97]. The algorithm is based upon the Bayes' theorem and assumes that the features are independent of each other, and in practical it is almost always untrue. Hence, naive. Bayes' theorem is stated as the following equation, mathematically:

$$P(X/Y) = \frac{P(Y/X) * P(X)}{P(Y)} \tag{1}$$

$P(X/Y)$	is the posterior probability of class (X, target) given predictor (Y, attributes).
$P(X)$	is the prior probability of class.
$P(Y/X)$	is the likelihood which is the probability of predictor given class.
$P(Y)$	is the prior probability of predictor.

8.6 *Deep Learning*

Deep learning is in fact, broadly speaking, a part of machine learning in which multi-layered networks are used to analyze patterns within the input data of raw images [114]. Deep neural networks, stacked auto encoders, deep Boltzmann machines and convolution neural networks (CNNs) are some deep learning algorithm tools. We'll concentrate on CNNs as these are most widely used for medical image data sets [113, 120].

A Convolution Neural Network is an algorithm that helps in identifying the features from the input provided to the system. Although the filters are hand-engineered in rudimentary methods, ConvNets are capable of learning these features with enough preparation. A CNN has several layers of convolutions and activation, mostly interspersed with pooling teams, and is trained as regular artificial neural networks using back-propagation and gradient descent [121]. Image convolution operations take place in the network by exchange in network weight values [122]. This way, it is not necessary for the network to learn different detectors for redundant objects in the image. This also significantly decreases the number of parameters that need to be learned (i.e. the number of weights are independent of input size [122]. A ConvNet is capable of capturing the Spatial and Temporary dependencies in an image by applying proper filters.

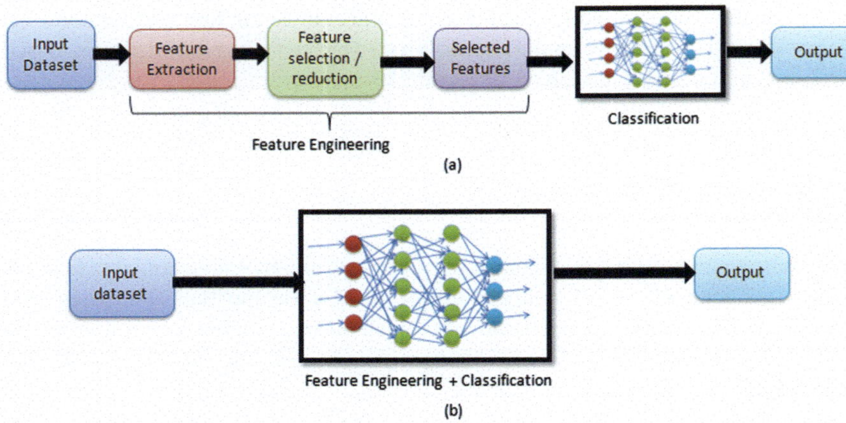

Fig. 10 **a** Conventional machine learning approach versus. **b** Deep learning workflow

Nevertheless, CNNs require very large and properly labeled data sets, as well as considerable computational resources to be equipped. Therefore, profound learning is applied in medical decision-making by using pre-trained CNNs [114].While deep learning enables computers to learn directly from image data, it takes millions of images for CNNs to be trained "from scratch" for each clinical function [114].

Deep learning algorithm can make diagnosis and care preparation simpler for clinicians and radiologists. Varieties of modalities are available for this purpose, X-Ray, CT, PET, MRI, fMRI, Ultrasound etc to detect the problem in various organs of human body. Deep learning also has advantages over conventional classifier. In conventional Machine learning approach, the vast majority of the applied features should be distinguished by a specialist and afterward hand-coded according to the area and information type. The performance of the most part of the Machine Learning algorithms relies upon how precisely the features are recognized and extracted. Deep Learning implemented the idea of end-to-end learning where the computer is given a data-set of images annotated with which object classes are present in each image [123, 124]. A Deep Learning model is thereby 'trained' on the data given, where NN discover the underlying trends in image classes and automatically evaluate the most concise and excellent features for each object's particular class [123]. As it is shown in Fig. 10 that a conventional machine learning approach (Fig. 10a) various steps for feature engineering are involved whereas in deep learning architecture the classifier skips those steps and does it automatically in one step. This feature of deep learning model (Fig. 10b) makes the model faster and accurate as the model is choosing the features by itself. When dealing with complex learning problems, deep architecture often has an advantage over shallow architectures. The stacking in layer-wise fashion of multiple linear and non-linear processing units gives the ability to learn complex representations at different abstraction levels [125].

Deep learning has seen massive growth in its prominence and effectiveness in recent years, mainly as a result of more mainframe devices, larger databases, and deeper-networking techniques.

9 Conclusion

In this article, the approaches taken in developing and designing computer aided diagnosis systems in the field of hematological disorders have been discussed. Techniques involved in the development of an efficient CAD system, namely image acquisition, pre-processing, segmentation and classification have been discussed in details. The chapter touches upon the common approaches that have been employed in a number of studies related to hematology and the general overview of said approaches. It covers a collection of state of the art techniques for disorders that can be identified from the analysis of blood cell images.

Acknowledgements The authors are thankful to Indian Institute of Technology, Kharagpur for providing the infrastructural support. We are thankful to Mr. Motiur Rahaman for his help with the rudimentary representative blood cell images. Special thanks to Ministry of Human Resource Development, Government of India and Ministry of Health and Family Welfare, Government of India for providing financial assistance through the IMPRINT-India initiative for project no. 5968, titled "Development of a portable decision support tool for screening high risk antenatal cases at household level by community health workers".

References

1. Doi, K.: Computer-aided diagnosis in medical imaging: historical review, current status and future potential. Comput. Med. Imaging Graph. **31**(4–5), 198–211 (2007)
2. Li, Q., Nishikawa, R.M.: Computer-Aided Detection and Diagnosis in Medical Imaging. Taylor & Francis (2015)
3. Jalalian, A., et al.: Foundation and methodologies in computer-aided diagnosis systems for breast cancer detection. In: EXCLI J. **16**, 113 (2017)
4. Balogh, E.P. et al.: The path to improve diagnosis and reduce diagnostic error. In: Improving Diagnosis in Health Care. National Academies Press (US), 2015
5. Dubey, R.B., Hanmandlu, M.: Integration of CAD into PACS. In: 2012 2nd International Conference on Power, Control and Embedded Systems, pp. 1–6. IEEE (2012)
6. Castellino, R.A.: Computer aided detection (CAD): an overview. Cancer Imaging **5**(1), 17 (2005)
7. Rogers, W., Ryack, B., Moeller, G.: Computer-aided medical diagnosis: literature review. Int. J. bio-med. Comput. **10**(4), 267–289 (1979)
8. Foran, D.J., Chen, W., Yang, L.: Automated image interpretation and computer-assisted diagnostics. Stud Health Technol Inform **185**, 77–108 (2013)
9. Kiszka, K., Haduch, J., Pasowicz, M.: Application of computer aided diagnosis (CAD) in clinical imaging. Presentation of the value and current directions in CAD development in various imaging techniques based on literature review. Przeglad lekarski **66**(8), 437–447 (2009)

10. Fleming, K.A., et al.: High-quality diagnosis: an essential pathology package. In: Disease Control Priorities: Improving Health and Reducing Poverty, 3rd edn. The International Bank for Reconstruction and Development/The World Bank (2017)
11. De la Salle, B.: Pre- and postanalytical errors in haematology. Int. J. Lab. Hematol. **41**, 170–176 (2019)
12. Rodak, B.F., Keohane, E.M., Fritsma, G.A.: Hematology-E-Book: Clinical Principles and Applications. Elsevier Health Sciences (2013)
13. Cahan, A., Cimino, J.J.: A learning health care system using computer-aided diagnosis. J. Med. Internet Res. **19**(3), e54 (2017)
14. Mar, C.D., Doust, C.D., Glasziou, P.: Clinical Thinking. Wiley Online Library (2007)
15. Kahneman, D., Klein, G.: Conditions for intuitive expertise: a failure to disagree. Am. Psychol. **64**(6), 515 (2009)
16. Hegde, R.B., et al.: Peripheral blood smear analysis using image processing approach for diagnostic purposes: a review. Biocybernetics Biomed. Eng. **38**(3), 467–480 (2018)
17. Shattil, S.J.: A (blood) smear campaign (2003)
18. Bain, B.J.: Diagnosis from the blood smear. N. Engl. J. Med. **353**(5), 498–507 (2005)
19. Nilsson, B., Heyden, A.: Segmentation of dense leukocyte clusters. In: Proceedings IEEE Workshop on Mathematical Methods in Biomedical Image Analysis (MMBIA 2001), pp. 221–227. IEEE (2001)
20. Tomari, R., et al.: Computer aided system for red blood cell classification in blood smear image. Proced. Comput. Sci. **42**, 206–213 (2014)
21. Ross, N.E., et al.: Automated image processing method for the diagnosis and classification of malaria on thin blood smears. Med. Biol. Eng. Comput. **44**(5), 427–436 (2006)
22. Dave, I.R., Kishor, P., Upla.: Computer aided diagnosis of Malaria disease for thin and thick blood smear microscopic images. In: 4th International Conference on Signal Processing and Integrated Networks (SPIN), vol. **2017**, pp. 561–565. IEEE (2017)
23. Adatrao, S., Mittal, M.: An analysis of different image preprocessing techniques for determining the centroids of circular marks using hough transform. In: 2016 2nd International Conference on Frontiers of Signal Processing (ICFSP), pp. 110–115. IEEE (2016)
24. Gonzalez, R.C., Woods, R.E., Eddins, S.L.: Digital image processing using MATLAB. Pearson Education India (2004)
25. Kaur, R., Kaur, S.: Comparison of contrast enhancement techniques for medical image. In: 2016 Conference on Emerging Devices and Smart Systems (ICEDSS), pp. 155–159. IEEE (2016)
26. Victor Haryanto, S.E., et al.: Malaria parasite detection with histogram color space method in Giemsa-stained blood cell images. In: 2017 5th International Conference on Cyber and IT Service Management (CITSM), pp. 1–4. IEEE (2017)
27. Sheeba, F., et al.: Detection of poor quality peripheral blood smear images used in detection of leukocytes and erythrocytes. In: 2017 Fourth International Conference on Image Information Processing (ICIIP), pp. 1–4. IEEE (2017)
28. Micheli-Tzanakou, E., Sheikh, H., Zhu, B.: Neural networks and blood cell identification. J. Med. Syst. **21**(4), 201–210 (1997)
29. Poon, S.S.S., Ward, R.K., Palcic, B.: Automated image detection and segmentation in blood smears. Cytometry: J. Int. Soc. Anal. Cytol. **13**(7), 766–774 (1992)
30. Harms, H., et al.: Segmentation of stained blood cell images measured at high scanning density with high magnification and high numerical aperture optics. Cytom.: J. Int. Soc. Anal. Cytol. **7**(6), 522–531 (1986)
31. Otsu, N.: A threshold selection method from gray-level histograms. IEEE Trans. Syst. Man, Cybern. **9**(1), 62–66 (1979)
32. Done, S.: Breast cancer: recent advances in biology. BoD-Books on Demand, Imaging and Therapeutics (2011)
33. Humaimi Mahmood, N., Asraf Mansor, M.: Red blood cells estimation using Hough transform technique. Sig. Image Proc. **3**(2), 53 (2012)

34. Ma, Y.-D., Dai, R.-L., Li, L.: A counting and segmentation method of blood cell image with logical and morphological feature of cell. Chin. J. Electron. **11**(1), 53–55 (2002)
35. Beucher, S., Meyer, F.: The morphological approach to segmentation: the watershed transformation. Math. Morphol. Image proc. **34**, 433–481 (1993)
36. Soille, P.: Morphological Image Analysis: Principles and Applications. Springer Science & Business Media (2013)
37. Dorini, L.B., Minetto, R., Leite, N.J.: White blood cell segmentation using morphological operators and scale-space analysis. In: XX Brazilian Symposium on Computer Graphics and Image Processing (SIBGRAPI 2007), pp. 294–304. IEEE (2007)
38. Wu, J., et al.: A novel color image segmentation method and its application to white blood cell image analysis. In: 2006 8th International Conference on Signal Processing, vol. 2. IEEE (2006)
39. Díaz, G., Manzanera, A.: Automatic analysis of microscopic images in hematological cytology applications. In: Clinical Technologies: Concepts, Methodologies, Tools and Applications. IGI Global, pp. 325–352 (2011)
40. Mohamed, M., Far, B., Guaily, A.: An efficient technique for white blood cells nuclei automatic segmentation. In: 2012 IEEE International Conference on Systems, Man, and Cybernetics (SMC), pp. 220–225. IEEE (2012)
41. Tosta, T.A.A.: Unsupervised segmentation of leukocytes images using thresholding neighborhood valley-emphasis. In: IEEE 28th International Symposium on Computer-Based Medical Systems, vol. **2015**, pp. 93–94. IEEE (2015)
42. Liao, Q., Deng, Y.: An accurate segmentation method for white blood cell images. In: Proceedings IEEE International Symposium on Biomedical Imaging, pp. 245-248. IEEE (2002)
43. Ongun, G. et al.: Feature extraction and classification of blood cells for an automated differential blood count system. In: IJCNN'01. International Joint Conference on Neural Networks. Proceedings (Cat. No. 01CH37222), vol. 4, pp. 2461–2466. IEEE (2001)
44. Sadeghian, F., et al.: A framework for white blood cell segmentation in microscopic blood images using digital image processing. Biol. Proced. Online **11**(1), 196 (2009)
45. Xing, F., Yang, L.: Robust nucleus/cell detection and segmentation in digital pathology and microscopy images: a comprehensive review. IEEE Rev. Biomed. Eng. **9**, 234–263 (2016)
46. Bala, A.: An improved watershed image segmentation technique using MATLAB. Int. J. Sci. Eng. Res. **3**(6), 1–4 (2012)
47. Di Ruberto, C., et al.: Analysis of infected blood cell images using morphological operators. Image Vis. Comput. **20**(2), 133–146 (2002)
48. Imroze Khan, M., et al.: Content based image retrieval approaches for detection of malarial parasite in blood images. Int. J. Biomet. Bioinf. (IJBB) **5**(2), 97 (2011)
49. Sio, S.W.S., et al.: MalariaCount: an image analysis-based program for the accurate determination of parasitemia. J. Microbiol. Methods **68**(1), 11–18 (2007)
50. Arslan, S., Ozyurek, E., Gunduz-Demir, C.: A color and shape based algorithm for segmentation of white blood cells in peripheral blood and bone marrow images. Cytometry Part A **85**(6), 480–490 (2014)
51. Vicar, T., et al.: Cell segmentation methods for label-free contrast microscopy: review and comprehensive comparison. BMC Bioinf. **20**(1), 360 (2019)
52. Rezatofighi, S.H., et al.: A new approach to white blood cell nucleus segmentation based on gram-schmidt orthogonalization. In: 2009 International Conference on Digital Image Processing, pp. 107–111. IEEE (2009)
53. Scotti, F.: Automatic morphological analysis for acute leukemia identification in peripheral blood microscope images. In: CIMSA. 2005 IEEE International Conference on Computational Intelligence for Measurement Systems and Applications, 2005, pp. 96–101 IEEE (2005)
54. Ravi Kumar, B., Joseph, D.K., Sreenivas, T.V.: Teager energy based blood cell segmentation. In: 2002 14th International Conference on Digital Signal Processing Proceedings. DSP 2002 (Cat. No. 02TH8628), vol. 2, , pp. 619–622. IEEE (2002)
55. Chassery, J.-M., Garbay, C.: An iterative segmentation method based on a contextual color and shape criterion. IEEE Trans. Pattern Anal. Mach. Intell. **6**, 794–800 (1984)

56. Ongun, G.: An automated differential blood count system. In: Conference Proceedings of the 23rd Annual International Conference of the IEEE Engineering in Medicine and Biology Society, vol. 3, pp. 2583–2586. IEEE (2001)
57. Madhloom, H.T., Kareem, S.A., Ariffin, H.: An image processing application for the localization and segmentation of lymphoblast cell using peripheral blood images. J. Med. Syst. **36**(4), 2149–2158 (2012)
58. Piuri, V., Scotti, F.: Morphological classification of blood leucocytes by microscope images. In: 2004 IEEE International Conference onComputational Intelligence for Measurement Systems and Applications, 2004. CIMSA, pp. 103–108. IEEE (2004)
59. Abdul Nasir, A.S., Mashor, M.Y., Rosline, H.: Unsupervised colour segmentation of white blood cell for acute leukaemia images. In: 2011 IEEE International Conference on Imaging Systems and Techniques, pp. 142–145. IEEE (2011)
60. Nor Hazlyna, H., et al.: Comparison of acute leukemia Image segmentation using HSI and RGB color space. In: 10th International Conference on Information Science, Signal Processing and their Applications (ISSPA 2010), pp. 749–752. IEEE (2010)
61. Mohapatra, S., Patra, D.: Automated cell nucleus segmentation and acute leukemia detection in blood microscopic images. In: 2010 International Conference on Systems in Medicine and Biology, pp. 49–54. IEEE (2010)
62. Yang, X., Li, H., Zhou, X.: Nuclei segmentation using marker-controlled watershed, tracking using mean-shift, and Kalman filter in time-lapse microscopy. IEEE Trans. Circ. Syst. I: Regular Papers **53**(11), 2405–2414 (2006)
63. Cheng, J., Rajapakse, J.C., et al.: Segmentation of clustered nuclei with shape markers and marking function. IEEE Trans. Biomed. Eng. **56**(3), 741–748 (2008)
64. Maitra, M., Gupta, R.K., Mukherjee, M.: Detection and counting of red blood cells in blood cell images using hough transform. Int. J. Comput. Appl. **53**(16) (2012)
65. Nasreen, N., Kumar, C., Nabeel, A.P.: Counting of RBC using circular hough transform with median filtering. In: Proceeding of Third National Conference on Emerging Trends in Engineering, pp. 150–153 (2015)
66. Mazalan, S.M., Humaimi Mahmood, N., Abdul Razak, M.A.: Automated red blood cells counting in peripheral blood smear image using circular hough transform. In: 2013 1st International Conference on Artificial Intelligence, Modelling and Simulation, pp. 320–324. IEEE (2013)
67. Bala, S., Doegar, A.: Automatic detection of sickle cell in red blood cell using watershed segmentation. Int. J. Adv. Res. Comput. and Commun. Eng **4**(6), 488–491 (2015)
68. Elsalamony, H.A.: Sickle anemia and distorted blood cells detection using hough transform based on neural network and decision tree. In: Proceedings of the International Conference on Image Processing, Computer Vision, and Pattern Recognition (IPCV). The Steering Committee of The World Congress in Computer Science, Computer, p. 1 (2014)
69. Hariharan, S., Parvathy, H.B., Aruna, S.N.: An overview of sickle cell anemia special emphasis on image processing on SEM images. Int. J. Appl. Eng Res **11**(1), 201–8 (2016)
70. Aruna, N.S., Hariharan, S.: Edge detection of sickle cells in red blood cells. Int. J. Comput. Sci. Inf. Technol. (IJCSIT) **5**(3), 4140–4144 (2014)
71. Punitha, S., et al.: Detection of malarial parasite in blood using image processing. Asian J. Appl. Sci. Technol. (AJAST) **1**(2), 211–213 (2017)
72. Dara, S., et al.: Feature extraction in medical images by using deep learning approach. Int. J. Pure Appl. Math. **120**(6), 305–312 (2018)
73. Umbaugh, S.E., Wei, Y.-S., Zuke, M.: Feature extraction in image analysis. A program for facilitating data reduction in medical image classification. IEEE Eng. Med. Biol. Mag. **16**(4), 62–73 (1997)
74. Merino, A., et al.: Optimizing morphology through blood cell image analysis. Int. J. Lab. Hematol. **40**, 54–61 (2018)
75. Bonilla, L.L., et al.: Progress in Industrial Mathematics at ECMI 2006, vol. 12. Springer (2007)

76. Eldar, S., et al.: Computer-assisted image analysis of small cell lymphoma of the thyroid gland: comparison of nuclear parameters of small lymphocytes in lymphomas and Hashimoto's thyroiditis. Comput. Med. Imaging Graphics **22**(6), 479–488 (1998)
77. Jahanmehr, S.A.H., et al.: Quantitation of cytological parameters of malignant lymphocytes using computerized image analysis. Int. J. Lab. Hematol. **30**(4), 278–285 (2008)
78. Kono, K., et al.: Quantitative distinction of the morphological characteristic of erythrocyte precursor cells with texture analysis using gray level co-occurrence matrix. J. Clin. Lab. Anal. **32**(1), e22175 (2018)
79. Patel, N., Mishra, A.: Automated leukaemia detection using microscopic images. Proced. Comput. Sci. **58**, 635–642 (2015)
80. Puigví, L., et al.: New quantitative features for the morphological differentiation of abnormal lymphoid cell images from peripheral blood. J. Clinic. Pathol. **70**(12), 1038–1048 (2017)
81. Zhou, X., et al.: Feature selection for image classification based on a new ranking criterion. J. Comput. Commun. **3**(03), 74 (2015)
82. Vogado, L.H.S., et al.: Leukemia diagnosis in blood slides using transfer learning in CNNs and SVM for classification. Eng. Appl. Artif. Intell. **72**, 415–422 (2018)
83. Andrew Hall, M.: Correlation-based feature selection for machine learning (1999)
84. Sun, Y., Lou, X., Bao, B.: A novel relief feature selection algorithm based on mean-variance model. J. Inf. Comput. Sci. **8**(16), 3921–3929 (2011)
85. Uğuz, H.: A two-stage feature selection method for text categorization by using information gain, principal component analysis and genetic algorithm. Knowl. Based Syst. **24**(7), 1024–1032 (2011)
86. Kezhi, Z.M.: Orthogonal forward selection and backward elimination algorithms for feature subset selection. IEEE Trans. Syst. Man, Cybern. Part B (Cybern.) **34**(1), 629–634 (2004)
87. Zhou, J., et al.: Streamwise feature selection. J. Mach. Learn. Res. **7**(Sep), 1861–1885 (2006)
88. Leardi, R.: Application of genetic algorithm-PLS for feature selection in spectral data sets. J. Chemometr. **14**(5–6), 643–655 (2000)
89. David, B.S.: Prototype and feature selection by sampling and random mutation hill climbing algorithms. In: Machine Learning Proceedings 1994. Elsevier, pp. 293–301 (1994)
90. Ahmad Ubaidillah, S.H.S.: Grey Relational Analysis Feature Selection for Cancer Classification Using Support Vector Machine. PhD thesis. Universiti Teknologi Malaysia (2014)
91. Jović, A., Brkić, K., Bogunović, N.: A review of feature selection methods with applications. In: 38th International Convention on Information and Communication Technology, Electronics and Microelectronics (MIPRO), vol. **2015**, pp. 1200–1205. IEEE (2015)
92. Farmer, M.E., Bapna, S., Jain, A.K.: Large scale feature selection using modified random mutation hill climbing. In: Proceedings of the 17th International Conference on Pattern Recognition, 2004. ICPR 2004. vol. 2, pp. 287–290, IEEE (2004)
93. Cintra, M.E., et al.: Feature subset selection using a fuzzy method. In: 2009 International Conference on Intelligent Human-Machine Systems and Cybernetics, vol. 2, pp. 214–217. IEEE (2009)
94. Kohavi, R., Sommerfield, D.: Feature subset selection using the wrapper method: over-fitting and dynamic search space topology. In: KDD, pp. 192–197 (1995)
95. Estévez, P.A., et al.: Normalized mutual information feature selection. IEEE Trans. Neural Networks **20**(2), 189–201 (2009)
96. Hong, Yi, et al.: Unsupervised feature selection using clustering ensembles and population based incremental learning algorithm. Pattern Recogn. **41**(9), 2742–2756 (2008)
97. Rish, I., et al.: An empirical study of the naive Bayes classifier. In: IJCAI 2001 Workshop on Empirical Methods in Artificial Intelligence, vol. 3. pp. 41–46 (2001)
98. Ma, Y., Guo, G.: Support Vector Machines Applications. Springer (2014)
99. Kramer, O.: K-nearest neighbors. In: Dimensionality Reduction with Unsupervised Nearest Neighbors. Springer, pp. 13–23 (2013)
100. Zhang, D., Chen, S., Zhou, Z.-H.: Learning the kernel parameters in kernel minimum distance classifier. Pattern Recognit. **39**(1), 133–135 (2006)
101. Hu, Y.H., Hwang, J.-N.: Handbook of neural network signal processing (2002)

102. Safavian, S.R., Landgrebe, D.: A survey of decision tree classifier methodology. IEEE Trans. Syst. Man Cybern. **21**(3), 660–674 (1991)
103. Tsuboka, E., Nakahashi, J.: On the fuzzy vector quantization based hidden Markov model. In: Proceedings of ICASSP'94. IEEE International Conference on Acoustics, Speech and Signal Processing, vol. 1, pp. I–637. IEEE (1994)
104. Wu, J.: Cluster analysis and K-means clustering: an introduction. In: Advances in K-means Clustering, pp. 1–16. Springer (2012)
105. Manjula, K., Vijayarekha, K., Vimaladevi, P.: Review on classification algorithms in image Processing. Int. J. Innovat. Trends Eng. Res. **2**(11) (2017)
106. Parui, S.: Emotion recognition from EEG signal using XGBoost algorithm. In: IEEE 16th India Council International Conference (INDICON). IEEE. **2019**, pp. 1–4 (2019)
107. Shah, T.: About train, validation and test sets in machine learning. In: Towards Data Science 6 (2017)
108. Li, Q.: Medical image classification with convolutional neural network. In: 13th International Conference on Control Automation Robotics & Vision (ICARCV), vol. **2014**, pp. 844–848. IEEE (2014)
109. Ge, J., et al.: Computer aided detection of clusters of microcalcifications on full field digital mammograms. Med. Phys. **33**(8), 2975–2988 (2006)
110. Lo, S.-C.B., et al.: Artificial convolution neural network for medical image pattern recognition. Neural Networks **8**(7–8), 1201–1214 (1995)
111. Nagel, R.H., et al.: Analysis of methods for reducing false positives in the automated detection of clustered microcalcifications in mammograms. Med. Phys. **25**(8), 1502–1506 (1998)
112. Dehmeshki, J., et al.: Classification of lung data by sampling and support vector machine. In: The 26th Annual International Conference of the IEEE Engineering in Medicine and Biology Society, vol. 2, pp. 3194–3197, IEEE (2004)
113. Erickson, B.J., et al.: Machine learning for medical imaging. Radiographics **37**(2), 505–515 (2017)
114. Giger, M.L.: Machine learning in medical imaging. J. Am. Coll. Radiol. **15**(3), 512–520 (2018)
115. Li, C., et al.: Using the K-nearest neighbor algorithm for the classification of lymph node metastasis in gastric cancer. In: Computational and Mathematical Methods in Medicine 2012 (2012)
116. Korn, F., et al.: Fast nearest neighbor search in medical image databases. Tech. rep. 1998
117. Paredes, R., et al.: Classification of medical images using local representations. In: Bildverarbeitung für die Medizin 2002, pp. 171–174 . Springer (2002)
118. Ramteke, R.J., Khachane Monali, Y.: Automatic medical image classification and abnormality detection using k-nearest neighbour. Int. J. Adv. Comput. Res. **2**(4), 190–196 (2012)
119. Wernick, M.N., et al.: Machine learning in medical imaging. IEEE Sig. Proc. mag. **27**(4), 25–38 (2010)
120. Lee, H. et al.: Face image retrieval using sparse representation classifier with gabor-lbp histogram. In: International Workshop on Information Security Applications, pp. 273–280. Springer (2010)
121. Suzuki, K.: Overview of deep learning in medical imaging. Radiol. Phys. Technol. **10**(3), 257–273 (2017)
122. Litjens, G., et al.: A survey on deep learning in medical image analysis. Med. Image Anal. **42**, 60–88 (2017)
123. O'Mahony, N., et al.: Deep learning versus traditional computer vision. In: Science and Information Conference, pp. 128–144. Springer (2019)
124. Koehn, P.: Combining genetic algorithms and neural networks: the encoding problem (1994)
125. Khan, A., et al.: A survey of the recent architectures of deep convolutional neural networks. In: arXiv preprint arXiv:1901.06032 (2019)

Magnets, Magnetism, and Magnetic Resonance Imaging: History, Basics, Clinical Aspects, and Future Directions

Shatadru Chakravarty and Erik M. Shapiro

Abstract Magnets and magnetism have played an intriguing and controversial role in human medicine. Undoubtedly, the most relevant use of magnetic phenomenon in modern clinics pertains to the diagnostic potential of Magnetic Resonance Imaging (MRI) that employs low-intensity radiofrequency electromagnetic radiation to study subjects placed in a strong magnetic field. The physical basis of MRI lies in its inherent ability to monitor the temporal and spatial distribution of tissue water protons, in the process taking into account local abnormalities to generate images with variable contrast. The contrast produced in MRI is further enhanced by the administration of paramagnetic entities called contrast agents that allow for superior spatial resolution in MRI. This chapter gives a glimpse into the history and development of MRI as a diagnostic imaging tool. The fundamentals of the MRI technique, contrast agent design, their current clinical status, and future directions are also discussed.

Keywords Magnetic resonance imaging · Contrast agents · Gadolinium · Nephrogenic systemic fibrosis

1 Introduction

Man's tryst with magnets and magnetism in general dates back to ancient times when the phenomenon of *aurora borealis* (or the northern lights) was first being observed [1]. It was, however, not until 1200 BC that an understanding of the general principles of magnetism began to develop with advances in iron smelting and the term 'magnetite' (coined to refer to iron oxide, Fe_3O_4) came into existence. The first exposition on the properties of magnetized needles was authored by Petrus Peregrinus in 1269 [2]. It was not until the year 1600 that a more enhanced and

S. Chakravarty (✉) · E. M. Shapiro
Department of Radiolgy, Michigan State University, 846 Service Road, East Lansing, MI, USA
e-mail: shatadru2006@gmail.com; shatadru@msu.edu

Institute for Quantitative Health Sciences and Engineering, Michigan State University, 775 Woodlot Dr, East Lansing, MI, USA

G. Dutta et al. (eds.), *Modern Techniques in Biosensors*, Studies in Systems, Decision and Control 327, https://doi.org/10.1007/978-981-15-9612-4_6

scientific treatise, including the discovery of earth's own magnetic properties was presented by the English physician, William Gilbert in his much-celebrated book *De Magnete* [3]. It is still widely considered to be a pioneering work in early scientific research. In medicine, magnetite or lodestone was considered to possess miraculous healing properties, stemming from the early belief that magnetite has a soul of its own, justified by its ability to move iron particles, just as the soul was believed to produce motion. The earliest proponent of this ancient theory was the Greek astronomer Thales of Miletus [4]. Magnetite's use as an effective agent to control bleeding and hemorrhage was championed by Hippocrates of Cos and his pupils [5]. In addition, there were references to using this ore for the treatment of burns, arthritis, gout, poisoning, and baldness. Together with these widespread external uses of magnetite, the Egyptian physician Avicenna (980–1037 AD) doctored its uptake with milk in order to facilitate the intestinal excretion of poisonous iron rust from the body of a patient [6]. The attracting prowess of a magnet was also put into use for surgical procedures, the earliest example being that of a Hindu surgeon Sucruta (600 BC) who employed it to remove an iron arrow tip. Other noted examples include its use as a cure for hernia and removal of iron dust or metal particles from the eye during the late sixteenth century [7]. However, most of these medical and surgical applications of lodestone were explicitly disqualified by Gilbert at the end of the sixteenth century who advocated its use only for the treatment of chlorosis [8].

Over the next 300 years, detailed practical accounts on the use of magnets for surgical procedures were being published. Most of these procedures put into use increasingly complex techniques employing either native magnets or artificial electromagnets for the removal of small iron or steel particles that had accidentally got into the eyes. The ophthalmic use of electromagnets was pioneered by two feuding compatriots, Dr. Julius Hirschberg from Germany who advocated the removal of small objects from the back of the eye, and the Swiss ophthalmologist Otto Haab, who insisted on removal from the front. Both these procedures were, however, rendered equally harmful in terms of consequent damage to the iris and related side effects. Subsequently, modern advancements in eye surgery techniques and instrumentation have rendered these techniques obsolete.

On a related note, the magnetic properties of lodestone and other artificial magnets was used as a cure for nervous disorders and moral conduct. Probably, the greatest champion of this treatment modality was Franz Anton Mesmer, a physician in Vienna who assumed that magnets worked by their ability to reorient the flow of universal 'fluidum' in a patient's body. His methods gained immense popularity, initially in Vienna and then in Paris, and were considered to be a clever mix of hypnotism and psychotherapy. Later on, several experiments conducted by Benjamin Franklin and Antoine Lavoisier conclusively indicated that *suggestibility and not magnetism* had a psychological impact on the patients who were subjected to this new treatment method and were subsequently cured because of their faith in it [9]. In modern times, magnets are still believed to have healing properties and are marketed in various size and shapes to be used as a suggested cure for headaches and even cancer [10].

2 Magnetism in Medicine

With the development of superconducting electromagnets at the Bell laboratories in 1961, followed by the evolution of strong permanent magnets [11], varying medical applications for these materials in increasingly diverse fields such as radiology, dentistry, cardiology, oncology, and neurosurgery began to appear. One of the most promising outcomes of these new techniques was the miniaturization of a magnet to an extent that it could be placed at the tip of a catheter and consequently used in recording electronic signals within the brain or treatment of arterial aneurysms [12] and related heart ailments [13]. The successful use of such catheter devices requires a strong magnetic navigation system that can guide them to the area of interest for site-specific delivery of radiation or therapeutic agents. Typically, such systems employ robust superconducting magnets [14] or more recently NdFeB [neodymium(Nd)-iron(Fe)-boron(B)] permanent magnets; used in the *Niobe*® system, developed by Stereotaxis Inc., in St. Louis, Missouri. Incidentally, this system was approved by the FDA in 2003 for multiple interventional cardiology and electro-physiology procedures.

Alternately, an external magnetic field has also been used for the localization of magnetic particles-such as nano- or micro-spheres in close vicinity to tumors or other target areas, such as blood vessels and arteries [15]. Such spheres can be encapsulated with therapeutic agents for a programmed release, triggered via irradiation at a specified frequency [16] or can act via the embolization (blockage) of vessels and capillaries or a combination of both [17]. An extension of this novel treatment modality, termed as 'Magnetic Fluid Hyperthermia' (MFH), involving the action of 'magnetite nanoparticles encapsulated in a dextran coating' on tumors was proposed by both Jordan et al. and Chang et al. in the year 1993 [18]. By 2003, Jordan had started phase II clinical trials of radiation therapy coupled with magnetic hyperthermia [19, 20] with a sample size of eight patients and achieved considerable success with minimal side effects [21].

The widespread use of magnets and magnetism in various treatment modalities is truly exceptional. However, the greatest medical boon of magnetism has been its applicability in the design of various methods for enhanced disease detection. In particular, the development of *Magnetic Resonance Imaging* or *MRI*, a Nuclear Magnetic Resonance (NMR) imaging technique as a clinical diagnostic tool has revolutionized the field of early disease diagnosis. Over the past three decades, MRI (Magnetic Resonance Imaging) has been the method of choice for in vivo imaging [22]. MRI employs low-intensity radiofrequency electromagnetic waves to study materials placed in a strong magnetic field. It is best suited for non-calcified tissue and has inherently superior contrast scale and better spatial resolution than X-Ray, Computed Tomography (CT), Single Photon Emission Computed Tomography (SPECT), or Positron Emission Tomography (PET). An inherent advantage of MRI over the other techniques is that it does not involve exposure to any harmful radiation. In some cases, MRI is the only way to do clear-cut diagnosis especially in the detection of cerebral abnormalities, multiple sclerosis, and in cases where bone artifacts are present in CT images [23]. In addition, MRI can also be used to monitor organ function using a relatively new technique called functional MRI or *f*MRI [24]. The development of MRI has also led to the extensive use of a potentially toxic and obscure lanthanide metal, *Gadolinium* (*Gd*) as a pharmaceutical agent that once injected into a patient can magnetically caress the water protons to produce startling effects in a magnetic resonance (*MR*) image. The *contrast* so produced aids the clinicians in alienating the healthy and normal tissues from the diseased ones and to indicate the status of organ function or blood flow. The chemical species that bring about such enhancements in MR images are termed as *Contrast Agents* (*CAs*). In addition to Gadolinium, contrast agents based on iron oxide nanoparticles and a manganese, Mn(II) complex have also been approved for clinical use, however, these have met with limited success commercially. Currently, all the eight contrast agents that are approved for clinical use and available commercially in the United States are small-molecule complexes with Gadolinium as the central metal ion. These *Gadolinium-based contrast agents (GBCAs)* are used in about 40% of all MRI exams and roughly 60% of all neural MRI exams, resulting in nearly 40 million uses of GBCAs worldwide [25]. In this chapter, we will discuss the various MRI contrast agents used in the clinics with recent examples, discuss their shortcomings and finally present a future outlook. However, to begin with, an interlude on the early developments of MRI as an imaging technique, its physical principles, and the logic behind the genesis of MRI CAs has been presented.

3 MRI-A Historical Perspective

On a strictly physical basis, MRI is an extension of the science underlying the nuclear magnetic resonance technique. The general experiment involves application of a time-varying magnetic field gradient to the subject of interest, followed by a spatial and temporal encoding of the resulting signals obtained from the NMR active nuclei (e.g., protons) present therein, thus generating an image [26]. The intensity of the images so obtained is augmented or curtailed by the nuclear relaxation times that in turn can be influenced by paramagnetic agents. The obvious chemical choice for a paramagnetic entity is a transition or a lanthanide metal ion, which have unpaired electrons that can influence the relaxation times of surrounding nuclei via dipolar interactions. Most of these metal ions are however extremely toxic in their native form and therefore can only be administered as complexes with multi-dentate metal chelator ligands. Such metal-ligand complexes must be water-soluble and extremely stable, both thermodynamically and kinetically, for in vivo use. A better understanding for successful application of the MRI technique would therefore require investigations into the physical behavior of aqueous solutions of such complexes in presence of a strong magnetic field with special emphasis on enhancing the relaxation rates of water protons. The following paragraph presents a small epilogue of the preliminary efforts carried out in this direction.

Initial experiments to augment the relaxation rates of nearby water protons using ferric nitrate, a paramagnetic salt was carried out by Bloch in 1948 [27]. The much-celebrated and now firmly established Solomon-Bloembergen-Morgan (SBM) theory and its various modifications aided immensely in underlining the basic physical principles of solvent proton relaxation rates for aqueous solutions of paramagnetic entities [28–32]. In 1961, Eisinger and co-workers showed that the binding of various transition metal ions such as Cu(II), Mn(II), and Cr(III) with exterior sites, such as phosphate groups within the DNA macromolecule can lead to enhanced water proton relaxation rates [33]. For decades, however, magnetic resonance was used to determine chemical structures and it was not until 1973 that the seminal work of Paul Lauterbur [34] and later improvements introduced by Peter Mansfield, led to the idea of using it for imaging the human body [35]. It was in 1977 when the first imaging studies on humans were performed [36, 37]. Once the idea of using magnetic resonance for human imaging found a strong footing, there was an increasing curiosity towards using it to alienate between healthy and diseased tissues. The answers arrived within a year when in 1978, Lauterbur and co-workers showed that normal and infarcted myocardial tissues can be differentiated based on the longitudinal proton relaxation rates ($1/T_1$) derived from tissue samples of a dog injected with a paramagnetic Mn(II) salt [38]. The idea of using a paramagnetic agent for MR imaging further gained strength with similar experiments carried out with excised dog hearts [39]. Interestingly, such contrast was only observed in the presence of the paramagnetic Mn(II) species. With considerable success in animals, Young and co-workers performed the first contrast agent assisted imaging study in humans in 1981, using ferric chloride to diagnose the gastrointestinal tract [40].

The field of CA development was, however, truly revolutionized with the arrival of CAs with a central paramagnetic gadolinium lanthanide metal ion. The idea was first proposed by Carr and co-workers who used the Gd(III) diethylenetriamine pentaacetate; $[Gd(DTPA)(H_2O)]^{2-}$ in patients with cerebral tumors and observed lesion enhancement in the images so obtained [41]. After multiple clinical trials, $[Gd(DTPA)(H2O)]^{2-}$ was approved for clinical use in 1988 [42], thus leading to an unprecedented and extensive foray into the development of novel contrast agents with an enhanced imaging profile. The majority of CAs approved for clinical use are Gd-chelates, that contain a central paramagnetic Gd(III) ion complexed using a *linear chain* or *macrocyclic* multi-dentate ligand. As discussed beforehand, CAs based on iron oxide nanoparticles and small-molecule Mn(II) chelate have also been developed and have been either used in human clinical trials or have received approval for clinical trials. However, all such agents are either no longer available commercially or their clinical development has been discontinued and they are no longer used as contrast agents for MRI [22]. In this chapter, our focus will only be on clinically approved GBCAs that are currently available and marketed.

Ongoing research aims at the generation of new and more effective class of CAs that have an enhanced MR image profile. This search clearly requires a deep understanding of the physics behind the MRI and answering one of the most pertinent questions; "*how the contrast is produced and how to improve on it…*" The following section provides a simple treatise as to how the MRI functions and the basic principles involved in generating the desired contrast. This section will also introduce various terms pertinent to the relaxation process that would enable a better understanding of the MR imaging phenomenon.

4 How the MRI Works

In a very general sense, MRI monitors the differential distribution of water within the subject of interest. It measures the spatial variation of *proton longitudinal* (T_1) and *transversal* (T_2) *magnetic relaxation times*. These terms will be explained later in the section in detail. Additionally, MRI takes advantage of the contrast produced within local sections of healthy and diseased parts of the same tissue. The proton relaxation times T_1 and T_2 are sensitive to biochemical conditions, such as pH, water concentration, and temperature. The essence of MRI is thus based on the *relaxation of tissue water protons subjected to an external magnetic field*.

Prior to understanding the functional principles of the MRI, let us look at its basic hardware components. A typical MRI scanner consists of the following parts; the magnet, gradient coils, radiofrequency transmitters, radiofrequency receivers, and finally a computer to process all the signals and produce an image for analysis by a trained radiologist. These components are shown in a schematic way in Fig. 1. The most important and expensive part of the MRI is the magnet at its core. These are mostly superconducting magnets, constructed of Niobium-Titanium alloy, though

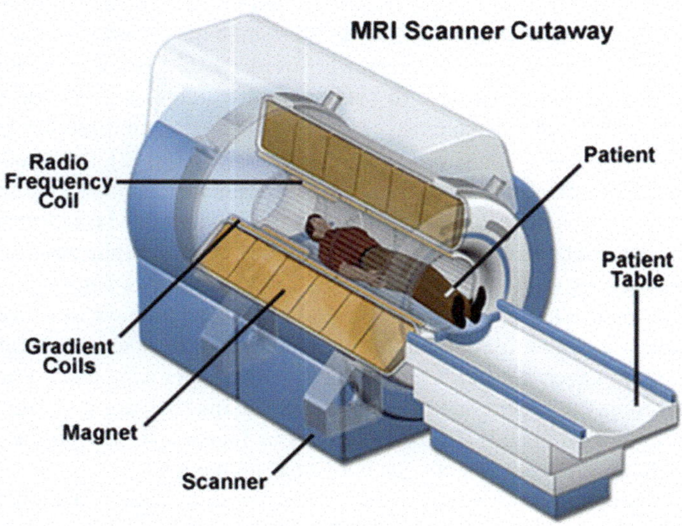

Fig. 1 A MRI scanner cutaway showing its various hardware components. Taken from Mahir Rashid & Fardin Kibria "MRI Scan-Components and Functions" https://snc2dmri.weebly.com/components--functions.html, 8 July 2020

other combination alloys such as Niobium-Tin can also be used. The strength of the MRI magnet is measured in units of Tesla (T); typical human MRI scanners range from 0.5 to 4.0 T while preclinical small animal scanners can be upto 7.0 T and 9.4 T. To establish a very stable field originating from such magnets, current flow within the superconducting coils is maintained indefinitely and the temperature is kept below the critical value by cooling them in an envelope of liquid helium. Resistive coils, also known as shim coils, and ferromagnetic blocks are housed within the magnet bore to even out field inhomogeneity. The shim coils are also used to generate fields that vary as a function of position.

The gradient coils are primarily used to produce a linear gradient in the magnetic field along one direction. Typically, the *z-axis* runs along the direction of the field and a *Maxwell coil* is used to generate a linear variation along the *z*-axis. To produce a gradient along the *x*- or *y-axes*, saddle coils, such as the *Golay coil*, are used. For producing gradients along directions other than the *x*-, *y*- or *z-axes*, currents are run along these axial coils in a proportionate manner. The gradient coils are also crucial for reducing the heat deposition and current requirements.

Another major component of the MRI system is the radiofrequency (RF) coils that are used to *transmit signals to* and *receive signals from* the subject of interest or the patient. Coils can be dual-purpose, which means that a single coil can be used to both transmit and receive signal or there can be separate, individual coils. RF coils are categorized as surface coils and volume coils. Surface coils rest on the surface of the object under consideration and can produce excellent images from the region of interest. Such images exhibit high signal-to-noise (*S*/*N*) ratio. Volume coils, on the other hand, are critical for imaging larger areas, such as the whole body or specific regions of interest such as the head or a limb. Thus, volume coils can be used to produce a homogenous field over a larger area and provide greater depth of penetration.

Finally, the entire scanning operation is controlled using a computer (not shown in Fig. 1) which also processes the signal obtained into a readout image on the MRI. For further details about specifics about each hardware component of the MRI, readers are directed towards various publications [43–45].

In order for a deeper insight into the process, let us consider a proton. A proton is a subatomic particle, which has a positive charge and a mass. What it additionally possesses is a term called "*spin*". The spin of the proton is denoted by the symbol I. As a result, the proton also has an angular momentum, characterized by its mass and charge as well as a magnetic moment, μ. What this means in a very general term is that the proton behaves as a tiny, subatomic magnet.

Now, let us consider this proton, a tiny magnet, in an external magnetic field, B_0. As the proton possesses a spin and has a charge, it undergoes a motion similar to that of a spinning top. This *precessional* motion of a proton under the influence of a field B_0 is characterized by a frequency, ν called the *Larmor frequency* and is defined by the relation; $\nu = \gamma B_0$; where γ is called the *gyromagnetic ratio* of the proton. For a proton, the value of the spin I is 1/2 and this means that in an external magnetic field (B_0) it can have a total of *two* ($2I + 1; I = 1/2$) possible orientations; one being aligned and the other being opposite to the direction of B_0. The alignment in the direction of B_0 is energetically more favorable and as a result, there is an excess population of protons, and consequently an excess number of nuclear spins, aligned with the external field. This introduces a net magnetization along the external field, termed as the *longitudinal* magnetization, M_0. However, the random distribution of spins ensures that there is *no transverse* magnetization, or in simpler terms, there is no net magnetization along the plane perpendicular to the direction of B_0.

Now, the net longitudinal magnetization, M_0 has a spin of its own. As a result, a small additional magnetic field, B_1 applied perpendicular or transverse to M_0, tends to rotate M_0 towards itself. The degree to which B_1 can rotate M_0 towards itself is dependent on its strength. This situation for a proton is very interesting and forms the basis of the NMR technique. The proton, characterized by a net longitudinal magnetization, M_0 is under the influence of two different and mutually perpendicular magnetic fields, B_0 and B_1. As a result, it experiences an effective magnetic field, characterized by B_{eff}. such that $B_{\text{eff.}} = B_0 + B_1$ ($B_{\text{eff.}}$ is, therefore a vector addition of the individual vectors B_0 and B_1). By virtue of its spin and under the influence of $B_{\text{eff.}}$, the longitudinal magnetization M_0 undergoes a precessional motion around this net effective magnetic field. In an exactly identical manner, the magnetic field vector of a radio frequency (RF) pulse generated in an NMR experiment acts as B_1, tipping over the longitudinal magnetization M_0 towards the transverse plane, in the process *exciting* the spins. This net magnetization, now directed towards the transverse plane is termed as the *transverse* magnetization, M_1. The precessional motion of M_1 in a NMR or MRI experiment induces an alternating voltage in a receiving coil with a frequency that is equal to the Larmor frequency, ν; in the process of generating the desired *MR signal* [46]. The signal so produced in a NMR or MRI experiment undergoes a multitude of simulations and computer processing to finally generate an image that can be used either for chemical species identification or for clinical purposes.

Once the RF pulse is switched off, the transverse magnetization M_1, eventually relaxes back to its original longitudinal counterpart, M_0. This apparent switch occurs either via a spin-lattice (T_1) relaxation or via a spin-spin (T_2) relaxation. In a *longitudinal relaxation process*, the transverse magnetization M_1 decreases with time and eventually the MR signal dies out. Consequently, the longitudinal magnetization, M_0 is slowly restored along the external magnetic field, B_0. In the process, energy is released within the surroundings and hence the overall process is termed as *spin-lattice relaxation.* It is characterized by a time constant, T_1, which is independent of B_0 and the internal motion of molecules and ranges within 0.5–5 s. Alternatively, in the *transverse relaxation phenomenon*, all spins that are initially *in phase*, undergo mutual energy exchange together with contributions from inhomogeneity in B_0 (with an additional time constant, T_2^*) resulting in *de-phasing* and consequent decrease in M_1. As the energy exchange occurs within the spins, the overall process is termed as *spin-spin relaxation.* It is characterized by a time constant T_2 and ranges within 100–300 ms [47].

With a general idea of the physics behind the MRI experiment firmly established, let us now look at the various factors that can affect the image contrast. One of the most important factors is the *proton density* that is defined by the number of excitable spins per unit volume. It determines maximum signal strength that can be emitted by tissue and can be tuned to obtain *proton-weighted* or *density-weighted* images. The *proton-weighted* images have a *higher signal-to-noise ratio* and are helpful to visualize bones and connective tissues. It is used to create high-resolution images and is extremely useful for clinical images of the brain, spinal cord as well as musculoskeletal system.

Other factors that contribute towards image contrast are the longitudinal relaxation time (T_1); a factor that decides how fast the spins would recover once disoriented by a small RF pulse. It can similarly be monitored to generate T_1-*weighted* images. The image contrast in a T_1-*weighted* experiment is influenced by a factor called the *repetition time*, T_R. T_R is defined as the time span between two successive excitations of the same slice and the experiment involves repeated excitation of the same tissue slice and consequent signal measurement to create an MR image. A short repetition time is characterized by a *high* T_1-*weighing* whereas long repetition times are characterized by *low* T_1-*weighing*. In a T_1-*weighted* image, tissues with *short* T_1 give *bright* images while tissues with *long* T_1 appear *dark*.

Another important factor that decides image contrast is the transversal relaxation time (T_2); a factor that is related to the time required for the decay of a NMR signal and can be modulated to generate T_2-*weighted* images. The influence of T_2 on image contrast is determined by a factor called the *echo time*, T_E. T_E is the time period between excitation and measurement of the MR signal. A short echo time is characterized by a *weak* T_2-*weighing* whereas long echo times are characterized by *strong* T_2-*weighing*. Consequently, with longer echo times, the contrast between tissues would be more pronounced. For a T_2-*weighted* image, tissues with short echo times give *dark* images while tissues with long echo times appear *bright*. Both the T_1 and T_2 parameters can be controlled to enhance the influence of proton density in MR images. All such factors are also influenced by *tissue type*, resulting in images with *distinct tissue to tissue contrast*.

In conclusion, *Image Contrast* in MRI is determined by the difference in signal intensity obtained from the tissues of interest. This contrast depends either upon *intrinsic* (body-related) or ***extrinsic*** (instrument related) factors. Signal intensity is influenced by factors such as proton density and T_1 or T_2 spin relaxation times. Contrast Agents are employed in MRI to improve the diagnostic information by increasing the signal intensity difference obtained from the diseased and healthy parts of ambient tissues of interest that are physiologically different [48].

5 MRI CAs: Mode of Action, Design, and General Considerations

The utility of MRI as a diagnostic tool is enhanced on administration of contrast agents prior to a MRI examination. Contrast agents are species that typically contain a paramagnetic entity (e.g., metal ion, iron oxides, labeled zeolites) that influence/reduce the relaxation times (T_1 or T_2) of ambient tissue water protons. Consequently, these are classified as either T_1 or T_2 agents, depending upon whether they reduce T_1 or T_2 for the tissue water protons. A Contrast agent (CA) can affect the image contrast in a number of ways. The CA being employed can *influence the spin or proton density* that in turn can lead to changes in signal intensity. For example, a reduction in spin or proton density can lead to consequent signal loss. CAs can also *reduce $T_1$and T_2 relaxation times*. For T_1-weighted images, the relaxation of nearby protons can be accelerated resulting in an increase in MRI signal strength. This leads to *positive enhancement in image contrast* and such CAs are termed as *positive CAs*. For *T_2-weighted* images greater de-phasing introduced by some CAs induce reduction in T_2 values. CAs can also bring about *changes in susceptibility* that can affect local field inhomogeneity leading to reduced signal intensity. This leads to a phenomenon called the *negative contrast*, wherein the image brightness is considerably reduced. Most T_2 agents that produce such effects are also termed as *negative CAs*. In addition, CAs can also *reduce the proton signal* by shifting the resonance frequency by several hundred ppm leading to drastic changes in image contrast. Thus the proton signal intensity of the MR image is either increased or decreased on administration of a T_1 or T_2 agent. The terms T_1 and T_2 are inversely related to *relaxivity*, r (r is inversely proportional to T) which is defined as the efficiency of a contrast agent in terms of its ability to reduce the relaxation time (T) or coherently, increase the *relaxation rate* (r) of the tissue water protons. Consequently, two terms, r_1 and r_2, are subsequently defined that refer respectively to the *longitudinal* and *transverse relaxation rates* of the concerned tissue water protons.

An ideal contrast agent must therefore possess some essential attributes [49]. First and foremost, it must be highly efficient in enhancing the relaxation rates of nearby aqueous protons. This requires the CA to possess high values of *relaxivity*, r_1*or* r_2; a higher r value ensuring better interaction of the contrast agent with the surrounding protons. Another important advantage of compounds with high relaxivity is related to their ease of detection at lower doses and the ability to provide greater contrast. Additionally, an ideal CA must preferentially localize within the target tissue, leading to a highly specific in vivo relaxivity. It is therefore important that the relaxation rates of the target tissue be enhanced in comparison to other ambient tissues. The CA must also exhibit sufficient in vivo stability and a consequent lack of toxicity to be put into human use. Finally, it should be able to display rapid tissue clearance within hours of its administration to a patient. Most of the clinically approved CAs are complexes of heavy transition and inner transition metals, which are toxic in their native form. The toxicity of the free chelating ligand is also a matter of concern, provided the CA undergoes in vivo dissociation in presence of a host of other endogenous metal ions such as Ca(II) and Zn(II).

A large number of contrast agents are available that are exceedingly diverse in their chemical composition [50–55]. They can be mono-nuclear or poly-nuclear ligands complexed with a paramagnetic metal ion of choice; polymeric or macromolecular carriers (bonded covalently or non-covalently with a paramagnetic transition or inner-transition metal ion); superparamagnetic iron oxides; metalloporphyrins

(HOPO class of ligands); ^{13}C labeled species; Chemical exchange saturation transfer (CEST) contrast agents; aerosols and gases to name just a few. The various contrast agents can also be classified based on their mode of administration viz. intravenous, oral, rectal, and a less common mode wherein the contrast agent is inhaled as a gas (CT agent for brain and lung imaging). Additionally, they can also be classified as extracellular, organ-specific, and blood pool contrast agents.

One of the most promising class of CAs constitutes polyaminocarboxylate ligands complexed with a paramagnetic metal ion of choice, usually Gadolinium (Gd^{3+}). Metal ions that contain one or more unpaired electrons possess a magnetic moment and are termed as *paramagnetic*. When complexes containing such paramagnetic metal ions are placed in aqueous solutions, a magnetic interaction between the magnetic moment of the unpaired electrons of the paramagnetic metal ion and that of the protons of surrounding water molecules ensues, which affects the proton longitudinal (T_1) and transversal (T_2) relaxation times. CAs that comprise a central Gd^{3+} ion augment both the transversal as well as the longitudinal relaxation rates. However, the percentage enhancement in the latter (r_1 or $1/T_1$) is much higher and as a result, such CAs are visualized best with T_1-*weighted* scans [56]. The metal ion Gd^{3+} is paramagnetic (7 unpaired electrons) and consequently has a very high magnetic moment (7.9 BM). In addition, Gd^{3+} has a symmetric ground state that results in longer electron spin relaxation times (10^{-8}–10^{-9} s) and allows for streamlined transfer of magnetic information to surrounding water molecules [32, 57]. With such favorable physical properties, the concerned species containing a paramagnetic metal ion are very effective in reducing the T_1 and T_2 relaxation time of the water protons, which can ultimately lead to better contrast in MR images. Gd^{3+} in its native form is toxic ($LD_{50} = 0.3$–0.5 mmol/kg for rats) and is known to be retained in various organs such as bones, liver, and spleen [58]. It is henceforth, administered in vivo as a complex with a suitable organic ligand in doses ranging from 0.1–0.3 mmol/kg of the body weight. On complexation with suitable ligands, the resulting Gd^{3+}-chelates possess minimal toxicity ($LD_{50} = 10$ mm mol/kg in rats). Once administered to a patient, issues pertaining to the in vivo thermodynamic and kinetic stability of these complexes need to be addressed. It is extremely important that the complex remains intact and is resistant to transmetallation/transchelation reactions to the various endogenous metal ions such as Zn(II) and Ca(II) and a host of endogenous ligands such as glutamate and citrate that are present in ample quantities inside the body. The stability of such complexes has been extensively studied and it is now understood that the in vivo kinetic stability of Gd^{3+} chelates are more important than their thermodynamic stability [59, 60].

Most of the concerned ligands that have been used for complexation with Gd^{3+} have one of two basic core structures. One of them is the open-chain diethylenetriaminepentaacetic acid (DTPA) while the other is the closed macrocyclic cyclen ring 1,4,7,10-tetraazacyclododecanetetraacetic acid (DOTA). The two core structures are shown in Fig. 2.

The first MR contrast agent to be approved for human use (1988) was the acyclic DTPA (diethylenetriaminepentaacetic acid) derivative, Gd-DTPA (Magnevist Schering). The first cyclen derivative to be put into human use was Gd-DOTA (Dotarem).

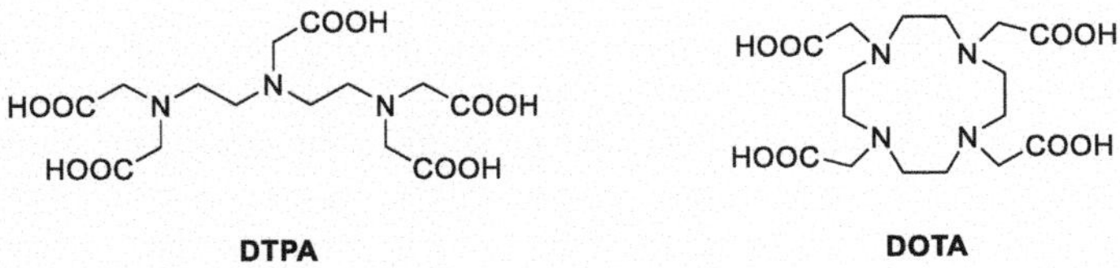

Fig. 2 The core structure of chelating ligands, DTPA and DOTA

Since then, a host of derivatives of both DTPA and DOTA have been studied as potential MR contrast agents. Both Magnevist and Dotarem along with a list of other CAs approved for clinical use in the United States are shown in Fig. 3.

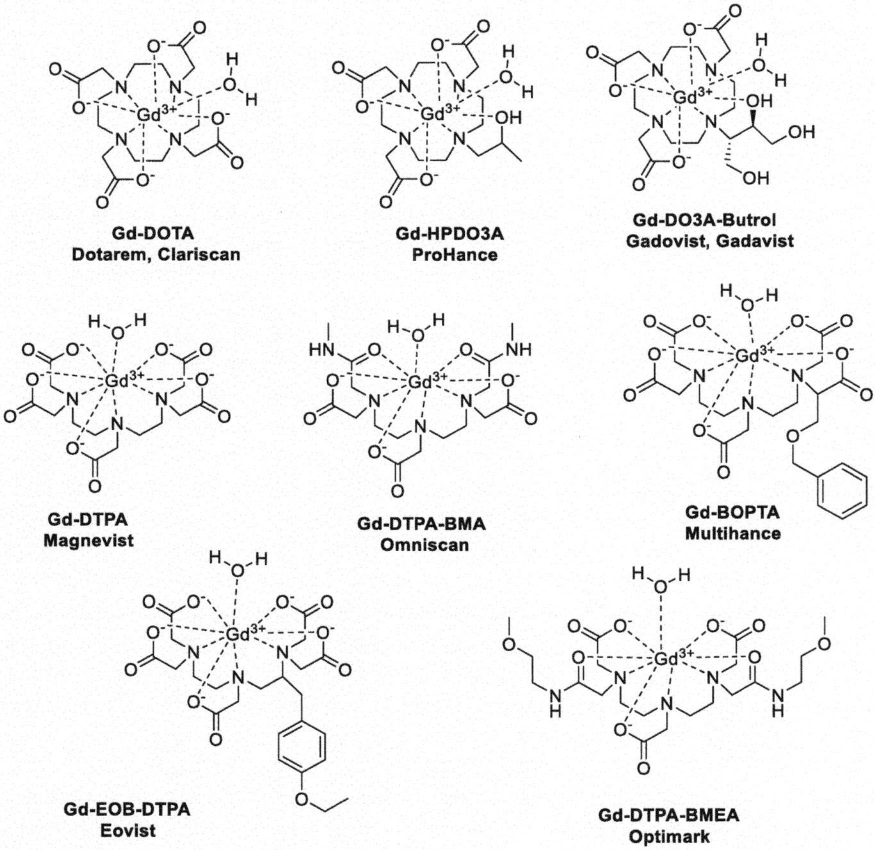

Fig. 3 MRI contrast agents approved for clinical use and currently sold in USA

About 40% of all MRI examinations performed around the world employ a *CA* such as a chelated paramagnetic Gd^{3+} ion. Such Gd^{3+} chelate systems must encompass a high thermo dynamic stability constant coupled with a miniscule metal dissociation rate, in addition to being hydrophilic and containing at least one coordinated water molecule available for rapid exchange with bulk water solvent. Enhanced T_1 relaxivity, r_1, is sought since the signal is brightened and the overall sensitivity is increased. Ongoing research involves efforts to enhance the relaxivity of contrast agents so as to obtain better resolution in the subsequent images, to achieve lower plasma clearance rates, and increase the specificity of these contrast agents. The successful design of a CA that ideally encompasses all the above attributes requires a thorough understanding of the behavior of such agents in aqueous solutions. As discussed beforehand, the Solomon-Bloembergen-Morgan (SBM) theory along with its various modifications provides a fundamental basis for such systems and considers all the possible interactions at interplay [61]. The following section presents a very basic treatise on the SBM theory, with special emphasis on the *points to remember* while designing such *ideal* CAs.

6 Contrast Agents and SBM Theory

The efficacy of a CA is determined in terms of its relaxivity, r_1 or r_2. The relaxivity of a CA is measured by its ability to enhance the relaxation rates of water protons at any given concentration, usually 1 mM. The observed relaxation rate of the water protons contains both a paramagnetic and a diamagnetic term. Mathematically, the relaxivity r_i ($i = 1, 2$) is defined by Eq. 1 and can be described as the slope of a plot of the observed relaxation rate $(1/T_i)_{obs.}$ and the concentration of the CA, represented by the term [Gd].

$$(1/T_i)_{\text{obs.}} = (1/T_i)_{\text{dia.}} + r_i[\text{Gd}] \quad (1)$$

In Eq. 1, the paramagnetic contribution is the second term on the right-hand side and is linearly related to the concentration of the concerned paramagnetic entity. The increase in paramagnetic relaxation can be attributed to both an *inner-sphere* as well as an *outer-sphere* contribution. The *inner-sphere* term consists of the contribution to the proton relaxation from a solvent molecule directly coordinated with the Gd^{3+} ion. Such solvent molecules belong to the first coordination sphere of the complex. Similarly, the *outer sphere* term relates to an identical contribution from solvent molecules in the second coordination sphere of the complex as well as to those in the bulk solvent. The ability of the central Gd^{3+} ion to influence the overall relaxivity depends on the strength of such interaction with nearby solvent species. This interaction is inversely proportional to the distance between the concerned species and thus the contribution of the solvent molecules closest to the Gd^{3+} ion, i.e., those within the first coordination sphere of the complex are extremely important. This contribution enhances the longitudinal, inner sphere relaxivity, r_{1p}. Another important factor relates to the *rate of solvent or water exchange* with the paramagnetic Gd^{3+} species. A faster exchange rate leads to a greater swap of magnetic information with the neighboring solvent molecules and this augments the resultant relaxivity. The exchange rate is characterized by the *mean water residence time*, τ_M and should have small values for a faster exchange rate. Equation 2 relates these factors mathematically to the paramagnetic proton relaxation enhancement;

$$1/T_1 = qP_m[1/(T_{1m} + \tau_M)] \tag{2}$$

In Eq. 2, the term $1/T_1$ refers to the longitudinal relaxation rate, q refers to the total number of solvent water molecules present in the first coordination sphere of the Gd-complex, while P_m and $1/T_{1m}$ refer to the mole fraction and the relaxation rate of water or solvent molecule coordinated to the Gd^{3+} center, respectively. According to the SBM theory and the related mathematical equations, the term $1/T_{1m}$ is governed by the dipole-dipole relaxation mechanism and is critically influenced by the *correlation time*, τ_{ic} $(i = 1, 2)$ of the molecular tumbling motion, defined mathematically by Eq. 3;

$$1/\tau_{ci} = 1/\tau_R + 1/T_{ie} + 1/\tau_M \tag{3}$$

In Eq. 3, τ_R refers to the *rotational correlation time* and signifies the extent of rotation or tumbling undergone by the Gd-complex. The term T_{ie} refers to the electronic *relaxation time.* The correlation time takes into account the alterations in the local magnetic field brought about by the tumbling motion of the complex and has to achieve optimum values for the desired high relaxivity of a CA.

In a nutshell, the SBM theory and its modifications relate to a number of factors that can influence the relaxivity of a CA. These factors can be colligated with ligand structure, design, and the overall motion of a CA when put into a solution. The three most important factors that affect the relaxivity attained by a complex are, the number of directly coordinated water molecules to the central metal ion (q); the residence time of a water molecule coordinated to the metal center (τ_M) and the rotational correlation time that measures the extent of rotation or tumbling undergone by a complex (τ_R). In addition, there are spates of electronic interactions especially with the water molecules in the outer coordination sphere of the complex that also affect the relaxivity for a contrast agent, although to a much lesser extent.

The current batch of CAs approved for clinical use are based on a central Gd^{3+} ion that is chelated to either an open-chain or a macrocyclic polyamino polycarboxylic acid ligand. For all such ligands, based on either the DTPA or DOTA framework (Fig. 2), the number of available ligation sites is eight. The central Gd^{3+} ion, however, has a coordination number of nine and hence for such clinically approved Gd-DTPA or a Gd-DOTA system, only one coordination site on Gd^{3+} is left for direct coordination with a water molecule. For these complexes, the value of q is 1. This coordinated water molecule is rapidly exchanged with the surrounding solvent water molecules and brings about the relaxation pertinent to these complexes. For complexes with larger q values, the extent of exchange of the coordinated water molecule with the surrounding solvent water molecules increases, and so does the

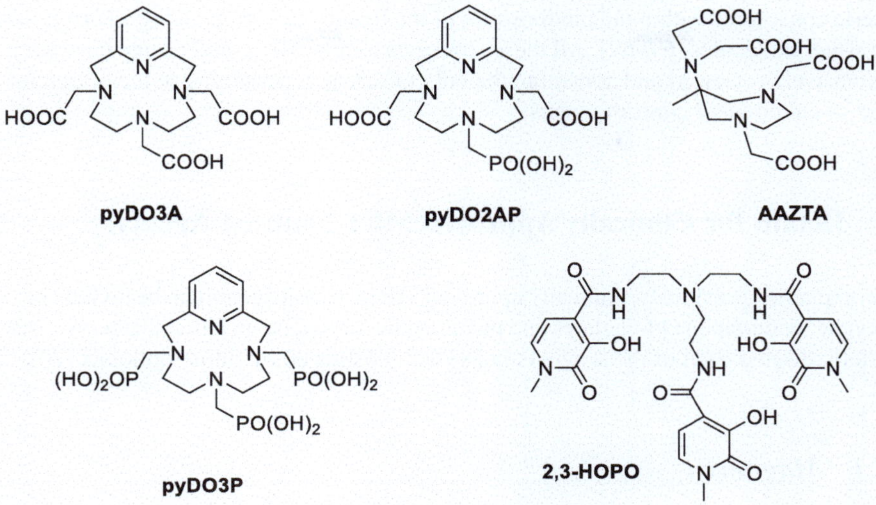

Fig. 4 Six and Seven coordinate ligands for conjugation with Gd^{3+} ion

relaxivity. However, with more water molecules directly coordinated with the central Gd^{3+} ion, the number of ligation sites left for the ligand attachment decreases and consequently the stability of the overall complex is drastically compromised. Ligands with seven and six coordination sites have been designed (Fig. 4) and corresponding Gd^{3+} complexes with $q = 2$ and 3 have been synthesized [62]. These complexes report higher relaxivity than Gd-DTPA or Gd-DOTA complexes, but their thermodynamic and kinetic stability profiles are not favorable and hence these have not been approved for clinical use, in spite of a promising relaxivity profile. Again, these complexes are also known to undergo dimer/trimer formation with the central metal ion [63, 64]. This decreases the q value for these complexes and consequently the gain in relaxivity is compromised.

The optimal values for τ_M and τ_R are both 10 ns at a magnetic field strength of 1.5 T [65]. For the various contrast agents put into clinical use, values attained for τ_M and τ_R are 100 ns and 0.1 ns, respectively. Clearly, there is an urgent need to optimize these values by designing complexes with a more favorable τ_M and τ_R. In principle, contrast agents with ligands having a more globular structure are bound to decrease the rotational tumbling of the resulting complex and therefore can bring about an increase in τ_R. Another strategy to enhance τ_R is to link up multiple DTPA/DOTA complexes with a suitable carrier, i.e., design a contrast agent with a *high pay load* [52, 66]. For optimization of τ_M values, various strategies can be undertaken. For example, it is known that complexes with an overall negative charge decrease the residence time of coordinated water molecules as compared to complexes that are neutral. Similarly, introduction of side chains in the pendant arms, especially replacement of acetate groups with bulkier phosphate groups, replacing carbon atoms in the core structure of DTPA and DOTA and introduction of greater steric constraints within the binding sites of the ligands can affect the residence time of water molecules [67–69]. All these strategies to improve ligand design have been extensively reviewed and constitute the bulk of research pertaining to the search for the so-called *ideal* contrast agents.

7 Details for Clinically Approved MRI Contrast Agents

As explained previously, clinically approved MRI contrast agents can be divided into acyclic, open-chain Gd-chelates and macrocyclic, closed chain Gd-chelates. We will briefly discuss each of these agents individually followed by future outlooks [70].

7.1 *Magnevist*

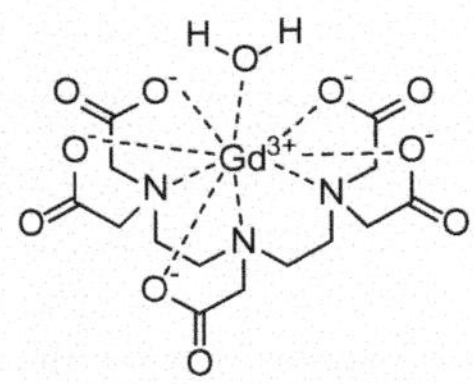

Gd-DTPA, Magnevist

Magnevist (Gd-DTPA, Gadopentate dimeglumine) was the first MRI CA approved for clinical use in 1988 as an extracellular fluid agent. It is an open chain, acyclic Gd-chelate that is used for detection of lesions with abnormal vascularity in the brain (intracranial lesions), spine, and associated tissues. It is also utilized for visualizing abnormalities in the body (excluding the heart), head, and neck. Upon IV injection, Magnevist undergoes non-specific biodistribution and is eliminated via the kidneys. It is marketed as single-dose vials, prefilled single-dose injections, and pharmacy bulk packages a clear, colorless to slightly yellow solution containing 0.5 mmol gadopentetate dimeglumine/mL (equivalent to 469.01 mg/mL of gadopentetate dimeglumine) for intravenous use.

As of September 2019, Magnevist will no longer be provided as an injection by its supplier (Bayer) to the United States. This is in response to the shift towards greater use of macrocyclic Gd-chelates in the clinics and a consequent reduction in use of open-chain agents [71].

7.2 *Omniscan*

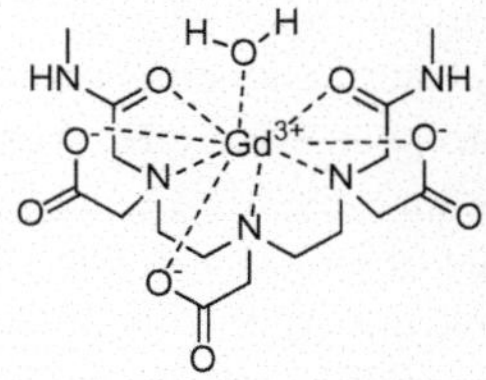

Gd-DTPA-BMA, Omniscan

Omniscan (Gd-DTPA-BMA, gadodiamide) was the second open chain, acyclic MRI CA approved for clinical use in 1993 as an extracellular fluid agent. It is approved as an intravenous injection for the detection of abnormalities in vasculature in the brain, spine, thoracic (non-related to heart), abdominal, pelvic cavities, retroperitoneal space and associated tissues. On IV injection, Omniscan undergoes non-specific biodistribution and is eliminated via the kidneys. It is marketed as single dose vials, prefilled injections and pharmacy bulk packages as a sterile aqueous solution for intravenous injection (287 mg/mL) bolus intravenous use.

As of July 20, 2017, the European Medicine Agency (EMA) has recommended the suspension of Omniscan for MRI exams in the European Union (EU) [72]. This is a direct consequence of concerns regarding the deposition of Gd in the brain tissues for patients who have undergone linear Gd-chelate assisted MRI. There are also additional concerns due to NSF in patients with impaired renal function.

7.3 *Optimark*

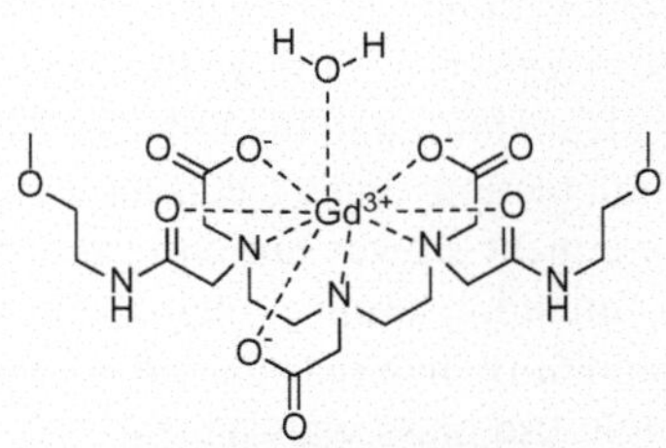

Gd-DTPA-BMEA, Optimark

Optimark (Gd-DTPA-BMEA, gadoversetamide) was the third open chain, acyclic MRI CA approved for clinical use in 1999 as an extracellular fluid agent. It was developed for MRI of the central nervous system and liver. It has also been approved for use in patients with abnormalities in vasculature of the spine, brain, the blood brain barrier (BBB), and associated tissues. It is a non-ionic species and henceforth has a lower osmolality than ionic agents such as Magnevist. Optimark undergoes non-specific biodistribution and is cleared through renal excertion (urine) on IV administration. It is marketed as glass vials and prefilled injections as a clear, colorless to slightly yellow solution for injection containing 330.9 mg gadoversetamide per mL (equivalent to 0.5 mmol/mL) sterile aqueous solution for intravenous injection.

As of July 20, 2017, the European Medicine Agency (EMA) has also recommended the suspension of Optimark for MRI exams in the European Union (EU) [72]. This is again a consequence of concerns regarding the deposition of Gd in the brain tissues for patients who have undergone linear Gd-chelate assisted MRI. There are also additional concerns due to NSF in subjects with renal dysfunction.

7.4 Multihance

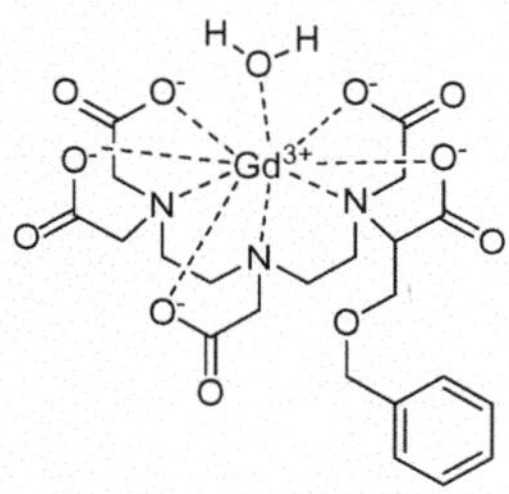

Gd-BOPTA, Multihance

Multihance (Gd-BOPTA, gadobenate dimeglumine) was the fourth open chain, acyclic MRI CA approved for clinical use in 2004 as an extracellular fluid agent. Multihance was originally developed for potential hepatobiliary distribution, however, it was realized that its biodistribution is species-specific. Thus, in mice and rats, Multihance undergoes about 50% clearance by the liver while in human subjects only 2–5% clearance is observed through the hepatobiliary route and the rest of the administered dose is cleared through the kidneys. Multihance also undergoes non-specific and transient interactions with macromolecules, resulting in increase in the overall size and consequent slower tumbling rate leading to higher r_1 and r_2 relaxivity in solutions containing serum proteins. MultiHance is a linear, ionic species and has a higher osmolality than plasma and is hypertonic under conditions of use. It is marketed in single-use glass vials a clear, colorless solution containing 529 mg gadobenate dimeglumine per mL.

Although there are concerns with free Gd-associated toxicity with Multihance use, EMA has suggested its continuation for liver MRI scans due to its somewhat liver-specific uptake and related importance in diagnosis of liver lesions in the EU [72].

7.5 *Eovist*

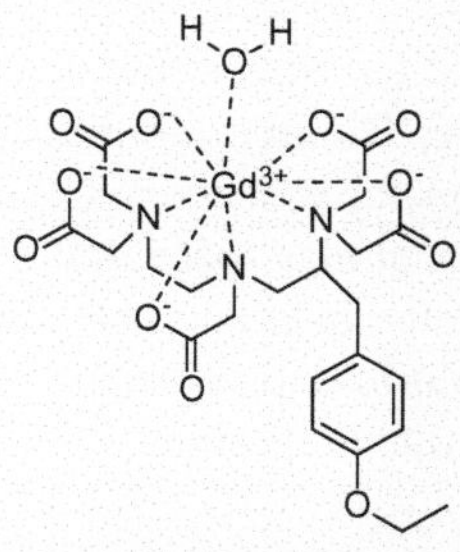

Gd-EOB-DTPA, Eovist

Eovist (Gd-EOB-DTPA, disodium gadoxetic acid) was the fifth open chain, acyclic MRI CA approved for clinical use in 2008 in the United States as a liver-specific MRI CA. Prior to this, the agent was already approved under the trade name Primovist in Europe in the year 2005 as a hepatobiliary contrast agent for MRI. The lipophilic ethoxybenzyl moiety within Gd-EOB-DTPA is unique to its structure and is responsible for liver-specific uptake. Once injected i.v., Eovist is rapidly taken up by both the liver and the kidneys and is subsequently excreted by the hepatobiliary (~50%) and renal (~50%) route. Interestingly, the hepatocyte cells in the liver drive the uptake of Eovist in it and this allows for delineation between healthy and tumorous liver tissues due to preferential uptake of Eovist by the neoplastic cells. Eovist is a linear, anionic charged species that exhibits low protein binding and has higher osmolality than plasma. It is marketed in single-use glass vials as a clear, colorless solution-containing 181.43 mg/mL of gadoxetate disodium, equivalent to 0.25 mmol/mL.

Similar to other open-chain Gd-chelates, free Gd-associated toxicity is a concern for Eovist. However, due to its specificity for delineation of liver lesions, it is used for analysis in patients with related disease scenarios.

7.6 *Dotarem*

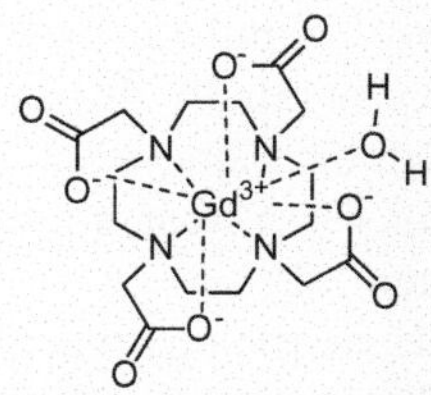

Gd-DOTA, Dotarem, Clariscan

Dotarem (Gd-DOTA, gadoterate meglumine) was the first macrocyclic, ionic MRI CA approved for clinical use in 1989 in Europe and was approved by FDA in 2013 for use in the United States for visualization of abnormalities in the brain, spine, and surrounding tissues. Clariscan is the generic version of Dotarem that was approved in 2019 for clinical use in the United States. In subjects with a compromised BBB and abnormalities in vasculature, Dotarem administration allows for detection of lesions such as neoplasm, infarcts, or abscesses. Upon i.v. administration, Dotarem is rapidly distributed in the extracellular space and is eliminated primarily through the kidneys and urine. It is marketed in single-use glass vials and pre-filled syringes a sterile, nonpyrogenic, clear, colorless to yellow, aqueous solution of 0.5 mmol/mL containing 376.9 mg/mL gadoterate meglumine.

A major advantage of macrocyclic MRI CAs as compared to their open-chain counterparts is their enhanced thermodynamic and kinetic stability. As a result, these are now the preferred agents for clinical use. Multiple reports have indicated their safety in terms of Gd release and they have emerged as a viable option for MRI contrast exams in patients with renal failure.

7.7 *ProHance*

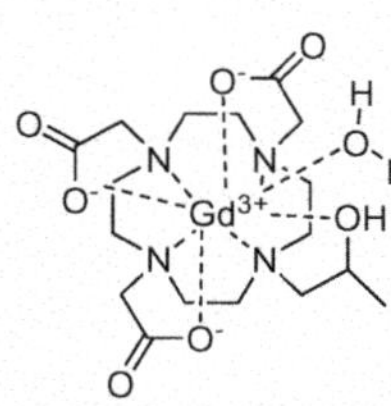

Gd-HPDO3A, ProHance

ProHance (Gd-HP-DO3A, gadoteridol) was the second macrocyclic MRI CA approved for clinical use in 1992 as an extracellular fluid agent for the non-targeted imaging of lesions in the CNS and extracranial/extraspinal tissues. ProHance is a non-ionic species that has a lower osmolality than agents such as Gd-DTPA but a higher Gd to solute particle ratio. The mechanism of action for ProHance pertains to the enhanced contrast observed in patients with a compromised BBB and perfusion deficiency in extracranial/extraspinal tissues as compared to patients with an intact BBB and associated tissue framework. Similar to the other macrocyclic CA Dotarem, upon i.v. administration, ProHance is rapidly distributed in the extracellular space and is eliminated primarily through the kidneys and urine. It is marketed in single-use glass vials and pre-filled syringes as a clear, colorless to slightly yellow solution containing 279.3 mg/mL of gadoteridol.

Again, similar to Dotarem, ProHance being a macrocyclic MRI CAs enjoys higher thermodynamic and kinetic stability as compared to their open-chain counterparts.

Thus, they have emerged as a viable option for MRI contrast exams in patients with renal failure.

7.8 *Gadovist*

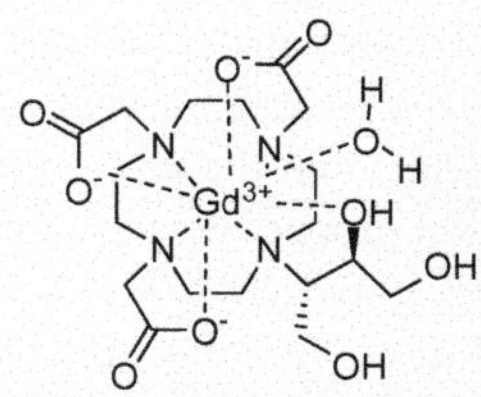

Gd-DO3A-Butrol, Gadovist, Gadavist

Gadovist (Gd-DO3A-Butrol, gadobutrol) was the third macrocyclic MRI CA approved for clinical use in 1998 in Europe and in 2011 in the United States (trade name Gadavist) as an extracellular fluid (ECF) agent for non-targeted imaging of the blood pool. Akin to other ECF agents, Gadovist does not cross an intact BBB. However, for subjects with a compromised BBB or the lack of it in the pituitary gland, meningiomas, or tumor margins, CNS lesions are clearly delineated following i.v. administration of Gadovist. The trihydroxybutyl group in Gadovist was introduced to enhance the overall hydrophilicity of the resulting chelate leading to lower protein binding and higher in vivo tolerance. Although a non-ionic species, the Gadovist formulation displays higher osmolality than the ionic Dotarem and the neutral ProHance. Following i.v. administration, Gadovist is rapidly distributed in the extracellular space and is excreted via kidneys. It is marketed in single-use glass vials and pre-filled syringes as a clear, colorless to pale yellow solution containing 1 mmol gadobutrol (equivalent to 604.72 mg gadobutrol) per mL as the active ingredient. Again, similar to Dotarem and ProHance, Gadovist exhibits higher thermodynamic and kinetic stability as compared to their open-chain counterparts and has emerged as a viable option for MRI contrast exams in patients with renal failure.

8 MRI-Clinical Applications

MRI has revolutionized the field of clinical imaging [73]. Conventional MRI is used in the clinics for functional and morphological characterization of a patient. For example, MRI has slowly developed as a method of choice for cardiovascular imaging. Several improvements in spin-echo techniques that provide access to various contrast mechanisms are now being used to delineate subtle abnormalities in cardiac function and flow, changes in cardiovascular morphology, myocardial viability and perfusion as well as coronary MR angiography. Another excellent use of MRI refers pertains to the mapping of clinical as well as experimental cerebral ischemia. Over the years, improvements in MRI scanner hardware, pulse sequence, and user interface have positioned it a technique of choice for image guidance in interventional procedures. MRI provides for unmatched soft tissue contrast even without the use of exogenous contrast media and this can aid tremendously in the delineation of target lesions. As the MRI does not use any harmful radiation, the damage to surrounding tissue anatomy is non-existent when compared to other techniques such as CT and PET. Further, by using various pulse sequences subtle differences in normal and damaged tissue can be visualized to obtain a plethora of functional information that is of critical importance to determine the end-point of a surgical intervention. MRI can also aid in thermal ablation procedures by monitoring temperature changes in tissues. Dynamic MR Mammography (MRM), a technique that combines a dedicated breast coil and a rapid 2D gradient-echo imaging sequence together with a bolus injection of GBCA is considered a prominent diagnostic tool for breast cancer imaging. It is an excellent technique to measure disease extent, presence of absence of malignancy, and evaluation of tumor response to neoadjuvant chemotherapy. Another interesting aspect of MRI relates to MR Spectroscopy (MRS) that can be used to study the metabolism of tissues and organs of interest, especially the tumor biochemistry that can be of utmost importance to determine prognosis and response to treatment for the subject of interest.

9 Future Directions and Conclusions

Despite several years of research and development, most clinically used low molecular weight MRI CAs suffer from some basic disadvantages, namely relatively low r_1 values (~4 s^{-1} mM^{-1} at higher magnetic field strength), lack of selectivity for tissues and extremely short intravascular half-lives (~20 min) [74]. Moreover, such low molecular weight CAs do not provide sufficient contrast at low concentrations, which is essential for biomedical and targeted imaging [75]. Till recently, GBCAs were considered to be the safest pharmaceutical agents barring very few short and long-term adverse effects [76]. However, this situation dramatically changed in 2006, when reports about a potentially fatal, new disease condition called Nephrogenic Systemic Fibrosis (NSF) was reported in patients with impaired renal function [77, 78]. Over the years, there have also been reports of long-term Gadolinium retention in the body following repeated administration of GBCAs [79]. Fortunately, the condition of NSF has only been reported for the usage of open chain, acyclic GBCAs that have with lower thermodynamic and kinetic stability than their macrocyclic, closed chain counterparts and are thus subject to Gd loss due to chelation with endogenous cations [Ca(II), Zn(II), etc.) and anions (citrate, glutamate, etc.) [80, 81]. This has led to the subsequent suspension of the use of open chain GBCAs, Magnevist, Omniscan, and Optimark by the European Medicines Agency (EMA) in the EU. Recently, Magnevist was also taken off from the market in the US due to potential toxicity concerns. Thus, there has been a renewed interest in the investigation of alternatives to GBCAs for MRI, which is now an active area of research. It must be emphasized that the design, synthesis, and potential human administration of a contrast agent requires a thorough understanding of the underlying principles of contrast generation in MRI, chelate chemistry, solution thermodynamics and kinetics, pharmacology, biodistribution, and the pertaining economics of the manufacturing process. Efforts are certainly underway in this direction and this would require active collaboration and exchange of ideas between chemists, biochemists, physicists, and clinicians.

References

1. Chakravarty, S.: Closomers at a click: a treatise on the design, synthesis and in vivo MRI of novel click closomers as high performance and efficacious contrast agents. University of Missouri, Columbia (2013)
2. Peregrinus, P.: Epistola Petri Peregrini de Maricourt ad Sygerum de Foucaucourt, Militem, De Magnete. Privately Published, Italy (1269)
3. Gilbert, W.: De Magnete, Magneticisque, Corporibus, et de Magno Magnete Tellure; Physiologica Nova (On the lodestone, magnetic bodies, and on the great magnet the earth). Dover (Paperback republication, 1991, Translation: Mottelay, P.F.), New York (1600)
4. Mourino, M.R.: From Thales to Lauterbur, or from the lodestone to MR imaging: magnetism and medicine. Radiology **180**, 593–612 (1991). https://doi.org/10.1148/radiology.180.3.1871268https://doi.org/10.1148/radiology.180.3.1871268
5. Mitchell, A.C.: Chapters in the history of terrestrial magnetism. Terr. Magn. Atmos. Electr. **44**, 77–80 (1939). https://doi.org/10.1029/TE044i001p00077https://doi.org/10.1029/TE044i001p00077
6. Häfeli, U.: The history of magnetism in medicine. In: Magnetism in Medicine, pp. 1–25 (2006)
7. Macklis, R.M.: Magnetic healing, quackery, and the debate about the health effects of electromagnetic fields. Ann. Int. Med. **118**, 376–383 (1993). https://doi.org/10.7326/0003-4819-118-5-199303010-00009https://doi.org/10.7326/0003-4819-118-5-199303010-00009
8. Lord Butterfield of Stechford: Dr Gilbert's magnetism. Lancet **338**, 1576–1579 (1991). https://doi.org/10.1016/0140-6736(91)92388-I
9. Shermer, M.: Mesmerized by magnetism. Sci. Am. **287** (2002)
10. Livingston, J.D.: Driving force: the natural magic of magnets, 1st edn. Harvard University Press, Cambridge (1996)
11. Asti, G., Solzi, M.: Permanent magnets. In: Gerber, R., Wright, C.D., Asti, G. (eds.) Applied Magnetism, pp. 309–375. Springer, Dordrecht (1994)
12. Hilai, S.K., Jost Michelsen, W., Driller, J., Leonard, E.: Magnetically guided devices for vascular exploration and treatment. Radiology **113**, 529–540 (1974). https://doi.org/10.1148/113.3.529https://doi.org/10.1148/113.3.529
13. Ram, W., Meyer, H.: Heart catheterization in a neonate by interacting magnetic fields: a new and simple method of catheter guidance. Cathet. Cardiovasc. Diagn **22**, 317–319 (1991). https://doi.org/10.1002/ccd.1810220412https://doi.org/10.1002/ccd.1810220412
14. McNeil, R.G., Ritter, R.C., Wang, B., et al.: Characteristics of an improved magnetic-implant guidance system. IEEE Trans. Biomed. Eng. **42**, 802–808 (1995). https://doi.org/10.1109/10.398641https://doi.org/10.1109/10.398641
15. Poznansky, M.J., Juliano, R.L.: Biological approaches to the controlled delivery of drugs: a critical review. Pharmacol. Rev. **36**, 277–336 (1984)
16. Rand, R.W., Snyder, M., Elliott, D., Snow, H.: Selective radiofrequency heating of ferrosilicone occluded tissue: a preliminary report. Bull. Los Angeles Neurol. Soc. **41**, 154–159 (1976)

17. Sako, M., Hirota, S., Morita, M., et al.: Clinical evaluation of ferromagnetic microembolization in the treatment of hepatoma. Nihon Gan Chiryo Gakkai Shi **20**, 1317–1326 (1985)
18. Chan, D.C.F., Kirpotin, D.B., Bunn, P.A.: Synthesis and evaluation of colloidal magnetic iron oxides for the site-specific radiofrequency-induced hyperthermia of cancer. J. Magn. Magn. Mater. **122**, 374–378 (1993). https://doi.org/10.1016/0304-8853(93)91113-L
19. Jordan, A., Scholz, R., Maier-Hauff, K., et al.: Presentation of a new magnetic field therapy system for the treatment of human solid tumors with magnetic fluid hyperthermia. J. Magn. Magn. Mater. **225**, 118–126 (2001). https://doi.org/10.1016/S0304-8853(00)01239-7
20. Gneveckow, U., Jordan, A., Scholz, R., et al.: Description and characterization of the novel hyperthermia- and thermoablation-system for clinical magnetic fluid hyperthermia. Med. Phys. **31**, 1444–1451 (2004). https://doi.org/10.1118/1.1748629https://doi.org/10.1118/1.1748629
21. Maier-Hauff, K., Rothe, R., Scholz, R., et al.: Intracranial thermotherapy using magnetic nanoparticles combined with external beam radiotherapy: results of a feasibility study on patients with glioblastoma multiforme. J. Neurooncol. **81**, 53–60 (2007). https://doi.org/10.1007/s11060-006-9195-0https://doi.org/10.1007/s11060-006-9195-0
22. Wahsner, J., Gale, E.M., Rodríguez-Rodríguez, A., Caravan, P.: Chemistry of MRI contrast agents: current challenges and new frontiers. Chem. Rev. **119**, 957–1057 (2019). https://doi.org/10.1021/acs.chemrev.8b00363https://doi.org/10.1021/acs.chemrev.8b00363
23. Pysz, M.A., Gambhir, S.S., Willmann, J.K.: Molecular imaging: current status and emerging strategies. Clin. Radiol. **65**, 500–516 (2010). https://doi.org/10.1016/j.crad.2010.03.011
24. Glover, G.H.: Overview of Functional Magnetic Resonance Imaging. Neurosurg. Clin. N. Am. **22**, 133–139 (2011). https://doi.org/10.1016/j.nec.2010.11.001
25. Runge, V.M.: Critical questions regarding gadolinium deposition in the brain and body after injections of the gadolinium-based contrast agents, safety, and clinical recommendations in consideration of the EMA's pharmacovigilance and risk assessment committee recommend. Invest. Radiol. **52**, 317–323 (2017). https://doi.org/10.1097/RLI.0000000000000374https://doi.org/10.1097/RLI.0000000000000374
26. Doan, B.-T., Meme, S., Beloeil, J.-C.: General principles of MRI. Chem. Contrast Agents Med. Magn. Reson. Imaging 1–23 (2013)
27. Bloch, F., Hansen, W.W., Packard, M.: The nuclear induction experiment. Phys. Rev. **70**, 474–485 (1946). https://doi.org/10.1103/PhysRev.70.474https://doi.org/10.1103/PhysRev.70.474
28. Solomon, I.: Relaxation processes in a system of two spins. Phys. Rev. **99**, 559–565 (1955). https://doi.org/10.1103/PhysRev.99.559https://doi.org/10.1103/PhysRev.99.559
29. Bloembergen, N.: Proton relaxation times in paramagnetic solutions. J. Chem. Phys. **27**, 572–573 (1957). https://doi.org/10.1063/1.1743771https://doi.org/10.1063/1.1743771
30. Lipari, G., Szabo, A.: Model-free approach to the interpretation of nuclear magnetic resonance relaxation in macromolecules. 1. Theory and range of validity. J. Am. Chem. Soc. **104**, 4546–4559 (1982a). https://doi.org/10.1021/ja00381a009https://doi.org/10.1021/ja00381a009
31. Lipari, G., Szabo, A.: Model-free approach to the interpretation of nuclear magnetic resonance relaxation in macromolecules. 2. Analysis of experimental results. J. Am. Chem. Soc. **104**, 4559–4570 (1982b). https://doi.org/10.1021/ja00381a010https://doi.org/10.1021/ja00381a010
32. Banci, L., Bertini, I., Luchinat, C.: Nuclear and Electron Relaxation. VCH, Weinheim (1991)
33. Eisinger, J., Shulman, R.G., Blumberg, W.E.: Relaxation enhancement by paramagnetic ion binding in deoxyribonucleic acid solutions. Nature **192**, 963–964 (1961). https://doi.org/10.1038/192963a0https://doi.org/10.1038/192963a0
34. Lauterbur, P.C.: Image formation by induced local interactions: examples employing nuclear magnetic resonance. Nature **242**, 190–191 (1973). https://doi.org/10.1038/242190a0https://doi.org/10.1038/242190a0
35. Mansfield, P., Maudsley, A.A.: Planar spin imaging by NMR. J. Magn. Reson. **27**, 101–119 (1977). https://doi.org/10.1016/0022-2364(77)90197-4
36. Hinshaw, W.S., Bottomley, P.A., Holland, G.N.: Radiographic thin-section image of the human wrist by nuclear magnetic resonance. Nature **270**, 722–723 (1977). https://doi.org/10.1038/270722a0https://doi.org/10.1038/270722a0

37. Pykett, I.L., Hinshaw, W.S., Buonanno, F.S., et al.: Physical principles of NMR imaging. Curr. Probl. Cancer **7**, 37–50 (1982). https://doi.org/10.1016/S0147-0272(82)80009-3
38. Lauterbur, P.C.: NMR zeugmatographic imaging in medicine. J. Med. Syst. **6**, 591–597 (1982). https://doi.org/10.1007/BF00995509https://doi.org/10.1007/BF00995509
39. Brady, T.J., Goldman, M.R., Pykett, I.L., et al.: Proton nuclear magnetic resonance imaging of regionally ischemic canine hearts: effect of paramagnetic proton signal enhancement. Radiology **144**, 343–347 (1982). https://doi.org/10.1148/radiology.144.2.6283594https://doi.org/10.1148/radiology.144.2.6283594
40. Young, I.R., Clarke, G.J., Baffles, D.R., et al.: Enhancement of relaxation rate with paramagnetic contrast agents in NMR imaging. J. Comput. Tomogr. **5**, 543–547 (1981). https://doi.org/10.1016/0149-936X(81)90089-8
41. Carr, D.H., Brown, J., Bydder, G.M., et al.: Intravenous chelated gadolinium as a contrast agent in NMR imaging of cerebral tumors. Lancet **323**, 484–486 (1984). https://doi.org/10.1016/S0140-6736(84)92852-6
42. Carr, D.H., Brown, J., Bydder, G.M., et al.: Gadolinium-DTPA as a contrast agent in MRI: initial clinical experience in 20 patients. Am. J. Roentgenol. **143**, 215–224 (1984). https://doi.org/10.2214/ajr.143.2.215https://doi.org/10.2214/ajr.143.2.215
43. Clare, S.: Functional MRI: Methods and Applications. University of Nottingham (1997)
44. Vlaardingerbroek, M.T., den Boer, J.A.: Magnetic Resonance Imaging. Springer, Berlin (2003)
45. Dale, B.M., Brown, M.A., Semelka, R.C.: Instrumentation. In: Dale, B.M., Brown, M.A., Semelka, R.C. (eds.) MRI Basic Principles and Applications, pp. 177–188 (2015)
46. Smith, R.C., Lange, R.C.: Understanding Magnetic Resonance Imaging, 1st edn. CRC Press, New York (1997)
47. Weishaupt, D., Köchli, V.D., Marincek, B.: How Does MRI Work? An Introduction to the Physics and Function of Magnetic Resonance Imaging, 2nd edn. Springer, Berlin (2006)
48. Geraldes, C.F.G.C., Laurent, S.: Classification and basic properties of contrast agents for magnetic resonance imaging. Contrast Media Mol. Imaging **4**, 1–23 (2009). https://doi.org/10.1002/cmmi.265https://doi.org/10.1002/cmmi.265
49. Hao, D., Ai, T., Goerner, F., et al.: MRI contrast agents: basic chemistry and safety. J. Magn. Reson. Imaging **36**, 1060–1071 (2012). https://doi.org/10.1002/jmri.23725https://doi.org/10.1002/jmri.23725
50. Tóth, É., Helm, L., Merbach, A.E.: Relaxivity of MRI contrast agents. In: Krause, W. (ed.) Contrast Agents I: Magnetic Resonance Imaging, pp. 61–101. Springer, Berlin (2002)
51. Muller, R.N., Vander Elst, L., Roch, A., et al.: Relaxation by Metal-Containing Nanosystems, pp. 239–292. Academic Press (2005)
52. Chan, K.W.-Y., Wong, W.-T.: Small molecular gadolinium(III) complexes as MRI contrast agents for diagnostic imaging. Coord. Chem. Rev. **251**, 2428–2451 (2007). https://doi.org/10.1016/j.ccr.2007.04.018
53. Werner, E.J., Datta, A., Jocher, C.J., Raymond, K.N.: High-relaxivity MRI contrast agents: where coordination chemistry meets medical imaging. Angew. Chem. Int. Ed. **47**, 8568–8580 (2008). https://doi.org/10.1002/anie.200800212https://doi.org/10.1002/anie.200800212
54. Viswanathan, S., Kovacs, Z., Green, K.N., et al.: Alternatives to gadolinium-based metal chelates for magnetic resonance imaging. Chem. Rev. **110**, 2960–3018 (2010). https://doi.org/10.1021/cr900284ahttps://doi.org/10.1021/cr900284a
55. Zhou, Z., Lu, Z.R.: Gadolinium-based contrast agents for magnetic resonance cancer imaging. WIREs Nanomed. Nanobiotechnol. **5**, 1–18 (2013). https://doi.org/10.1002/wnan.1198https://doi.org/10.1002/wnan.1198
56. Caravan, P., Ellison, J.J., McMurry, T.J., Lauffer, R.B.: Gadolinium(III) chelates as MRI contrast agents: structure, dynamics, and applications. Chem. Rev. **99**, 2293–2352 (1999). https://doi.org/10.1021/cr980440xhttps://doi.org/10.1021/cr980440x
57. Koenig, S.H., Brown, R.D., III., Spiller, M., Lundbom, N.: Relaxometry of brain: why white matter appears bright in MRI. Magn. Reson. Med. **14**, 482–495 (1990). https://doi.org/10.1002/mrm.1910140306https://doi.org/10.1002/mrm.1910140306

58. McDonald, R.J., Levine, D., Weinreb, J., et al.: Gadolinium retention: a research roadmap from the 2018 NIH/ACR/RSNA workshop on gadolinium chelates. Radiology **289**, 517–534 (2018). https://doi.org/10.1148/radiol.2018181151https://doi.org/10.1148/radiol.2018181151
59. Hermann, P., Kotek, J., Kubíček, V., Lukeš, I.: Gadolinium(iii) complexes as MRI contrast agents: ligand design and properties of the complexes. Dalt. Trans. 3027–3047 (2008). https://doi.org/10.1039/B719704G
60. Clough, T.J., Jiang, L., Wong, K.L., Long, N.J.: Ligand design strategies to increase stability of gadolinium-based magnetic resonance imaging contrast agents. Nat. Commun. **10**, 1420 (2019). https://doi.org/10.1038/s41467-019-09342-3https://doi.org/10.1038/s41467-019-09342-3
61. De León-Rodríguez, L.M., Martins, A.F., Pinho, M.C., et al.: Basic MR relaxation mechanisms and contrast agent design. J. Magn. Reson. Imaging **42**, 545–565 (2015). https://doi.org/10.1002/jmri.24787https://doi.org/10.1002/jmri.24787
62. Yang, C.T., Chuang, K.H.: Gd(iii) chelates for MRI contrast agents: from high relaxivity to "smart", from blood pool to blood–brain barrier permeable. Med. Chem. Commun. **3**, 552–565 (2012). https://doi.org/10.1039/C2MD00279Ehttps://doi.org/10.1039/C2MD00279E
63. Kang, S.I., Ranganathan, R.S., Emswiler, J.E., et al.: Synthesis, characterization, and crystal structure of the gadolinium(III) chelate of (1R,4R,7R)-.alpha., alpha′., alpha″.-trimethyl-1,4,7,10-tetraazacyclododecane-1,4,7-triacetic acid (DO3MA). Inorg. Chem. **32**, 2912–2918 (1993). https://doi.org/10.1021/ic00065a019https://doi.org/10.1021/ic00065a019
64. Chang, C.A., Francesconi, L.C., Malley, M.F., et al.: Synthesis, characterization, and crystal structures of M(DO3A) (M = iron, gadolinium) and Na[M(DOTA)] (M = Fe, yttrium, Gd). Inorg. Chem. **32**, 3501–3508 (1993). https://doi.org/10.1021/ic00068a020https://doi.org/10.1021/ic00068a020
65. Aime, S., Botta, M., Fasano, M., Terreno, E.: Lanthanide(III) chelates for NMR biomedical applications. Chem. Soc. Rev. **27**, 19–29 (1998). https://doi.org/10.1039/A827019Zhttps://doi.org/10.1039/A827019Z
66. Caravan, P.: Strategies for increasing the sensitivity of gadolinium based MRI contrast agents. Chem. Soc. Rev. **35**, 512–523 (2006). https://doi.org/10.1039/B510982Phttps://doi.org/10.1039/B510982P
67. Aime, S., Barge, A., Batsanov, A.S. et al.: Controlling the variation of axial water exchange rates in macrocyclic lanthanide(iii) complexes. Chem. Commun. 1120–1121 (2002). https://doi.org/10.1039/B202862J
68. Thompson, A.L., Parker, D., Fulton, D.A., et al.: On the role of the counter-ion in defining water structure and dynamics: order, structure and dynamics in hydrophilic and hydrophobic gadolinium salt complexes. Dalt. Trans. 5605–5616 (2006). https://doi.org/10.1039/B606206G
69. Caravan, P., Farrar, C.T., Frullano, L., Uppal, R.: Influence of molecular parameters and increasing magnetic field strength on relaxivity of gadolinium- and manganese-based T1 contrast agents. Contrast Media Mol. Imaging **4**, 89–100 (2009). https://doi.org/10.1002/cmmi.267https://doi.org/10.1002/cmmi.267
70. Ibrahim, M.A., Hazhirkarzar, B., Dublin, A.B.: Magnetic Resonance imaging (MRI) gadolinium. In: StatPearls Publ. Treasure Isl. https://www.ncbi.nlm.nih.gov/books/NBK482487/. Accessed 5 May 2020
71. Magnevist® (gadopentetate dimeglumine) injection 0.5 mmol/mL. https://www.radiologysolutions.bayer.com/sites/g/files/kmftyc641/files/MVEOSLetter-GPOPDFR8v1.pdf. Accessed 8 July 2020
72. Gadolinium-Containing Contrast Agents. https://www.ema.europa.eu/en/medicines/human/referrals/gadolinium-containing-contrast-agents. Accessed 8 July 2020
73. Li, D., Larson, A.C., Speck, O., et al.: Modern applications of MRI in medical sciences. Magn. Med. 343–476 (2006)
74. Caravan, P.: Protein-targeted gadolinium-based magnetic resonance imaging (MRI) contrast agents: design and mechanism of action. Acc. Chem. Res. **42**, 851–862 (2009). https://doi.org/10.1021/ar800220phttps://doi.org/10.1021/ar800220p
75. Laurent, S., Vander, E.L., Muller, R.N.: Comparative study of the physicochemical properties of six clinical low molecular weight gadolinium contrast agents. Contrast Media Mol. Imaging **1**, 128–137 (2006). https://doi.org/10.1002/cmmi.100https://doi.org/10.1002/cmmi.100

76. Prince, M.R., Zhang, H., Zou, Z., et al.: Incidence of immediate gadolinium contrast media reactions. Am. J. Roentgenol. **196**, W138–W143 (2011). https://doi.org/10.2214/AJR.10.4885https://doi.org/10.2214/AJR.10.4885
77. Grobner, T.: Gadolinium—a specific trigger for the development of nephrogenic fibrosing dermopathy and nephrogenic systemic fibrosis? Nephrol. Dial. Transplant. **21**, 1104–1108 (2006). https://doi.org/10.1093/ndt/gfk062https://doi.org/10.1093/ndt/gfk062
78. Idée, J.-M., Port, M., Medina, C., et al.: Possible involvement of gadolinium chelates in the pathophysiology of nephrogenic systemic fibrosis: a critical review. Toxicology **248**, 77–88 (2008). https://doi.org/10.1016/j.tox.2008.03.012
79. Kanda, T., Fukusato, T., Matsuda, M., et al.: Gadolinium-based contrast agent accumulates in the brain even in subjects without severe renal dysfunction: evaluation of autopsy brain specimens with inductively coupled plasma mass spectroscopy. Radiology **276**, 228–232 (2015). https://doi.org/10.1148/radiol.2015142690https://doi.org/10.1148/radiol.2015142690
80. Thomsen, H.S., Morcos, S.K., Almén, T., et al.: Nephrogenic systemic fibrosis and gadolinium-based contrast media: updated ESUR Contrast Medium Safety Committee guidelines. Eur. Radiol. **23**, 307–318 (2013). https://doi.org/10.1007/s00330-012-2597-9https://doi.org/10.1007/s00330-012-2597-9
81. Brücher, E., Tircsó, G., Baranyai, Z., et al.: Stability and toxicity of contrast agents. In: The Chemistry of Contrast Agents in Medical Magnetic Resonance Imaging, pp. 157–208. John Wiley & Sons Ltd., Chichester (2013)

Current Methods and Future of Tuberculosis (TB) Diagnosis

Sanchit Sood, Rakesh Arya, Nirmita Dutta, Abhishek Paul, Rajendra Kumar Behera, Ranjan Kumar Nanda, and Gorachand Dutta

Abstract Nearly, 2.1 billion people are supposed to be infected with a deadly disease of tuberculosis (TB) today. Besides this, there is an additional 8 million emerging TB cases every year, indicating a more serious threat. This catastrophic effect of tuberculosis is also accelerated by its deadly cooperation with HIV. Thus, quick and precise diagnosis is highly needed to fight against TB and TB-HIV/AIDS confection especially in developing nations like India. The review describes several already in use and novel diagnostic methods beside their advantages and limitations for establishing a standard TB treatment in resource limited areas. As per WHO policy decision, development of some novel methods have been made available at affordable prices for health care sector in TB endemic countries. The future looks encouraging with many new upcoming tools specifically molecular based diagnosis, next-generation-based technologies and many new tests which are approved by WHO for investment and adoption. But still the developments of new weapons for diagnosis are yet to realize its power in order to reach out the maximum patients possible. Many international and national workers and donors are synergistically working toward the development of diagnosis methods that can be easily installed and used in point-of-care (POC) settings.

S. Sood and R. Arya are contributed equally.

S. Sood
Department of Agricultural and Food Engineering, Indian Institute of Technology, Kharagpur, West Bengal, India

R. Arya · R. K. Nanda
Translational Health Group, International Centre for Genetic Engineering and Biotechnology (ICGEB), New Delhi, India

N. Dutta · A. Paul · G. Dutta (✉)
School of Medical Sciences and Technology, Indian Institute of Technology, Kharagpur, West Bengal, India
e-mail: g.dutta@smst.iitkgp.ac.in

R. Arya · R. K. Behera
School of Life Sciences, Sambalpur University, Sambalpur, Odisha 768019, India

G. Dutta et al. (eds.), *Modern Techniques in Biosensors*, Studies in Systems, Decision and Control 327, https://doi.org/10.1007/978-981-15-9612-4_7

Keywords Tuberculosis · Diagnosis · Technologies · Resource limited areas · Point-of-care

1 Introduction

Tuberculosis (TB) is a communicable disease observed due to bacterial infection of *Mycobacterium tuberculosis* (MTB). Lungs are the main site of infection,system (CNS), spinal cord, bone, ovary etc. which is called extra-pulmonary TB. Approximately, 5–10% of total world's population is believed to develop TB in their lifetime [1]. In 2018, approximately, 10.0 million people (57% adult men, 32% adult women, 11% children, 8.6% bearing HIV co-infection) reported to be infected with TB, globally [1]. People suffering from Human Immunodeficiency Virus (HIV)/AIDS and exposed to other risk factors like under-nutrition, diabetes or alcohol/drug abuse are prone to TB disease [1]. The fear due to TB is way more severe than HIV/AIDS and it stands at 10th leading cause of death [1]. As per 2019s WHO report, nearly, 1.2 million HIV negative people died due to TB infection (the score being 1.7 million in 2000) while an additional death scored 0.25 million among HIV-positive people (down from 0.62 million in 2000) in 2018 [1]. TB disease is transmitted by aerosol produced from infected patients to healthy subjects. The driving forces behind many nations holding high TB endemic rate in population are due to the unavailability of sensitive point-of-care diagnostic tests [2]. As per data, it was pointed out that a significantly less than 30 percent of people were actually diagnosed by a proven method between 2012 and 2017. The major reason behind this that initial TB diagnosis was dependent on the test accuracy, complexity, accessibility, cost, and sensitivity and also regulated by political will and investment by funder. Therefore, it becomes clear that precise and immediate diagnosis of TB is critical in breaking the transmission of MTB [3, 4]. Since centuries, TB infections have waxed and waned throughout the world, mankind is continuously struggling to regulate it and be prepared against the adversity of the disease. Today, many cofactors like HIV, multi-drug resistant (MDR) and over population have continuously threatened and decreased the control of the disease [5]. The high rate of increase in co-infections is majorly due to nonspecific signs, with respect to HIV negative people, causing the detection rate to fall and allowing MTB to transmit over the community. Fast and accurate detection of infectious agents and its extent of drug susceptibility is crucial for timely initiation of treatment, and disease control and transmission [6]. Commonly used Bacilli Calmette-Guérin (BCG) vaccine is a century old method and protects pediatric TB patients to develop severe form of meningeal TB but ineffective in adults. According to WHO, 12 TB vaccine candidates are in clinical trials [7]. However, there is no vaccine available to control TB (especially in adults) till date. Antiretroviral Treatment (ART) for TB and HIV/AIDS coinfected patients can eliminate an additional 9 million death [7]. TB can take mainly three prominent forms which show associated severity differently (as shown in Fig. 1). Technical advancements in diagnostics resulted in increased improved tools; some are appropriate for TB endemic areas

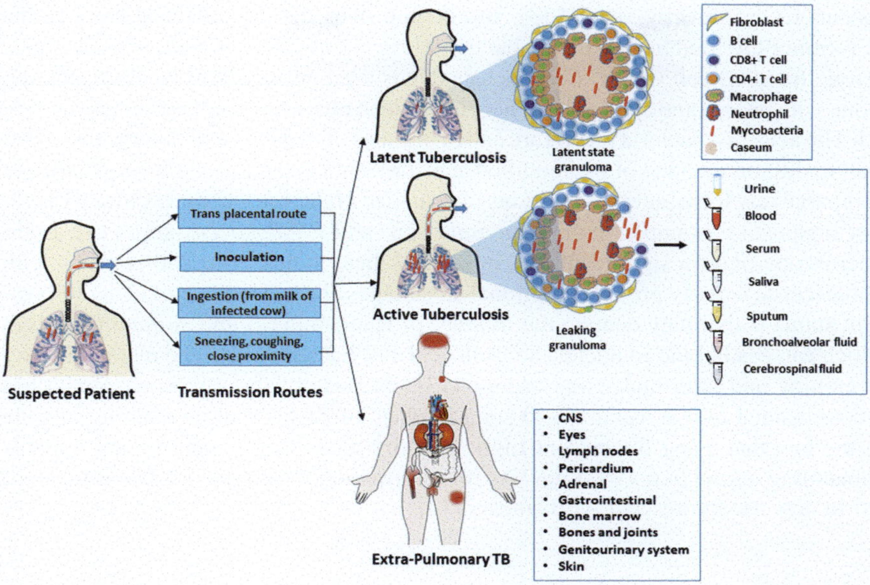

Fig. 1 Types of TB and associated severity

but focus on advanced technology should be adopted for proper diagnosis. Much work is still needed to integrate new methods of diagnosis in preventive programs of TB endemic countries. In this article, we have summarized presently available TB diagnosis methods and their application in diverse clinical settings.

2 Methods

Several diagnostic methods that are new in addition to already used are described in detail.

2.1 *Microscopy*

Sputum microscopy is the most common method to diagnose TB. It is the inexpensive, quick, and relatively easy to perform. Though detection of gram-positive bacteria is possible by gram stain due to their cell wall composition (peptidoglycan), MTB cannot be detected by the same. One of the unique features of MTB is their hydrophobic cell wall. Due to which it withstands penetration of dyes like crystal violet (aniline dye). Thus, they become invisible or clear zones of bacilli shaped when specimens are directly stained [8]. But certain dyes like arylmethane are capable of forming thermodynamically stable complexes in association with mycolic acids across the MTB cell wall. Carbol fuchsia dye is used during Ziehl-Neelsen staining along with phenol and applied heat, here methylene blue serves as the counter stain [9, 10] The acid-fast bacteria (AFB) are called because of resistance of complexes in cell wall to destaining. One of the modified acid-fast stain is the Fite stain which utilizes comparatively lower acidic conditions to visualize *M. leprae* and other partially acid-fast organisms, examples being *Nocardia* spp. and *Rhodococcus* spp. [11]. Light microscopy plays a significant role in each of these stains, while other stains with fluorescence activity provide additional advantages. The advantage of using fluorescent stains is that they contain the mixture of Rhodamine B and Auramine A dyes which can easily bind to nucleic acids in acid-fast bacteria [12]. The method is easy to employ and scan under microscope [13]. Because of these reasons, WHO has recommended to use Auramine O stain in place of Ziehl-Neelsen staining in spite of the fact that using fluorescent Light emitting diode (LED) microscopy remains demanding option in poor set ups. As for information 2014 only 7% laboratories of world now use fluorescent AFB smears [14].

2.2 *Mycobacterial Culture: Solid Media*

Clinical samples can be readily cultured for *M. tuberculosis* complex (MTBC) and non-tuberculous mycobacteria (NTM) growth. The most common egg-based culture media is Lowenstein-Jensen (L-J) media, which contains components such as malachite green, glycerol, egg proteins, salts and potato starch. L-J media is useful in growing many species of MTB but it is not the apt choice for the whole range of mycobacterial species, such as *M. genavense* and *M. bovis* [15, 16]. Though the L-J media is still in use by many researchers but many laboratories have now started opting for chemically defined media which are optimized for MTB growth. One such example is the Middlebrook 7H10 or 7H11 agars which allows one to observe MTBC cultured colonies in 10–12 days, as opposed to a duration of 18–24 days in case of L-J media [17]. One of the disadvantages of agar media is that concerning stability and sensitivity toward damage by external means. For example, heat or light accelerates the releasing of formaldehyde which is toxic enough to inhibit mycobacterial growth. The optimum temperature for growth of most Mycobacterium species, including MTB, is 35–37 °C. As for aerobic bacteria, mycobacterial growth is increased when incubated with 5–10% CO_2. The minimum time for doubling required by MTB and NTM species is 12–24 h. Because of this fact, it takes at least 2–3 weeks for MTB colonies to be visible to the open eye on petri dish. Generally cultures require an incubation time of 6–8 weeks [18]. Irrespective of sluggish growth, MTB culture is more sensitive than AFB smear, only 10–100 CFU/mL of sample (clinical) is sufficient for good growth [17].

2.3 Mycobacterial Culture: Liquid Media

MTB can be efficiently diagnosed from clinical samples by combining both solid and liquid culture media [19]. Researchers have observed that the growth of MTB from clinical specimens require about 10 days on an average, assisted by automated broth systems, while the same sums up to 20–25 days for solid media [19–21]. Till date, FDA has passed three platforms for commercial usage to obtain the semi-automated broth-based culture of mycobacteria: the BACTEC MGIT 960 system (Becton Dickinson Microbiology Systems), the VersaTREK system (Trek Diagnostic Systems), and the MB/BacT Alert 3D (bioMérieux). The individual tube in the system is equipped with a modified Middlebrook 7H9 broth and a fluorescent indicator. This indicator is so designed that it is quenched by the presence of oxygen within the tube. It works as a function of MTB growth, i.e., as the oxygen is consumed over time the fluorescent indicator signals positive at threshold value. The instrument has shown promising results as it helps scientists to keep a note of positive tubes and can easily identify mycobacteria in samples. The MGIT culture system is the more advanced form when compared with the previous BACTEC machines which employed radiometric measurement as the parameter for growth assessment, and called for assigning someone with the task of placing bottles on the machine, periodically and manually, say once or twice per day.

Apart from faster growth, there are other findings that indicate toward the superior capability of the MGIT broth system over solid culture [22–26]. Consider the example of a collection of 1381 strains from 14,745 clinical specimens from a meta-analysis of 10 published studies, which revealed that MGIT system bore a sensitivity of 81.5% (99.6% specificity) while the value was only 67% in case of L-J solid media [21]. The VersaTREK system employs a change in pressure in the headspace of the culture bottle as an indicator of growth of cultured specimen in broth medium. Finally, the MB/BacT Alert 3D system uses a colorimetric carbon dioxide sensor as an indicator of growth. Myco/F Lytic culture bottles, as compared to standard BACTEC aerobic bottles, are more preferred for growth of MTB [27, 28]. The Myco/F Lytic culture system has been shown to give positive MTBC colonies from blood specimen within 20 days, while the growth period for BacT/Alert culture system is 16.4 days [29]. Whatever the broth system, sub-culturing in solid media comes next for all specimens that test positive. This helps one to analyze and ascertain the presence of mixed cultures and determine the correlation between colony morphologies, if present.

2.4 Nucleic Acid Hybridization Probes

The hybridization probe is a fragment of DNA that can be efficiently used to check the prevalence of MTB. Even many commercially available probes can easily be employed to detect various clinically significant mycobacterium species. FDA approved hybridization probes can be accessed from Hologic Gen-Probe to detect

M. tuberculosis complex, *M. avium* complex, *M. kansasii* and *M. gordonae*. Extraction of nucleic acids from cultured microorganisms has been done by subjecting them heat treatment and ultrasonic frequencies. A species-specific DNA probe conjugated with a chemiluminescent label is added next. The probe hybridizes with the 16S ribosomal RNA target on account of its sequence complementarity. This generates chemiluminescent signal from the DNA-rRNA heteroduplex, which can now be identified and quantified. While using these probes, no PCR amplification steps are followed. So this demands for increased amount of nucleic acid to be present in a sample. This also poses a limitation of having to use probes on cultured specimen rather than using them directly on urine or sputum specimens collected from the patient. However, culture hybridization probes detection claim high sensitivity and specificity, call for a simpler protocol,, and deliver results within about two hours [30–32]. Various studies have shown minor cross reactivity with infrequent species of NTM; however, this can be removed by manipulation time and temperature set for hybridization. Only few reports have shown cross reactivity between the MAC probe (to *M. palustre, M. saskatchewanense, M. paraffinicum, M. vulneris M. arosiense, M. colombiense, M. chimaera* and *M. nebraskense*) and the MTBC probe (to *M. hol-saticum* and *M. celatum*) [30, 31]. Hybridization probes are an efficient tool, but the one main disadvantage is that they are only limited to the detection of four mycobacterial species or complexes, which means that the identification of other mycobacterial species available in the clinical laboratory would call for additional molecular probes, and newer, optimized strategies.

2.5 Next-Generation Sequencing (NGS)

One of the most trusted and acceptable method till date for identifying most mycobacterial species is the Sanger dideoxy sequencing method. To this date, about 40 species of MTB have been sequenced fully; however, sequencing of targets such as the heat shock protein hsp65 gene, 16S or 23S ribosomal DNA genes or the rpoB gene encoding the beta subunit of the bacterial RNA polymerase is a more common practice [32]. The procedure involves using PCR primers complementary to conserved sequences, and polymerases which amplify DNA from all MTB species. The concerned genes might also be investigated for flanking variable regions, which can act as a marker of species differentiation. The next step is sequencing of amplified DNA and comparing the results with already established sequence databases. There are public databases available, such as the NCBI GenBank (within the International Nucleotide Sequence Database Collaboration; http://www.ncbi.nlm, https://www.nih.gov/genbank) which can be queried using a BLAST search, but the amplified genes may contain novel sequences which have not been reported or entered into the databases yet [33]. Medical diagnostics of TB usually use commercial and curated databases such as the SmartGene IDNS (SmartGene), MicroSeq (Thermo Fisher Scientific), or RipSeq (Pathogenomix, Inc.) libraries. These in silico databases are reliable tools for identifying and comparing entries. DNA sequencing can be a potential tool for the identification of many clinical species including MTB. Also, a fairly rapid TAT identification is possible within either the same day or the day after, at the species level. Sequencing reports generally arrive within 8–24 h after the growth of the organism in culture. Major drawbacks of these methods are the labor-intensive procedures and requirement of highly trained technicians. Sequencing often comes in use when cheaper and relatively faster methods such as hybridization probes fail to yield accurate analysis.

2.6 Tuberculin Skin Test (TST)

The tuberculin skin test (TST), or Mantoux test, in the field of TB diagnostics has proven to be the most trustful and accurate method for screening [34]. In this test, a concoction of purified protein derivative of MTB-secreted proteins is intradermally injected (100 μL, PPD) into the palmar surface of the forearm. This may or may not induce a delayed-type or type-IV hypersensitivity reaction (as millimeters of induration), based on which the test is deemed either positive or negative after 48–72 h [30]. Though the technique is relatively simple and cost effective with precise definitions for interpretation, it requires the training and experience of professional clinicians [35]. A follow up by patients to the diagnostic center is required within 48–72 h for the final interpretation of the test [36]. This test suffers from a limitation of reduced sensitivity and specificity, displaying false results in patients with a previous history of the Bacilli Calmette-Guérin (BCG) vaccination or exposure to non-tuberculous mycobacteria (NTM) [37, 38]. In few cases, a standard assessment is needed for TST. Sometimes, it may so happen that individuals who undergo routine TST testing (e.g., health care workers) may produce a reaction upon TST exposure. This would simply signify an immunological response mounted due to the stimulus provided by the last positive infection state and not necessarily recent infection [38, 39]. Several studies have also reported low reproducibility of test causing false negatives in some populations comprising of immunocompromised patients with HIV infection, chronic renal disease, and systemic infections; people who are malnourished; individuals with early gastrointestinal surgical procedures; who have undergone live vaccination within the previous 2 months; or people taking systematic immunosuppressive medication [39, 40].

2.7 Antigen-Based Serological Tests

This diagnostic technique is based on mycobacterial antigens following MTB exposure which can be detected in various amounts in different fluids [41]. The most commonly found antigens from extra-pulmonary TB (EPTB) specimens are non-protein cell wall antigen, antigen 5 (14 kDa), antigen A60 (45/47 kDa), cord factor (trehalose-6, 6-dimycolate), haemoglycolipid-lipid antigen, and lipoarabinomannan

(LAM). But none of them shows accurate sensitivity or specificity especially to pauci-bacillary EPTB. In recent times, one of the antigen-based assay designed to identify mycobacterial LAM in urine has acquired momentum as it holds a promise for POC test. But it is confined to active TB cases due HIV and not for pauci-bacillary EPTB. Meta-analysis of antigen-based tests by researchers have cocncluded that still there is a lack of much study in this field and more work is needed in order to develop proficient antigen-based tests [42].

2.8 Immunohistochemistry (IHC) and Immunocytochemistry (ICC) Based Antigen Detection Tests

Immunological staining can work as a better option with respect to conventional acid-fast staining [43, 44]. The increased sensitivity of immune-based assay can be attributed to the fact that degraded bacilli can also be detected unlike intact bacilli in case of Z-N staining. Various kinds of mycobacterial antigens have been identified in tissue aspirate, formalin-fixed paraffin-embedded (FFPE) tissue, and body fluids, for example, *M. tuberculosis* antigen 85, MPT64, BCG, Tb8.4, LAM, HspX, Antigen 5 (38 kDa), ESAT6, and the phospholipase C encoding A (PlcA) protein. Research has pointed out that immunostaining can be used for extra-pulmonary clinical specimens with reasonably high sensitivity (70–100%) and specificity (65–100%). Even it shows comparable results in HIV-coinfected TB cases with typical histological features. It is a powerful technique and result available within 24 h and insensitive to contamination. However, this works more accurately in high TB prone areas. One of the common limitations and pre-requisite of IHC/ICC is that it involves invasive sample collection and preparation before the application of test [42].

2.9 Histopathology

Histopathology is regarded as the method of choice for diagnosis. Frequent histological examination combined with excision biopsy, Z-N staining and culture for MTB play an almost indispensable tool at the hands of the pathologist for diagnostic purposes. In TB endemic, inflammation of granulomatous accompanied with or without caseation is generally regarded as the important symptom of TB. Though the most common etiological agent of EPTB is MTB, there has been a hike in the spread of NTM regardless of HIV, and can roughly fall between 4% and 50% in various parts of the world. The histological features of NTM lesions look similar under the microscope as that of tuberculous MTB. But the accurate determination and differentiation between the two is crucial on account of different courses of treatment prescribed. The scope of biopsy is restricted for non-reachable sites of EPTB [42]. The scarcity of biopsy facilities in primary medical centers and its invasive characteristics prevents its use in all cases of EPTB. Even with incision biopsy, there is a limitation as it involves sinus tract and fistula formation which hinders screening of accessible lesions. Histopathology is currently followed only for patients having negative fine needle aspiration cytology (FNAC) or with high clinical suspicion, and is enacted as the first line investigation in nearly all accessible mass lesions [42].

2.10 Cytology

Cytological examination specifically FNAC is one of the simple and cost-effective diagnostic measure [45]. This technique is an outpatient procedure with high sensitivity [46]. As an initial diagnostic measure, this procedure is applied in TB-suspected patients as cytology evidence for TB is well established. However, with 57.8% (95% C.I.) [47] specificity, this may give false positive results and can lead to differentiation of lymphocyte-predominant "tuberculosis" cytology pattern from non-tuberculous lesions. Another challenge that it may pose is the lack of accurate distinction between infected (tuberculous) lesions from other granulomatous condition which may be non-tuberculous or lesions that appear during HIV disease. Although the technique is easy and affordable, the precise diagnosis is possible only on bacteriological confirmation in body fluids which may reach up to 20–25% [42].

2.11 Antibody-Based Serological Test

Various serological tests rely on identification of specific antibodies in clinical samples are routinely used for TB diagnosis. These tests can be used in resource-compromised settings for they are simple to handle, cost-effective, and fulfill the criteria of point-of-care (POC) diagnosis. Antibody-based assay works as an additional measure for confirmation of MTB in the patient beside the routine tests for inaccessible sites. Superoxide dismutase, a secretory protein of MTB, has been assessed for serological detection with ELISA. In low prevalence countries, this test bears a 93–94% positive predictive value (PPV), but in high prevalence countries, it falls down to 75% [48]. The positive test may perhaps help to "rule in" the possibility of infection, but a negative test does not necessarily "rule out" this possibility of TB infection. Because of shortage of non-reproducible tests, the readings of antibody-based assay vary according to geographical place. Besides this, in BCG vaccinated patients, delayed-type hypersensitivity kicks in and leads to a delayed response, giving the false measurement of an antibody even after the sub-clinical or clinical disease subsides, and it is also sensitive to exposure of environmental NTM and HIV susceptibility. Now, newly purified antigens and monoclonal antibodies have shown high specificity and sensitivity and are also reproducible. As per WHO guidelines sero-diagnosis is not the preferred method in low- and high-income countries due to compromised sensitivity and specificity [49, 50].

2.12 Interferon-γ Release Assays (IGRAs)

This technology makes use of mononuclear cells residing in peripheral blood which are stimulated in-vitro using specific mycobacterial antigens such as CFP-10 and ESAT-6. This immunological stimulus induces the secretion of interferon-ɣ (Cytokine), which is then measured. Commercially, this test is marketed as QuantiFERON-TB Gold (Cellestis, Australia) (QFT-TB) and T-SPOT TM TB assay (Oxford Immunnotec, UK). Developed countries generally make use of it to diagnose latent TB condition. While the T-SPOT TM. TB assay is relatively more tedious to conduct in a diagnostic laboratory, the QFT-TB may prove to be better adapted for routine diagnosis. IGRAs show that the test has a sensitivity of 75% [51] in active TB patients (including pleuritis) and specificity of 94% in non-tuberculous infected patient. Recent evidence indicates that enzyme-linked immunospot (ELISPOT) assay is more efficient than hematologic malignancy-associated immunosuppression and some types of iatrogenic immunosuppression, including corticosteroids, and azathioprine. The success rate of test is found in HIV co-infection, malnutrition, young children with either latent or active TB. But these assays cannot differentiate between latent- and active-stage infection, thereby posing a major challenge in their use in TB endemic countries. Measurement of both IL-2 and IFN-ɣ secretion by MTB-specific T-cells is directly proportional to the treatment. So, a next-generation T-cell-based test which is capable of dual cytokines enumeration may prove to be promising in giving more clinical information. WHO guidelines have shown negative recommendations with regard to its usage in low-income and middle-income countries

2.13 Adenosine Deaminase (ADA) Activity

ADA an enzyme which is responsible for transformation of adenosine to inosine. The sensitization of immune cells when exposed to mycobacterial antigen drives ADA activity. Two isoenzymes of ADA, i.e., ADA-1 and ADA-2 are available. Most cells of body express ADA-1, whereas ADA-2 is exclusively expressed by the macrophages and monocytes. ADA test can be a convenient method for diagnosing in endemic and poor settings. Dinnes et al. have reviewed ADA activity procured from NHS Health Technology Assessment Program and shown restricted applicability of test in EPTB diagnosis [52]. But there is also evidence supporting the later used for EPTB in literature. It has been found that ADA-2 levels to be increased during TB infection. The two factors which regulate the increased ADA activity in body fluids are identification of particular ADA isoenzyme and proportion of different isoenzymes present. However, there is still a dearth of information on usage of isoenzymes of ADA for TB diagnosing.

2.14 Line Probe Assays (LPA)

These probes are a type of hybridization probe used for enumerating MTB from culture isolates. Line probe assay uses genus- and species-specific probes fixed in nitrocellulose membrane. By using this technology, the DNA from lysed cells hybridizes to complementary probes. Currently, the three most commonly used line probe assays are INNO-LiPA Mycobacteria assay (Fujirebio), the GenoType MTBC test (Hain Lifescience), and the Speed-oligo Mycobacteria (Vircell) [53]. The key advantage of using this method is that it can distinguish between nucleotides with 16S and/or 23S subunits of RNA [54]. Line probe assays are capable of identifying MTB complex members beside other NTM species. The reported sensitivity and specificity levels are >90% and the performer of the test can note the readings within 4–8 h post administration [55, 56]. However, there are reports which point toward cross reactivity of these probes among certain uncommon NTM members [57]. The results of this assay are concordant with nucleic acid hybridization probes. But what poses a major hurdle is that they are not cleared for usage in clinical settings in USA. At present, two-line probe assay is commercially available INNO-LiPA Rif. TB and Genotype MTBDR. Literature evidence suggests that the sensitivity and specificity of INNO-LiPA test are fairly reliable in identifying rifampicin-resistance strains in culture. But the probe is seen to shown decreased specificity when used directly on clinical samples [58].

2.15 Matrix-Assisted Laser Desorption Ionization-Mass Spectrometry (MALDI-TOF MS)

Matrix-assisted laser desorption ionization-time-of-flight mass spectrometry (MALDI-TOF MS) can identify microorganisms by analyzing their protein profile either from disrupted cells [59] or whole bacterial cells [60].This technique works by isolating bacterial colonies from culture plates, followed by mixing with UV absorbing matrix and finally drying on steel plates. The dried samples are exposed to laser which leads to transmission of energy from matrix to non-volatile molecules with analyte desorption into a gas phase. The resulting molecules in ionized state are moved across an electric potential along a flight tube until it reaches mass detector, which ultimately identifies molecules on the basis of their mass/charge ratio (m/z). The resulting profile can then be collated and compared to the already characterized entries in a data base [61]. Many studies supports promising use of MALDI-TOF for MTB diagnosis [33]. This technology over last few years has become a routine procedure for analyzing wide pathogens. Various modifications in the original procedure have been made to suit the purpose. El et al. [33] have used this technique for evaluating heat-inactivated MTB following its segregation in Tween-20, break free of cell wall and finally protein extraction using acetonitrile and formic acid. The results achieved by this are highly fascinating as they were able to achieve to nearly

105 colony-forming units isolates of MTB, *M. avium*, and 20 other MTB species. Also, they were successful in achieving species-specific mass spectra for formulating a local database. Today, their protocol is widely followed for clinical purposes. It has allowed discerning nearly 87 MTB, 25 *M. avium* and 12 NTM isolates with identification in less than 2.5 h. Therefore, as a first line approach MALDI-TOF MS has proved to be a portable and promising technology for pinpointing large variety of bacteria. Being highly sensitive, i.e., 72.4% and specific, i.e., 92.4% [62] MALDI requires stringent conditions to be maintained during its operation including both bacterial culture and pre-sample processing for contaminant-free output [63].

2.16 Gas Chromatography-Mass Spectrometry (GC-MS)

A recent method of TB diagnosis using gas chromatography–mass spectrometry (GC–MS) has been developed for efficient screening [64]. Unlike nucleic acid amplification test (NAAT) which are expensive though accurate but unsuitable for resource limited countries. GC-MS is a novel technique with high potential to translate into a precise, sensitive, and more specific method for diagnosing of TB especially in urine specimens. Collection of urine samples is far easier than sputum collection and this also avoid harmful exposure to health care workers. Many crucial biomarkers have been identified with the help of GC-MS providing high hopes for future diagnosis. One of the applications is the detection of D-arabinose biomarker from urine for confirming TB infection. Owing to unique character of TB cell wall, it is mainly composed of lipids, lipoglycans, polysaccharides, and other constituents. The principal component of the MTB wall is lipoarabinomannan (LAM). As per research, the D-arabinose acts in a similar manner as LAM and can be used as a replacement to detect MTB infection [65]. D-arabinose is not easily confirmed by LC-MS or HPLC columns due to its polar nature and small size. Other methods such as immunological based assays suffer from poor sensitivity and high pre-processing steps. Scientists have distinguished D-arabinose from L-arabinose by using chiral GC column [6]. The results achieved have a detection limit of 10 ng/mL for D-arabinose which gave high hopes to use this method for clinical samples. GC-MS/MS analysis of D-arabinose and other potential biomarkers is an effective means of confirming TB and it gives benefits over conventional methods that rely on multi-step procedures and cannot be carried out in resource limited settings (Table 1). This method holds promise to treat TB and give more cost effective and rapid treatment options to communities around the globe [66]. Figure 2 depicts all the diagnostic methods discussed briefly.

Table 1 Techniques and methods used for the TB diagnosis, its advantages, and limitations

S. no.	Techniques/methods	Specificity/sensitivity	Advantages	Limitations	Ref.
1.	Ziehl-Neelsen staining	>100%/57%	• Rapid, simple method with treatment monitoring	• Tedious, less sensitive • Presumptive diagnosis of MTB • Species differentiation is not possible	[67]
2.	Culture (Solid/liquid)	100%/69.2%	• One test to confirm definite TB • More sensitive than Z-N staining	• Very slow growth • Less accessible and more expensive than microscopy	[68]
3.	Nucleic acid hybridization	100%/100%	• Simultaneous analysis of multiple genes • High throughput • Allow comparison of gene expressions (disease vs control) • Can categorize diseases into sub-groups.	• Confusing to first time users • Result not always reproducible • Still too expensive • Gene expression dictated by qualitative data	[69]
4.	Sequencing (Maxam Gilbert and Sanger)	100%/66.7%	• Produces maximum read lengths, single molecule produces large number of reads	• High capital cost • Accuracy and longevity remain questionable	[70]
5.	Tuberculin skin test (TST)	97.8%/76.7%	• Easily available • Useful in high prevalence areas	• Overestimation of disease in healthy individual and underestimation in immunocompromised individuals • Previous BCG vaccination may interfere with reports	[71]

(continued)

Table 1 (continued)

S. no.	Techniques/methods	Specificity/sensitivity	Advantages	Limitations	Ref.
6.	Interferon-γ release Assays (IGRAs)	92.0%/89.3%	• Independent of reader • Reproducible • No booster effects • No interference with previous vaccination	• Not freely available • Expensive	[71]
7.	Adenosine Deaminase Activity (ADA)	92.7%/89.0%	• Highly sensitive	• Depends on the specimen • Variable sensitivity	[72]
8.	Histopathology	92%/42%	• Confirmative/definitive diagnosis	• Invasive	[73]
9.	Ag–Ab-based serology test	47.4%/87.4%	• Quick diagnostic test • Cost effective • usage of little number of reagents	• Discrepancies in interpreting results • Serum albumin titer difficult to determine	[74]
10.	ImmunoHistoChemistry (IHC)	100%/97%	• Routinely performed • Low technology requirements • Presence of histological information	• Semi-quantitative • Subjective interpretation of results • Poor visual morphological analysis	[75]
11.	Cytology	57.8% (95% C.I.)/80.08% (95% C.I.)	• Non-invasive and painless, minimal instrumentation, cost effective, easily done at patient's side	• False negative results • Contamination, inadequate sampling • Low sensitivity	[76]

(continued)

Table 1 (continued)

S. no.	Techniques/methods	Specificity/sensitivity	Advantages	Limitations	Ref.
12.	Line Probe Assay (LPA)	98.7%/98.1%	• Identifies MTB and genetic mutants associated with INF and RIF resistance, has the potential to reduce turnaround time of DST results	• Available LPAs are recommended only on smear positive sputum specimens, sensitivity of LPAs is lower than culture methods (~70–75%)	[77]
13.	Matrix-Assisted Laser Desorption Ionization-Mass Spectrometry (MALDI-TOF MS)	92.4%/72.4%	• Rapid identification (1 min per isolate), reduce cost per test and hands-on time, flexible	• Sequencing information not as complete as that provided by LC-MS • Limited success in identification of bacterial proteins that are not in database	[62]
14.	Gas Chromatography-Mass Spectrometry (GC-MS)	67%/75%	• Non-invasive, low cost, easy to handle • No risk of disease transmission as in blood sampling • Can detect both recent and long-term exposure unlike blood, urine, or saliva	• Instrument maintenance • Range of chemicals that instrument scans before actually performing • Host proteins and normal flora might overlap mass spectra	[78]

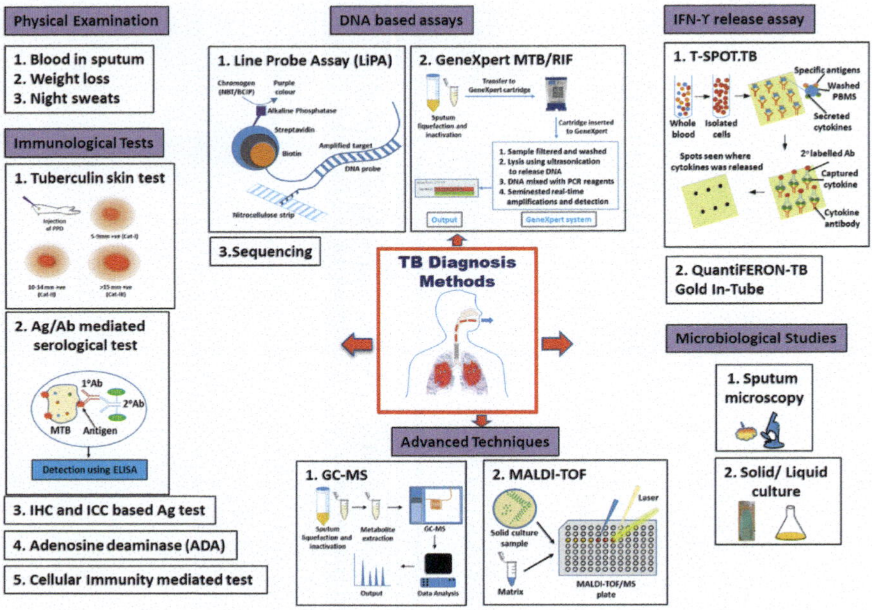

Fig. 2 Different methods of tuberculosis (TB) diagnosis

3 Concluding Remarks

High TB endemic countries rely on conventional techniques like spear microscopy and culturing for TB diagnosis which are time consuming and require repeated laboratory visits. The incoming of novel diagnostic methods is beginning to change the scenario. TB epidemiology highlights that targeted sensitization, quick, low cost and time saving methods are needed to generate the importance of this disease. Being aware about TB infection is the primary step in this direction, as early diagnosis and instant decision of treatment are important to achieve best results. Though the development of rapid molecular techniques stimulated the rate of diagnosis in part, there is still growing interest in more efficient technologies. New and rapid diagnostic methods are yet to reach its full potential and there is a need to develop balance between development of new diagnosis methods and comparatively smaller drug treatment in order to avoid MDR-TB. One of the positive hopes in this direction is POC devices which hold the promise of accurate and ultrasensitive inspection of multiple key components of body fluids, which assert indispensable information regarding the state of human health. This paves the path for personalized medicines. In developing nations like India, these devices are capable of bringing laboratory to homes, thus facilitating the ease of TB treatment and control its transmission worldwide.

References

1. World Health Organization: Global tuberculosis report. https://www.who.int/tb/publications/global_report/en/ (2019)
2. Frieden, T. (ed.): Toman's Tuberculosis: Case Detection, Treatment, and Monitoring: Questions and Answers, 2nd edn. World Health Organization, Geneva (2004)
3. Domínguez, J., Boettger, E.C., Cirillo, D., Cobelens, F., Eisenach, K.D., Gagneux, S., et al.: Clinical implications of molecular drug resistance testing for *Mycobacterium tuberculosis*: a TBNET/RESIST-TB consensus statement. Int. J. Tuberc. Lung Dis. **20**(1), 24–42 (2016). https://doi.org/10.5588/ijtld.15.0221
4. McNerney, R., Cunningham, J., Hepple, P., Zumla, A.: New tuberculosis diagnostics and rollout. Int. J. Infect. Dis. **32**:81–86 (2015). https://doi.org/10.1016/j.ijid.2015.01.012
5. Steingart, K.R., Flores, L.L., Dendukuri, N., Schiller, I., Laal, S., Ramsay, A., et al.: Commercial serological tests for the diagnosis of active pulmonary and extrapulmonary tuberculosis: an updated systematic review and meta-analysis. PLoS Med. **8**(8), e1001062 (2011). https://doi.org/10.1371/journal.pmed.1001062
6. Weyer, K., Mirzayev, F., Migliori, G.B., Van Gemert, W., D'Ambrosio, L., Zignol, M., et al.: Rapid molecular TB diagnosis: evidence, policy making and global implementation of Xpert MTB/RIF. Eur. Respir. J. **42**(1), 252–271 (2013). https://doi.org/10.1183/09031936.00157212
7. World Health Organization: Global tuberculosis report (2017)
8. Trifiro, S., Bourgault, A.M., Lebel, F., Rene, P.: Ghost mycobacteria on gram stain. J. Clin. Microbiol. **28**(1), 146–147 (1990)
9. Ziehl, F.: Zur Färbung des tuberkelbacillus. DMW-Deutsche Medizinische Wochenschrift. **8**(33), 451 (1882)
10. Neelsen, F.: Ein casuistischer Beitrag zu Lehre von der Tuberkulose. Centralblatt für die medizinischen Wissenschaften **28**, 497–501 (1883)
11. Fite, G.L., Cambre, P.J., Turner, M.H.: Procedure for demonstrating lepra bacilli in paraffin sections. Arch. Pathol. (Chic) **43**(6), 624 (1947)
12. Hanscheid, T., Ribeiro, C.M., Shapiro, H.M., Perlmutter, N.G.: Fluorescence microscopy for tuberculosis diagnosis. Lancet Infect. Dis. **7**(4), 236–237 (2007)
13. Kommareddi, S., Abramowsky, C.R., Swinehart, G.L., Hrabak, L.: Nontuberculous mycobacterial infections: comparison of the fluorescent auramine-O and Ziehl-Neelsen techniques in tissue diagnosis. Hum. Pathol. **15**(11), 1085–1089 (1984)
14. World Health Organization: Global tuberculosis report, 20th ed. Geneva, Switzerland. http://www.who.int/tb/publications/global_report/en/ (2015)
15. Realini, L., De Ridder, K., Hirschel, B., Portaels, F.: Blood and charcoal added to acidified agar media promote the growth of *Mycobacterium genavense*. Diagn. Microbiol. Infect. Dis. **34**(1), 45–50 (1999)
16. Robbe-Austerman, S., Bravo, D.M., Harris, B.: Comparison of the MGIT 960, BACTEC 460 TB and solid media for isolation of *Mycobacterium bovis* in United States veterinary specimens. BMC Vet. Res. **9**, 74 (2013)
17. Pfyffer, G.E.: *Mycobacterium:* general characteristics, laboratory detection, and staining procedures. In: Manual of Clinical Microbiology, 11th ed., pp. 536–569. ASM Press, Washington, DC (2015)
18. Simner, P.J., Doerr, K.A., Steinmetz, L.K., Wengenack, N.L.: *Mycobacterium* and aerobic actinomycete culture: are two medium types and extended incubation times necessary? J. Clin. Microbiol. **54**(4), 1089–1093 (2016)
19. Clinical Laboratory Standards Institute (CLSI): Laboratory detection and identification of mycobacteria. Document M48-A. Wayne, PA (2008)
20. Pfyffer, G.E., Welscher, H.M., Kissling, P., Cieslak, C., Casal, M.J., Gutierrez, J., et al.: Comparison of the mycobacteria growth indicator tube (MGIT) with radiometric and solid culture for recovery of acid-fast bacilli. J. Clin. Microbiol. **35**(2), 364–368 (1997)

21. Cruciani, M., Scarparo, C., Malena, M., Bosco, O., Serpelloni, G., Mengoli, C.: Meta–analysis of BACTEC MGIT 960 and BACTEC 460 TB, with or without solid media, for detection of mycobacteria. J. Clin. Microbiol. **42**(5), 2321–2325 (2004)
22. Chihota, V.N., Grant, A.D., Fielding, K., Ndibongo, B., van Zyl, A., Muirhead, D., et al.: Liquid versus solid culture for tuberculosis: performance and cost in a resource-constrained setting. Int. J. Tuberc. Lung Dis. **14**(8), 1024–1031 (2010)
23. Somoskovi, A., Magyar, P.: Comparison of the mycobacteria growth indicator tube with MB redox, Lowenstein-Jensen, and Middlebrook 7H11 media for recovery of mycobacteria in clinical specimens. J. Clin. Microbiol. **37**(5), 1366–1369 (1999)
24. Lu, D., Heeren, B., Dunne, W.M.: Comparison of the automated mycobacteria growth indicator tube system (BACTEC 960/MGIT) with Lowenstein-Jensen medium for recovery of mycobacteria from clinical specimens. Am. J. Clin. Pathol. **118**(4), 542–545 (2002)
25. Srisuwanvilai, L.O., Monkongdee, P., Podewils, L.J., Ngamlert, K., Pobkeeree, V., Puripokai, P., et al.: Performance of the BACTEC MGIT 960 compared with solid media for detection of *Mycobacterium* in Bangkok. Thailand. Diagn. Microbiol. Infect. Dis. **61**(4), 402–407 (2008)
26. Hasan, M., Munshi, S.K., Momi, S.B., Rahman, F., Noor, R.: Evaluation of the effective- ness of BACTEC MGIT 960 for the detection of mycobacteria in Bangladesh. Int. J. Mycobact. **2**, 214–219 (2013)
27. Fuller, D.D., Davis Jr., T.E., Denys, G.A., York, M.K.: Evaluation of BACTEC MYCO/F Lytic medium for recovery of mycobacteria, fungi, and bacteria from blood. J. Clin. Microbiol. **39**(8), 2933–2936 (2001)
28. Vetter, E., Torgerson, C., Feuker, A., Hughes, J., Harmsen, S., Schleck, C., et al.: Comparison of the BACTEC MYCO/F Lytic bottle to the isolator tube, BACTEC Plus Aerobic F/bottle, and BACTEC Anaerobic Lytic/10 bottle and comparison of the BACTEC Plus Aerobic F/bottle to the isolator tube for recovery of bacteria, mycobacteria, and fungi from blood. J. Clin. Microbiol. **39**(12), 4380–4386 (2001)
29. Crump, J.A., Morrissey, A.B., Ramadhani, H.O., Njau, B.N., Maro, V.P., Reller, L.B.: Controlled comparison of BacT/Alert MB system, manual Myco/F lytic procedure, and isolator 10 system for diagnosis of *Mycobacterium tuberculosis* bacteremia. J. Clin. Microbiol. **49**(8), 3054–3057 (2011)
30. Simner, P.J., Stenger, S., Richter, E., Brown-Elliot, B.A., Wallace, R.J. Jr., Wengenack, N.L.: Mycobacterium: laboratory characteristics of slowly growing mycobacteria. In: Manual of Clinical Microbiology, 11th ed. ASM Press, Washington, DC (2015)
31. Lebrun, L., Espinasse, F., Poveda, J.D., Vincent-Levy-Frebault, V.: Evaluation of non-radioactive DNA probes for identification of mycobacteria. J. Clin. Microbiol. **30**(9), 2476–2478 (1992)
32. Bull, T.J., Shanson, D.C.: Evaluation of a commercial chemiluminescent gene probe system 'AccuProbe' for the rapid differentiation of mycobacteria, including 'MAIC X', isolated from blood and other sites, from patients with AIDS. J. Hosp. Infect. **21**(2), 143–149 (1992)
33. Tortoli, E., Pecorari, M., Fabio, G., Messino, M., Fabio, A.: Commercial DNA probes for mycobacteria incorrectly identify a number of less frequently encountered species. J. Clin. Microbiol. **48**(1), 307–310 (2010)
34. Christiansen, D.C., Roberts, G.D., Patel, R.: *Mycobacterium celatum,* an emerging pathogen and cause of false positive amplified *Mycobacterium tuberculosis* direct test. Diagn. Microbiol. Infect. Dis. **49**(1), 19–24 (2004)
35. Turenne, C.Y., Tschetter, L., Wolfe, J., Kabani, A.: Necessity of quality-controlled 16S rRNA gene sequence databases: identifying nontuberculous *Mycobacterium* species. J. Clin. Microbiol. **39**(10), 3637–3648 (2001)
36. Andersen, P., Doherty, T.M., Pai, M., Weldingh, K.: The prognosis of latent tuberculosis: can disease be predicted? Trends Mol. Med. **13**, 175–182 (2007)
37. Campos-Outcalt, D.: Tuberculosis: old problem, new concerns. J. Fam. Pract. **52**, 792–798 (2003)
38. Trajman, A., Steffen, R.E., Menzies, D.: Interferongamma release assays versus tuberculin skin testing for the diagnosis of latent tuberculosis infection: an overview of the evidence. Pulm. Med. **2013**, 601737 (2013)

39. Fact sheet: trends in tuberculosis, 2012. Atlanta, GA: Centers for Disease Control and Prevention (CDC); 2013 http://www.cdc.gov/tb/publications/factsheets/statistics/TBTrends.htm. Accessed 20 July 2014
40. Madariaga, M.G., Jalali, Z., Swindells, S.: Clinical utility of interferon gamma assay in the diagnosis of tuberculosis. J. Am. Board Fam. Med. **20**, 540–547 (2007)
41. Schluger, N.W.: Advances in the diagnostics of latent tuberculosis infection. Semin. Respir. Crit. Care Med. **34**, 60–66 (2013)
42. Purohit, M., Mustafa, T.: Laboratory diagnosis of extra-pulmonary tuberculosis (EPTB) in resource-constrained setting: state of the art, challenges and the need. J. Clin. Diagn. Res JCDR **9**(4), EE01 (2015)
43. Hwang, L.Y., Grimes, C.Z., Beasley, R.P., Graviss, E.A.: Latent tuberculosis infections in hard-to-reach drug using population-detection, prevention and control. Tuberculosis (Edinb) **89**(Suppl 1), S41–S45 (2009)
44. Andersen, P., Munk, M.E., Pollock, J.M., Doherty, T.M.: Specific immune-based diagnosis of tuberculosis. Lancet **356**, 1099–1104 (2000)
45. ATS. Diagnostic standards and classification of tuberculosis in adults and children. Joint statement of the American Thoracic Society and the Centers for Disease Control and Prevention, July 1999. Endorsed by the Council of the Infectious Disease Society of America, September 1999. Am. J. Respir. Crit. Care. Med. **161**(4 Pt 1), 1376–1395 (2000)
46. Hauck, F.R., Neese, B.H., Panchal, A.S., El-Amin, W.: Identification and management of latent tuberculosis infection. Am. Fam. Physician **79**, 879–886 (2009)
47. WHO: Treatment of tuberculosis: Guidelines for national programmes. pp 11–25 (2003)
48. Purohit, M.R., Mustafa, T., Wiker, H.G., Morkve, O., Sviland, L.: Immunohistochemical diagnosis of abdominal and lymph node tuberculosis by detecting Mycobacterium tuberculosis complex specific antigen MPT64. Diagn. Pathol. **2**, 36 (2007)
49. Tadesse, M., Abebe, G., Abdissa, K., Aragaw, D., Abdella, K., Bekele, A., Apers, L., de Jong, B.C., Rigouts, L.: GeneXpert MTB/RIF assay for the diagnosis of tuberculous lymphadenitis on concentrated fine needle aspirates in high tuberculosis burden settings. PLoS One **10**(9), e0137471 (2015)
50. Mustafa, T., Wiker, H.G., Mfinanga, S.G., Morkve, O., Sviland, L.: Immunohistochemistry using a Mycobacterium tuberculosis complex specific antibody for improved diagnosis of tuberculous lymphadenitis. Mod. Pathol. **19**(12), 1606–1614 (2006)
51. Samaila, M.O., Oluwole, O.P.: Extrapulmonary tuberculosis: fine needle aspiration cytology diagnosis. Nigerian J. Clin. Prac. **14**(3), 297–299 (2011)
52. Ahmad, A., Afghan, S., Raykundalia, C., Catty, D.: Diagnosis of tuberculosis by using enzyme-linked immunosorbent assay (ELISA) to detect anti-mycobacterial superoxide dismutase in the patients. J. Islamic Acad. Sci. **11**, 1 (1998)
53. Meuzelaar, H.L., Kistemaker, P.G.: A technique for fast and reproducible fingerprinting of bacteria by pyrolysis MS. Anal. Chem. **45**(3), 587–590 (1973)
54. Caulfield, A.J., Wengenack, N.L.: Diagnosis of active tuberculosis disease: from microscopy to molecular techniques. J. Clin. Tuberc Other Mycobact. Dis. **4**, 33–43 (2016)
55. Global tuberculosis control: key findings from the december 2009 WHO report. Wkly. Epidemiol. Rec. **85**(9), 69–80 (2010)
56. World Health Organization: Policy Statement: Commercial serodiagnostic tests for diagnosis of tuberculosis. WHO, Geneva, Switzerland WHO/HTM/TB/2011. http://www.who.int/tb/laboratory/policy_statement/en (2011)
57. Pai, M., Menzies, D.: Interferon-gamma release assays: what is their role in the diagnosis of active tuberculosis? Clin. Infect. Dis. **44**(1), 74–77 (2007)
58. Ling, D.I., Zwerling, A.A., Pai, M.: Rapid diagnosis of drug-resistant TB using line probe assays: from evidence to policy. Expert Rev. Resp. Med. **2**(5), 583–588 (2008)
59. Dinnes, J., Deeks, J., Kunst, H., Gibson, A., Cummins, E., Waugh, N., et al.: A systematic review of rapid diagnostic tests for the detection of tuberculosis infection. Health Technol. Assess. **11**(3), 1–196 (2007)

60. Russo, C., Tortoli, E., Menichella, D.: Evaluation of the new GenoType *Mycobacterium* assay for identification of mycobacterial species. J. Clin. Microbiol. **44**(2), 334–339 (2006)
61. Tortoli, E., Mariottini, A., Mazzarelli, G.: Evaluation of INNO-LiPA MYCOBACTERIA v2: improved reverse hybridization multiple DNA probe assay for mycobacterial identification. J. Clin. Microbiol. **41**(9), 4418–4420 (2003)
62. Su, K.Y., Yan, B.S., Chiu, H.C., Yu, C.J., Chang, S.Y., Jou, R., Liu, J.L., Hsueh, P.R., Yu, S.L.: Rapid sputum multiplex detection of the *M. tuberculosis* complex (MTBC) and resistance mutations for eight antibiotics by nucleotide MALDI-TOF MS. Sci. Rep. **7**, 41486(66) (2017)
63. Kemptner, J., Marchetti-Deschmann, M., Mach, R., et al.: Evaluation of matrix-assisted laser desorption/ ionization (MALDI) preparation techniques for surface characterization of intact Fusarium spores by MALDI linear time-of-flight mass spectrometry. Rapid Commun. Mass Spectrom. **23**, 877–884 (2009)
64. Prithwiraj, D. et al., PLoS One **10**(12). https://doi.org/10.1371/journal.pone.0144088 (2015)
65. De, P., Amin, A.G., Valli, E., Perkins, M.D., McNeil, M., Chatterjee, D.: Estimation of D-arabinose by gas chromatography/mass spectrometry as surrogate for mycobacterial lipoarabinomannan in human urine. PLoS One **10**(12) (2015)
66. Kind, T., Fiehn, O.: Advances in structure elucidation of small molecules using mass spectrometry. Bioanal Rev **2**(1–4), 23–60 (2010)
67. Saeed, M., Iram, S., Hussain, S., Ahmed, A., Akbar, M., Aslam, M.: GeneXpert: a new tool for the rapid detection of rifampicin resistance in mycobacterium tuberculosis. J. Pak. Med. Assoc. **67**(2), 270–274 (2017)
68. Kulkarni, S., Singh, P., Memon, A., Nataraj, G., Kanade, S., Kelkar, R., Rajan, M.G.R.: An in-house multiplex PCR test for the detection of *Mycobacterium tuberculosis*, its validation & comparison with a single target TB-PCR kit. Indian J. Med. Res. **135**(5), 788 (2012)
69. Tevere, V.J., Hewitt, P.L., Dare, A., Hocknell, P., Keen, A., Spadoro, J.P., Young, K.K.: Detection of *Mycobacterium tuberculosis* by PCR amplification with pan-mycobacterium primers and hybridization to an *M. tuberculosis*-specific probe. J. Clin. Microbiol. **34**(4), 918–923 (1996)
70. Chen, Y., Zhang, L., Hong, L., Luo, X., Chen, J., Tang, L., Chen, J., Liu, X., Chen, Z.: Rapid diagnosis of pulmonary tuberculosis and detection of drug resistance by combined simultaneous amplification testing and reverse dot blot. J. Clin. Pathol. (2017)
71. Lopes, L.B., Alves, T.M., Stynen, A.P.R., Mota, P.M., Leite, R.C., Lage, A.P.: Parameter estimation and use of gamma interferon assay for the diagnosis of bovine tuberculosis in Brazil. Pesquisa Veterinária Brasileira **32**(4), 279–283 (2012)
72. Garcia-Zamalloa, A., Taboada-Gomez, J.: Diagnostic accuracy of adenosine deaminase and lymphocyte proportion in pleural fluid for tuberculous pleurisy in different prevalence scenarios. PLoS One **7**(6), e38729 (2012)
73. Jacomelli, M., Silva, P.R.A.A., Rodrigues, A.J., Demarzo, S.E., Seicento, M., Figueiredo, V.R.: Bronchoscopy for the diagnosis of pulmonary tuberculosis in patients with negative sputum smear microscopy results. Jornal Brasileiro de Pneumologia **38**(2), 167–173 (2012)
74. Kochak, H.E., SeyedAlinaghi, S.A., Zarghom, O., Hekmat, S., Jam, S., Sabzvari, D., Abdi, Z.: Evaluation of serological tests using A60 antigen for diagnosis of tuberculosis. Acta. Medica. Iranica **48**(1), 21–26 (2010)
75. Goel, M.M., Budhwar, P., Jain, A.: Immunocytochemistry versus nucleic acid amplification in fine needle aspirates and tissues of extrapulmonary tuberculosis. J. Cytol. Indian Acad. Cytol. **29**(3), 157 (2012)
76. Tadesse, M., Abebe, G., Abdissa, K., Aragaw, D., Abdella, K., Bekele, A., Bezabih, M., Apers, L., de Jong, B.C., Rigouts, L.:GeneXpert MTB/RIF assay for the diagnosis of tuberculous lymphadenitis on concentrated fine needle aspirates in high tuberculosis burden settings. PLoS One **10**(9), e0137471(53) (2015)
77. Shah, A., Rodrigues, C.: The expanding canvas of rapid molecular tests in detection of tuberculosis and drug resistance. Astrocyte **4**(1), 34 (2017)
78. Sethi, S., Nanda, R., Chakraborty, T.: Clinical application of volatile organic compound analysis for detecting infectious diseases. Clin. Microbiol. Rev. **26**(3), 462–475 (2013)

Dielectrically Modulated Bio-FET for Label-Free Detection of Bio-molecules

Amit Bhattacharyya, Manash Chanda, and Debashis De

Abstract An inclusive survey of dielectric modulated (DM) biosensor field-effect transistors (Bio-FETs) has been discussed in this chapter. Biosensor field-effect transistors (Bio-FETs) have been appeared akin to essentially important contender as label-free biosensor that directs towards a considerable improvement in this area during the last ten years. This chapter proposes a common synopsis on Bio-FET, mentioning the fundamental physics of DM transduction and charge modulation methods. Subsequently, state-of-the-art improvements in DM Bio-FET have been talked about, as well as the consequences of device downscaling over transduction presentation has been usually highlighted. Later, various DM Bio-FET proposals for optimized transducer presentation have been examined depends on device electrostatical approach and mechanisms of carrier transportation. The relative advantage/disadvantage of this promising alternative of DM Bio-FET is further specified in this chapter. Here, an apparent conception of DM Bio-FET-based biosensors and the state-of-the-art research framework and future declaration with research confronts of this area are suitably detailed for concerned researchers.

Keywords Dielectric modulation · Bio-FET · MOSFET · Tunnel FET · Biomolecules · Sensitivity

A. Bhattacharyya
Department of Electronics and Communication Engineering, Haldia Institute of Technology, Haldia 721657, India
e-mail: amit_elec06@yahoo.com

M. Chanda (✉)
Department of Electronics and Communication Engineering, Meghnad Saha Institute of Technology, Kolkata 700150, India
e-mail: manash.bst@gmail.com

D. De
Department of Computer Science and Engineering, Maulana Abul Kalam Azad University of Technology, Kolkata 700064, India
e-mail: dr.debashis.de@gmail.com

G. Dutta et al. (eds.), *Modern Techniques in Biosensors*, Studies in Systems, Decision and Control 327, https://doi.org/10.1007/978-981-15-9612-4_8

1 Introduction

The satisfying advancement of molecular biology has shown to a superior recognization of various categories of biomolecular relations with their impact on the physico-chemical features for biological arrangement. The constantly developing biosensing tools utilize this reality on behalf of consistently recognizing a broad variety of target bio-analytes inside a heterogeneous mixture. Usually, the recognition and assessment of definite bio-analytes within body fluidic test have turned into of supreme attention for untimely analysis of different infections. The microbiology and ecological research, along with food technology, have been appeared likely further possible functional regions of biosensing mechanism. Consequently, the popularity of biosensing mechanism is quickly growing, and scientists from various streams of science and engineering are incessantly doing their work for relocating the research-based awareness into industrially feasible outcomes. The International Union of Pure and Applied Chemistry (IUPAC) has described the biosensor like—"Biosensors are devices which use a biological detection component preserved in direct spatial contact with transduction system" [1]. Hence, biosensor transfers the bio-signals that instigated from the variation in various physico-chemical features of the biological scheme following biomolecular relations towards an assessable electric-signal separate from unintentional and surrounding signals. After that, the signal may be moreover electronically progressed and symbolized in an expedient variety to the ending client. Nowadays, a wide range of biosensor is accessible because of the continual improvements of various biosensing methods. Aforesaid biosensors may be segregated upon the origin of the bio-detection component (bioreceptor) that carefully combines itself with explicit analyte and its transduction methods.

The receptor-oriented categorization generally comprises protein, cell, enzyme, DNA as well as bacteriophage. In every class, a common variety of target/bioreceptor relations may be examined. Conversely, the transducer supported arrangement specifies the physico-chemical feature of the target/bioreceptor binding schemes that have been developed in favour of bio-analyte recognition. The biotransducer system may be largely categorized into optical, mechanical and electrochemical natures [2–4]. Conversely, every biosensor may additionally be categorized likely label-based and label-free, based on whatever every label which acquires different physico-chemical feature is connected through the local target/bioreceptor molecules. In label-based biosensor, the recognition is appreciated upon the supports of features for label, while the intrinsic feature of target/bioreceptor molecules has been developed for recognition in label-free biosensing. The label-based biosensor suggests the benefit of connecting different and enviable physico-chemical features through explicit biomolecules via connecting suitable label to them. Conversely, label-based biosensor forever engages extra labelling and sanitization procedure stages and irregular change for inhabitant features of target/bioreceptor molecules which might contrarily influence their biological actions throughout conjugate mechanism. Hence, the label-free biosensing is favoured in support of examining of biomolecular relations and faster with cheaper recognition practices [5, 6]. Here, the biosensor field-effect transistor (Bio-FET) has been appeared akin to significant contender for label-free electrochemical biosensor that directs to a considerable growth in the area for last few years.

The execution of Bio-FET, ensuing the traditional CMOS procedure flow, suggests the advantage of system-on-chip assimilation and regular miniaturized process for perceiving affinity through a broad variety of analyte. The possibility of system-on-chip assimilation is mostly hopeful for point-of-care (POC) interpretation, where the rapid as well as prior recognition of different infections is accomplished via incorporating the digital microfluidics through Bio-FET upon a lab-on-a-chip (LOC) framework. In these proposals, each droplet holding bio-samples is conveyed to a collection of Bio-FETs with the consequent electrical response which is handled upon the alike chip via employing the large-scale integration [6–9].

The beginning of ion-sensitive field-effect transistor (ISFET) by Bergveld for perceiving concentration of inorganic ion with pH within a solution of electrolyte is extensively accepted as the beginning of Bio-FET supported label-free electrochemical biosensor [8]. Primarily, Bio-FETs utilize the occurrence of unbalanced charges within the functional groups of various bio-analytes in favour of the label-free recognition. Consequently, these Bio-FETs are categorized into the class of charge modulated (CM) Bio-FETs. Newly substitute recognition methods employ the dielectric modulation (DM) of target/bioreceptor bonding scheme throughout conjugation. This has significantly broadened the appropriateness of DM Bio-FET, which can constantly perceive a huge quantity of charged and charge-neutral bio-analytes. Afterwards, here, widespread surveys of DM Bio-FETs have been offered along with the transducing arrangement feature of these biosensors which is usually highlighted.

2 Bio-FET: A General Survey

The Bio-FETs are recognized by removing the metal-polysilicon gate electrode along with placing a sheet of bioreceptor over the gate insulator section of traditional MOSFET. Here, the existence of charged bio-molecule within a mixture changes density of charges close to gate insulator. This ultimately directs to an alteration inside gating effect of Bio-FET along with the conduction effect through channel. Consequently, bio-analytes is able to perceive as well as computed from modulations of electrical factors of Bio-FET [8, 9]. The bio-analyte recognition in Bio-FET is recognized by a progression of common steps. At first, a proper bioreceptor is physically or chemically appended with the surface of gate insulator, and this procedure is described as functionalization.

The functionalization stage is pursued by the incubation stage, where the nano-biosensor is exposed to a solution consisting of the molecule for a definite instance. The molecule disperses approaching sensor surface with chemically bounds itself by receptor, which is recognized as conjugation. The biological action of receptor film and thus the efficacy of Bio-FETs greatly depend upon the selection of bioreceptor along with its functionality. Generally, the bioreceptor has easy connection method, steady connection through the surface of the sensor, large density throughout surface area and reliability of biological performances with realistic self-life. Consequently, the bioreceptor-oriented categorization of Bio-FETs is explained later.

A. Enzyme-FET

The primary exertions on Bio-FETs utilized the biomolecular response of molecule in existence of enzyme. These categories of Bio-FETs are usually recognized as enzyme-FET or ENFET. Enzyme is typically proteins or in various conditions ribonucleic acid (RNA) analytes those are able of performing alike catalyst towards a certain biomolecular relations with controlling the possibility of almost all of the metabolic actions inside a cell. For ENFET, certain enzymes are hybridized over the biosensing surface with ion derivatives of the consequent enzymatic effect which changes the measurement of pH of surface for Bio-FETs with the electrical reaction [10].

B. Gene-Modified FET

Deoxyribonucleic acid (DNA) recognition is an additional possible functional area of Bio-FET that is usually identified as gene-modified FET or GenFET. DNA is an essential element of living species that bears genetic statistics which regulates the features along with operation of almost existing creature. DNA is compiled of two polynucleotide chains organized within the double-helical shape, in which one of the four sorts of nucleobases is connected through every deoxyribose phosphate backbone of separate chains. Moreover, every type of nucleobase on single chain makes binding with only one complementary nucleobase upon the further chain. Following partition, the separate polynucleotide chain is known as single-stranded DNA (ss-DNA). During proper action of gate/insulator sheet of GenFET, a probe ss-DNA is functionalized over its surface that carefully bound itself through its complementary ss-DNA strand in heterogeneous surroundings. These procedures are identified likely DNA-hybridization, with because of the negative charge beard inside phosphate group near the DNA backbone, the DNA-hybridization adjusts the potential along the surface of GenFET. On the other hand, the DNA recognitions have also been appreciated from the ion derivatives of biomolecular relations throughout the assimilation of nucleotide in a probe ss-DNA. The ionic by-products adjust the measurement of pH of surface and thus potential along the surface of GenFET. Separately from nucleotides, rest biospecies have been further identified via the GenFET immobilization through probe DNA succeeding whichever of the recognition methods examined formerly [11].

C. Immuno-FET

The usual category of Bio-FET which is applied for antibody recognition is identified as immuno-FET. All utilities of alive biospecies are correlated with proteins which are engaged in various bio-mechanisms likely metabolism, cell signalling and immune response. Immune interface engages greatly specific interfaces among antigen and antibody. Antigen as well as antibody is a protein which is a linear sequence of amino acid organized in an extremely systematized configuration. The biological purpose of definite antigen-antibody relies on the series of amino acids and in general architectural association of the analyte. Relying on the pH intensity of buffer mixtures, the functional groups exist in amino acids is frequently initiated to be electrically charged. Afterwards, antibody-antigen recognition is generally identified via the alteration of charge condition in an immune contact which may ultimately perform potential modulation on the surface and thus electrical features of immuno-FET [12].

D. Cell-FET

The Bio-FET-dependent cell examination is a growing functional region that can scrutinize the impact of exterior stimulus on a living organism, and such alternative of Bio-FET is normally denoted as cell-FET. The cell-FET can utilize the extra-cellular potentials happens in bio-electrogenic cells, viz. neuron cell, myocyte, etc. The succeeding modulation of surface potential is perceived from the electrical features of cell-FET. On contrary, the cell-FET can supervise the actions of alive biospecies through various situations commencing the cell metabolism by-products that usually occupy pH or ion concentration alterations of electrolyte [13].

3 DM Bio-FET: Working Perception and Categorizations

The main drawbacks of any charge modulated (CM) Bio-FET are its failure to perceive charge/neutral analytes that bounds its function for a few quantity of biomolecular relations. This will require a label-free electrochemical recognition procedure for charge-neutral analytes well suited by Bio-FET and will direct to the beginning of DM Bio-FET. The DM Bio-FETs include a nano-cavity formed either within the gate electrode or within the gate insulator section of a FET configuration, as exposed in Fig. 1a. The cavity functionalization can be obtained with certain bioreceptor which directs to bioreceptor/target bond after inserting the solution including analyte. This contact changes the effectual permittivity (k) within the cavity [14, 15]. In existence of a useful electric field (E_X) through gate electrode, bioreceptor/target analytes initiate a total polarized charge density, and hence, a polarized field exists inside the nano-cavity. Moreover, this polarized field alters the effectual E_X inside the nano-cavity, and in case of charged analytes, the charge persuaded E_X also supplies in this procedure. This changes the gating effect of DM Bio-FET and thus the carrier insertion element inside the channel. Successive modulation in the electrical factors of DM Bio-FET is used to perceive the existence of analytes as depicted in Fig. 1b. This must be recognized that although an advanced E_X through gate electrode increases the polarized field inside nano-gap, the subsequent superior inversion charges throughout channel efficiently displayed the impact of electrostatic modulation of nano-gap from channel. Consequently, the utmost current modulation in drain section of DM Bio-FET is generally examined in subthreshold region. Furthermore, this specifies to a lesser channel inversion near comparatively superior bias across gate electrode which is usually enviable for DM Bio-FET arrangement. Here, various alternatives of DM Bio-FET are explored elaborately in the subsequent sections.

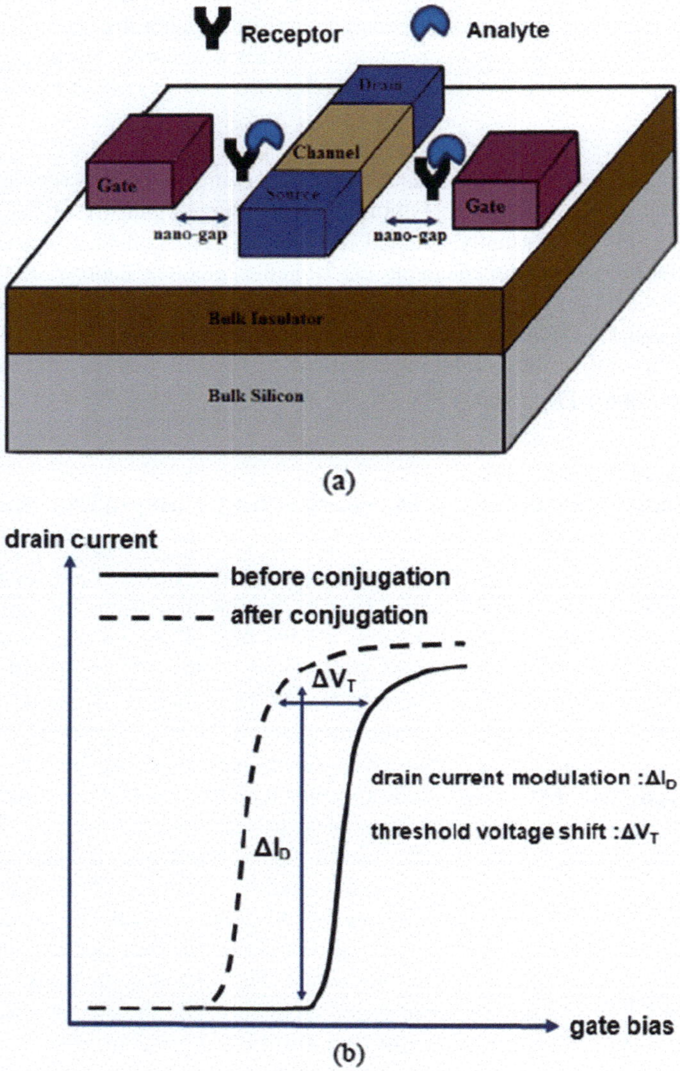

Fig. 1 **a** Cross-sectional outlook of DM Bio-FET, **b** ideal transfer characteristics after and before conjugation

A. Dielectric Modulated MOSFET (DM-MOSFET)

The perceptions of DM supported transduction system for Bio-FET have been earliest initiated by Im et al. [16]. Here, the nano-gap of DM-MOSFET has been recognized by carefully removing a gate metal from the dual-metal gate-stack region of tri-gate MOSFET design [16]. On the other hand, the cavity can be initiated by removing the thick gate/insulator section of double-gate MOSFET construction. The later method is generally favoured because of the elevated nano-gap possession and simpler specimen initiation in favour of the dual-sided opened walls nano-gap configuration [17]. A suspension chromium gate DM-MOSFET structural design is moreover suggested which moreover promotes fluid dynamics of specimen release within the nano-gap section under the suspension gate [18]. Since examined, bioreceptor/target bond modifies the electrostatic coupling through gate to channel of DM-MOSFET and thus the barrier potential for thermionic insertion, as exposed in Fig. 2. Consequently, an important modulation of threshold potential and current through drain may be examined following bioreceptor/target bonding. The DM-MOSFET can be recognized for n-category as well as p-category channels, and the option relies on the charge-status of the analyte. This has been established that p-category DM-MOSFET is appropriate for negative-charged analytes, since the existence of this analyte in the close to channel advances injection of hole element, i.e. incidence of analyte performs equivalent to enhancing negative gate/source potential. Correspondingly, the n-category DM-MOSFET is appeared to be further suitable for positive-charged analyte recognition. Though the existence of charge-neutral analytes enhances injection elements for p-category as well as n-category DM-MOSFETs, generally the last is favoured in favour of comparatively superior mobility of electrons. This is significant to declare

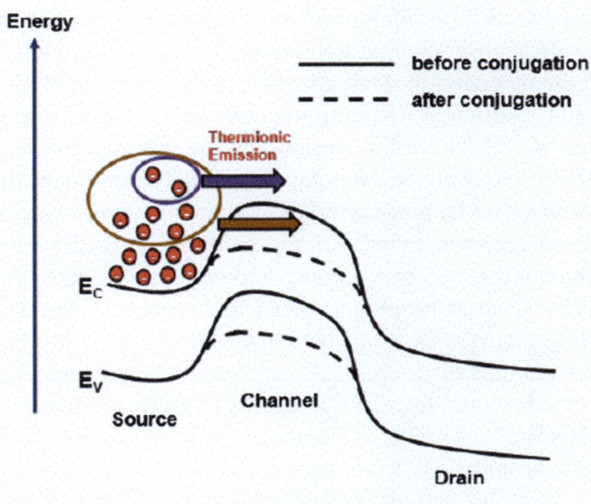

Fig. 2 Energy-band profile of n-category DM-MOSFET [20]

that if essential CM and DM are related by any definite analyte, the option of DM-MOSFET must be such that CM and DM have a general impact on gating effect. Contrarily, the total sensitivity of the device might be significantly imperfect [19].

Presently, the DM-MOSFET biosensing effectively perceived various categories of biomolecular interfacing, namely protein ligand bonding, neural-immune interaction and nucleic acid hybridization [16–22]. In addition, the stated sensitivity of DM Bio-FET is also equivalent to the CM Bio-FET complements. Conversely, this has been established that within aqueous surroundings, the Debye charge-screening consequence of ion species which exist inside buffer mixture can significantly reduce the signal instigated from biomolecular interface inside nano-gap [23]. Afterwards, in almost all of the practical works, electrical demonstration of DM-MOSFET is regarded as in air surroundings that may be appreciated via succeeding de-ionized water for final wash before dry nitrogen blow-off stages, devoid of adjusting the biological functionalization of bioreceptor/target analytes [16–20]. The primary published works demonstrate that a superior conditional conformity among practical and simulation can be realized for the design of DM-MOSFET [16, 17, 19, 20]. Because of the reason, different researchers have executed device simulation supported examination on DM-MOSFETs in favour of developing the electrical reactions and describing the main searching supported on theoretical modelling. In aforesaid numerical simulation persuaded details, generally appropriate substances have been included inside the nano-gap and the features are consequently corrected to characterize the occurrence/nonexistence of definite analytes within the nano-gap regime [16–24]. Afterwards, quantity of assuring DM-MOSFET constructions has been commenced that illustrates a major presentation enhancement in contrast with their traditional complements. Furthermore, gates all around (GAA) DM-MOSFETs with vacuum gate dielectric have been recommended while no gate insulator is included among gate/channel section. A configuration of transducer illustrates a higher electrical modulation through conjugation contrasted to its traditional DM-MOSFET equivalent, since the inversion through channel is relatively smaller in nonexistence of gate insulator [24]. Moreover, these appreciations have been developed to initiate a novel alternative of DM-MOSFET, somewhere the transduction method relies on gate-persuaded fringe effect inside nano-gap rather than straight gate field. The transducing effect includes a better gate underlap section, and comparatively feeble electric field coupling of gate-persuaded fringing field maintains the inversion through channel and consequently outcomes a higher modulation in electrical factors of this DM-MOSFET [25]. Contrarily, plenty of research papers illustrated junctionless (JL) DM-MOSFET which is essentially beneficial above the traditional complements because of the less manufacture complexity and minor short-channel effects (SCEs) in nonexistence of substantial intersections [26, 27]. Here, the presentation of every DM Bio-FET robustly relies on the character of cavity status, i.e. total gate/channel coupling inflection through analyte conjugate. Consequently, these have been originated that the arrangements and fill factor of bio-analytes within a moderately filled nano-gap regime can importantly influence the presentation of DM-MOSFET and thus emerge to be a vital architectural deliberation for this transducing effect [28, 29]. The rest proficient organizational architectural factors of DM-MOSFET usually comprise cavity length and height, length of gate region, height of channel and work-function of gate metal (φ_M) [24–29].

Here, an entire POC analysis stage has been practically recognized by a collection of DM-MOSFET that comprises a little-noise read-out arrangement. The current through drain of chosen DM-MOSFET is imitated with current-mirror arrangement, and the equivalent charge is accumulated in an integration capacitor, where output potential is buffered with a voltage-following arrangement [30]. Moreover, inverter-oriented logic-stage recognitions/read-out approaches have been initiated in favour of DM-MOSFET. This has been observed that charged as well as charge-neutral bio-analytes need dissimilar inverter arrangements, where either pull-up or pull-down components of traditional inverter are restored via DM-MOSFETs. This type of bioinverter architecture demonstrates an apparent output logic-stage conversion following conjugation in favour of a specified input potential [31]. Figure 3 shows the deterioration of electric response through variation of length of channel for DM-MOSFET which is a major constraint for this transducer. Moreover, this figure illustrates such deterioration becomes apparently less than 1 μm length of channel, where SCEs of traditional MOSFET are generally insignificant. These features can be recognized from the reality that the inclusion of nano-gap and thus excessive larger value of effectual oxide thickness require a major decrease in gate/channel electrostatic coupling of DM-MOSFET. Consequently, at comparatively longer length of channel, the drain-to-channel coupling begins the controlling of channel electrostatics [32] and finally directs to superior channel inversion at comparatively inferior gate potential. This strictly restricts channel electrostatic modulations through bio-analyte conjugation and deteriorates succeeding current response through drain. On contrary, a shorter

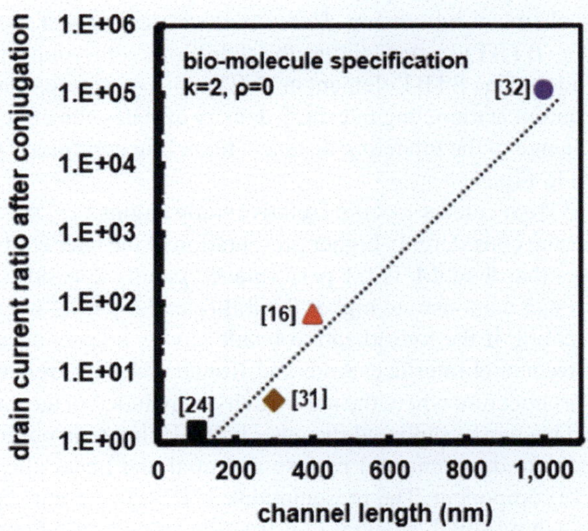

Fig. 3 Graphs of surveyed drain current modulations of DM-MOSFET for various lengths of channel

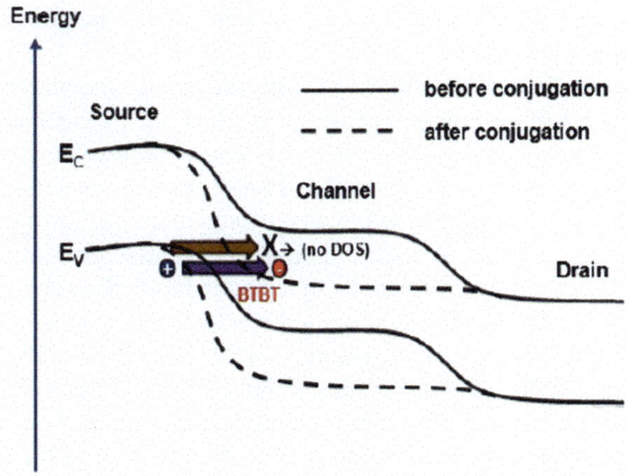

Fig. 4 Energy-band contour of n-category DM-TFET [35]

length of channel for DM Bio-FET is not merely essential in favour of the efficient on-chip integration other than moreover for realizing inferior detection limit (DL). This obstruct has mainly directed examination of other promising transistor constructions and metals for dielectric modulated biosensing functions that may possibly surpass DM-MOSFET in shorter dimension of device as shown in Fig. 4.

B. Dielectric Modulated Tunnel FET (DM-TFET)

The DM-TFETs have presently appeared akin to a possible transducer in favour of label-free electrochemical biosensing. Here, main carrier transfer method is band-to-band tunneling (BTBT) in opposite to the thermionic-emission-diffusion in the MOSFET contenders. The BTBT element in TFET relies on various parameters, akin to energy band contour at tunneling interface, density of states and effectual tunneling mass. With the change of the tunneling distance, tunneling current of DM-TFET will modulate, shown in Fig. 4.

Narang et al. [33] executed a specified relative study among the DM-FET and DM-TFET that creates the considerably higher presentation of the later configuration. The work also exposes that the DM-TFET performance greatly depends upon conjugation profile and it can demonstrate higher flexibility under partial hybridization and channel length scaling if the hybridization inside cavity is specially located in the vicinity of source/channel interface. Abdi and Kumar [34] have recommended a p-type DM-TFET architecture where the nano-cavity is included in the gate-drain overlapping section. This work employed the gate/drain electrostatic coupling for transduction that modulate drain/channel reverse potential and hence the band-to-band tunneling (BTBT) component. The presentations of DM-TFET may additionally be developed by pertaining suitable device construction procedures akin to shorter gate single metal configurations by Kanungo et al. [35]. Consequently, a shorter gate insulator-less (SGIL) DM-TFET constructions have been recognized by removing the gate-stack regime that illustrates important presentation enhancement over both its completely filled and partially filled cavity equivalents by the similar group [36]. Also, related proposal has been approved to progress the presentation of DM-TFET through dual-metal gate-engineering, i.e. introducing metals of dissimilar φ_M above cavity and gate-stack sections [37]. Alternatively, it has been originated that a source-pocket doped and source-overlapping gate is initiated into a DM-TFET configuration that directs to a best possible vertical tunneling blockade among source and pocket section. The consequent vertical tunneling factors have guided to a greater concert evaluated to those of lateral tunneling-based DM-TFET [38]. This has been efficiently accepted that the BTBT possibility of TFET may be significantly improved by introducing a lesser energy bandgap material. Consequently, the functions of bandgap engineering have been investigated by changing the germanium (Ge) mole fraction of a silicon-germanium source (SiGe) DM-TFET to illustrate when raising the tunneling possibility; a lesser bandgap too disgraces the electrostatic modulations via enhancing inversion through channel at a comparatively inferior potential across gate electrode. Hence, optimal transduction effectiveness may be realized at a comparatively inferior Ge mole fraction of SiGe-DM-TFET [39]. Moreover, electrically doped (ED) DM-TFET employed the perception of dopingless for electrically recognizing extremely doped source as well as drain sections by placing metals with suitable φ_M above the insulating layers of source and drain areas. The ED DM-TFET has exposed equivalent presentation to the traditional DM-TFET contenders [40]. Furthermore, the presentations of ED DM-TFET may further be developed through concerning proper device construction procedures akin to short-gate dual-metal configurations [41]. Dwivedi and Kranti [42] have suggested and estimate the feasibility of transconductance-to-current ratio (gm/Ids) as a sensing factor for a DM-TFET-based biosensor. Recently, Wadhwa and Raj [43] have proposed the functioning of charge-plasma-based gate underlap DM-JL-TFET (CPB DM-JLTFET) based biosensor. Goswami and Bhowmick [44] integrated non-ideal factors likely, issues of steric hindrance and probe positioning in DM-TFET, sustaining the analysis of realistic concerns of the devices. Moreover, the authors have exposed the fundamental expectations of circular gate (CG) TFET as biosensor with partially filled cavity. Dwivedi and Kranti [45] detailed a feasibility evaluation of tunneling and accumulation mode p-type transistor configurations for usage as DM biosensor with partial hybridization (PH) of bio-analytes within cavity. Recently, Bhattacharyya et al. [46] illustrated a new dual-pocket (DP), DM heterostructured TFET (DM-HTFET) based biosensor considering steric hindrance issue together with various nonuniform step patterns such as decreasing, increasing, concave and convex inside the cavity. Furthermore, Goswami and Bhowmick [47] performed a comparison elaborately between CG TFET and uniform gate Heterojunction (HJ) TFET as label-free biosensors based on DM.

C. Dielectric Modulated Impact Ionization FET (DM-IFET)

DM-IFET has presently drawn significant research consideration for dielectric modulated biosensing relevance that may probably surpass its DM-MOSFET equivalent at nano-scale sizes, as specified in Fig. 5. The main carrier transportation systems of these transducers have been impact ionization in which the amplitude of impact ionization may be modulated via bio-analyte conjugation. In DM-IFET, shorter gate constructions through nano-gap section have been included in which the analyte conjugation may change gate-to-channel capacitive coupling as well as the inversion through channel under cavity region. Consequently, while a high potential across drain is affected, an advanced inversion through channel under gated region directs to a huge lateral E_X in the non-gated section of channel, as exposed in Fig. 6. For an adequately superior lateral E_X, major impact ionization may be examined in the non-gated section that abruptly increases the magnitude of terminal current [48, 49]. Later,

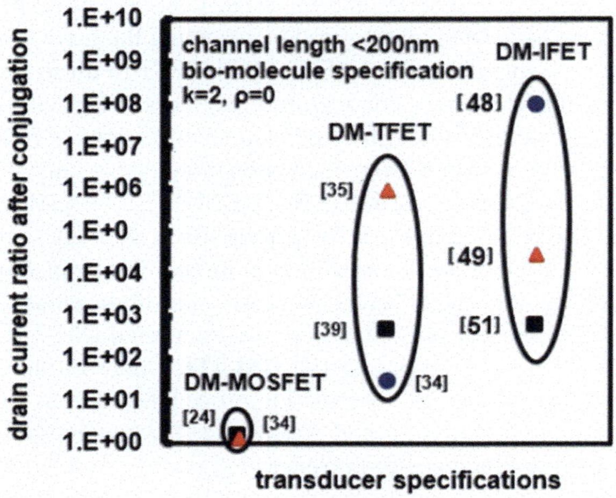

Fig. 5 Graphs of surveyed drain current modulations of short-channel length for DM-MOSFET, DM-TFET and DM-IFET, respectively

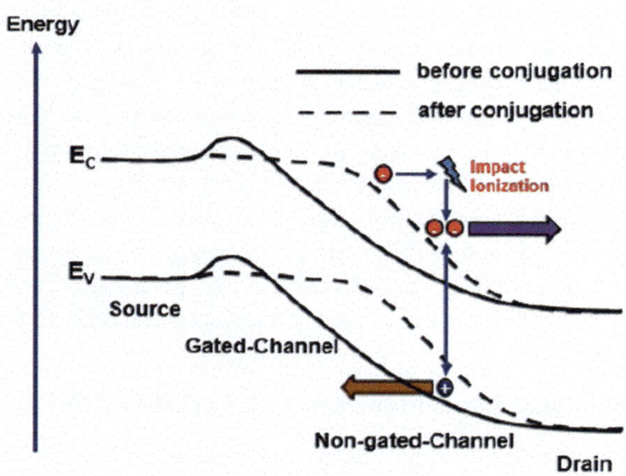

Fig. 6 Energy-band contour of n-category DM-IFET [49]

the initiation of DM-IFET, little possible alternatives has moreover been commenced, specifically JL-DM-IFET and electrically doped (ED) DM-IFET that incorporates the primary benefits of junctionless construction and charge-plasma-oriented electrically doped by means of DM-IFET [50]. Here, remarkable DM-IFET constructions have been suggested that appreciate Schottky barrier tunneling insertion method within the channel rather than thermionic emission or diffusion module. Hence, this transducer construction has exposed assurance in terms of higher transduction effectiveness and comparatively less potential procedure [51].

4 Current Framework and Future Expectations

The aforementioned survey obviously specifies that the DM Bio-FET mechanism has been speedily describing research interest of label-free electrochemical biosensing area. The promising material as well as device configurations is regularly gaining significance for DM Bio-FET proposal. In this chapter, it must be messaged that despite probability exposed through DM-IFET, the foremost examine confronts of these transducers comprise less potential operation with reliability issues in nano-scaling. On the other hand, the DM-TFET has revealed assertion for recognition of superior sensitivity and inferior LOD under nano-scale architecture sizes. However, the DM biosensing mechanism is yet far away from being mature and thus needs widespread examination for accomplishing comprehensive analytical and design-level realizing for their booming industrial comprehension. This highlights a stimulating prospect for associates to devote in the progress of DM Bio-FET in its pre-manufacture phase, since the simulation-oriented transducer configuration and numerical analysis or analytical modelling become an essential branch in the expansion of these skills. This assists the possibility of effectual involvements of associates in this field from various branches of world devoid of the prerequisite of having wide investigated amenities.

References

1. Thevenot, D.R., Tóth, K., Durst, R.A., Wilson, G.S.: Electrochemical biosensors: recommended definitions and classification. Pure Appl. Chem. **71**(12), 2333–2348 (1999)
2. Chambers, J.P., Arulanandam, B.P., Matta, L.L., Weis, A., Valdes, J.J.: Biosensor recognition elements. Mol. Biol. **10**, 1–12 (2008)
3. Grieshaber, D., MacKenzie, R., Voros, J., Reimhult, E.: Electrochemical biosensors—sensor principles and architectures. Sensor **8**(3), 1400–1458 (2008)
4. Turner, A.P.F.: Biosensors: sense and sensibility. Chem. Soc. Rev. **42** (2013)
5. Syahir, A., Usui, K., Tomizaki, K., Kajikawa, K., Mihara, H.: Label and label-free detection techniques for protein microarrays. Microarrays **4**, 228–244 (2015)
6. Sang, S., Wang, Y., Feng, Q., Wei, Y., Ji, J., Zhang, W.: Progress of new label-free techniques for biosensors: a review. Crit. Rev. Biotechnol. **36**(3), 465–481 (2016)

7. Schöning, M.J., Poghossian, A.: Recent advances in biologically sensitive field-effect transistors (BioFETs). Analyst **127**, 1137–1151 (2002)
8. Lee, C.S., Kim, S.K., Kim, M.: Ion-sensitive field—effect transistor for biological sensing. Sensors **9**, 7111–7131 (2009)
9. Pachauri, V., Ingebrandt, S.: Biologically sensitive field-effect transistors: from ISFETs to NanoFETs. Essays Biochem. **60**, 81–90 (2016)
10. van der Schoot, B.H., Bergveld, P.: ISFET based enzyme sensors. Biosensors **3**(3), 161–186 (1988)
11. Veigas, B., Fortunato, E., Baptista, P.V.: Field effect sensors for nucleic acid detection: recent advances and future perspectives. Sensors **15**, 10380–10398 (2015)
12. Schasfoort, R.B.M., Kooyman, R.P.H., Bergveld, P., Greve, J.: A new approach to immunoFET operation. Biosens. Bioelectron. **5**, 103–124 (1990)
13. Wang, P.: Cell-Based Biosensors: Principle and Applications, Bioinformatics & Biomedical Imaging, edited volume. Artech House, Norwood (2010)
14. Chen, X., Guo, Z., Yang, G.M., Li, J., Li, M.Q., Liu, J.H., Huang, X.J.: Electrical nanogap devices for biosensing. Mater. Today **13**(11), 28–41 (2010)
15. Choi, Y.K., Kim, C.H., Ahn, J.H., Kim, J.Y., Kim, S.: Dielectric detection using biochemical assays. In: Point-of-Care Diagnostics on a Chip, edited volume, pp. 97–123. Springer, Berlin (2012)
16. Im, H., Huang, X.J., Gu, B., Choi, Y.K.: A dielectric-modulated field-effect transistor for biosensing. Nat. Nanotechnol. **2**, 430–434 (2007)
17. Ahn, J.H., Kim, J.Y., Im, M., Han, J.W., Choi, Y.K.: A nanogap-embedded nanowire (NW) field effect transistor for sensor applications: immunosensors and humidity sensors. In: Proceedings of 14th International Conference on Miniaturized Systems for Chemistry and Life Sciences, pp. 1301–1303. Groningen, Netherlands (2010)
18. Gu, B., Park, T.J., Ahn, J.H., Huang, X.J., Lee, S.Y., Choi, Y.K.: Nanogap field-effect transistor biosensors for electrical detection of avian influenza. Small **5**(21), 2407–2412 (2009)
19. Kim, C.H., Jung, C., Park, H.G., Choi, Y.K.: Novel dielectric-modulated field-effect transistor for label-free DNA detection. Biochip J. **2**(2), pp. 127–134 (2008)
20. Ahn, J.H., Im, M., Choi, Y.K.: Label-free electrical detection of PSA by a nanogap field effect transistor. In: Proceedings of 12th International Conference on Miniaturized Systems for Chemistry and Life Sciences, pp. 979–981. California, USA (2008)
21. Tang, X., Jonas, A.M., Nysten, B., Demoustier-Champagne, S., Blondeau, F., Prévot, P.P., Pampin, R., Godfroid, E., Iñiguez, B., Colinge, J.P., Raskin, J.P., Flandre, D., Bayot, V.: Direct protein detection with a nano-interdigitated array gate MOSFET. Biosens. Bioelectron. **24**, 3531–3537 (2009)
22. Kim, C.H., Ahn, J.H., Kim, J.Y., Choi, J.M., Park, T.J., Choi, Y.K.: Improvement of sensitivity and limit of detection in a nanogap biosensor by controlling surface wettability. BioNanoSci. **3**, 192–197 (2013)
23. Kim, J.Y., Ahn, J.H., Choi, S.J., Im, M., Kim, S., Duarte, J.P., Kim, C.H., Park, T.J., Lee, S.Y., Choi, Y.K.: An underlap channel-embedded field-effect transistor for biosensor application in watery and dry environment. IEEE Trans. Nanotechnol. **11**(2), 390–394 (2012)
24. Gautam, R., Saxena, M., Gupta, R.S., Gupta, M.: Numerical model of gate-all-around MOSFET with vacuum gate dielectric for biomolecule detection. IEEE Electron. Device Lett. **33**(12), 1756–1758 (2012)
25. Kanungo, S., Chattopadhyay, S., Sinha, K., Gupta, P.S., Rahaman, H.: A device simulation-based investigation on dielectrically modulated fringing field-effect transistor for biosensing applications. IEEE Sens. J. **17**(5) (2017)
26. Ajay, Narang, R., Saxena, M., Gupta, M.: Investigation of dielectric modulated (DM) double gate (DG) junctionless MOSFETs for application as a biosensors. Superlattices Microstruct. **85**, 557–572 (2015)
27. Ahangari, Z.: Performance assessment of dual material gate dielectric modulated nanowire junctionless MOSFET for ultrasensitive detection of biomolecules. RSC Adv. **6**, 89185–89191 (2016)

28. Kalra, S., Kumar, M.J., Dhawan, A.: dielectric-modulated field effect transistors for DNA detection: impact of DNA orientation. IEEE Electron. Device Lett. **37**(11), 1485–1488 (2016)
29. Rahman, E., Shadman, A., Khosru, Q.D.M.: Effect of biomolecule position and fill in factor on sensitivity of a dielectric modulated double gate junctionless MOSFET biosensor. Superlattices Microstruct. **13**, 49–54 (2017)
30. Im, M., Ahn, J.H., Han, J.W., Park, T.J., Lee, S.Y., Choi, Y.K.: Development of a point-of-care testing platform with a nanogap-embedded separated double-gate field effect transistor array and its readout system for detection of avian influenza. IEEE Sens. J. **11**(2), 351–360 (2011)
31. Kanungo, S., Mondal, S.A., Chattopadhyay, S., Rahaman, H.: Design and investigation on bio-inverter and bio-ring-oscillator for dielectrically modulated biosensing applications. IEEE Trans. Nanotechnol. **16**(6), 974–981 (2017)
32. Choi, J.M., Han, J.W., Choi, S.J., Choi, Y.K.: Analytical modeling of a nanogap-embedded FET for application as a biosensor. IEEE Trans. Electron. Devices **57**(12), 3477–3484 (2010)
33. Narang, R., Saxena, M., Gupta, M.: Comparative analysis of dielectric-modulated FET and TFET-based biosensor. IEEE Trans. Nanotechnol. **14**(3), 427–435 (2015)
34. Abdi, D.B., Kumar, M.J.: Dieletric modulated overlapping gate-on-drain tunnel-FET as a label-free biosensor. Superlattices Microstruct. **86**, 198–202 (2015)
35. Kanungo, S., Chattopadhyay, S., Gupta, P.S., Rahaman, H.: Comparative performance analysis of the dielectrically modulated full-gate and short-gate tunnel FET-based biosensors. IEEE Trans. Electron. Devices **62**(3), 994–1001 (2015)
36. Kanungo, S., Chattopadhyay, S., Rahaman, H.: Investigating the performance of short gate insulator less dielectrically modulated tunnel field effect transistor based bio-sensors. In: Proceedings of 6th International Conference on Computers and Devices for Communication, Kolkata, India (2015)
37. Verma, M., Sharma, D., Pandey, S., Nigam, K., Kondekar, P.N.: Performance comparison of single and dual metal dielectrically modulated TFETs for the application of label free biosensor. Superlattices Microstruct. **101**, 219–227 (2017)
38. Verma, M., Tirkey, S., Yadav, S., Sharma, D., Yadav, D.S.: Performance assessment of a novel vertical dielectrically modulated TFET-based biosensor. IEEE Trans. Electron. Devices **64**(9), 3841–3848 (2017)
39. Kanungo, S., Chattopadhyay, S., Gupta, P.S., Sinha, K., Rahaman, H.: Study and analysis of the effects of SiGe source and pocket-doped channel on sensing performance of dielectrically modulated tunnel FET-based biosensors. IEEE Trans. Electron. Devices **63**(6), 2589–2596 (2016)
40. Venkatesh, P., Nigam, K., Pandey, S., Sharma, D., Kondekar, P.N.: A dielectrically modulated electrically doped tunnel FET for application of label free biosensor. Superlattices Microstruct. **109**, 470–479 (2017)
41. Sharma, D., Singh, D., Pandey, S., Yadav, S., Kondekar, P.N.: Comparative analysis of full-gate and short-gate dielectric modulated electrically doped tunnel-FET based biosensors. Superlattices Microstruct. **111**, 767–775 (2017)
42. Dwivedi, P., Kranti, A.: Applicability of transconductance-to-current ratio (gm/Ids) as a sensing metric for tunnel FET biosensors. IEEE Sens. J. **17**(4), 1030–1036 (2017)
43. Wadhwa, G., Raj, B.: Parametric variation analysis of symmetric double gate charge plasma JLTFET for biosensor application. IEEE Sens. J. **18**(15), 6070–6077 (2018)
44. Goswami, R., Bhowmick, B.: Dielectric-modulated TFETs as label-free biosensors. In: D. Vikraman (ed.) Design, Simulation and Construction of Field Effect Transistors, vol. 1, ch. 2, pp. 17–35. IntechOpen, London (2018). https://doi.org/10.5772/intechopen.76000
45. Dwivedi, P., Kranti, A.: Dielectric modulated biosensor architecture: tunneling or accumulation based transistor? IEEE Sens. J. **18**(8), 3228–3235 (2018)
46. Bhattacharyya, A., Chanda, M., De, D.: Performance assessment of new dual-pocket vertical heterostructure tunnel FET-based biosensor considering steric hindrance issue. IEEE Trans. Electron. Devices **66**(9), 3988–3993 (2019). https://doi.org/10.1109/TED.2019.2928850. (Early Access)

47. Goswami, R., Bhowmick, B.: Comparative analyses of circular gate TFET and heterojunction TFET for dielectric-modulated label-free biosensing. IEEE Sens. J. (2017). https://doi.org/10.1109/jsen.2019.2928182. (Early Access)
48. Kannan, N., Jagadesh Kumar, M.: Dielectric-modulated impact-ionization MOS transistor as a label-free biosensor. IEEE Electron. Device Lett. **34**(12), 1575–1577 (2013)
49. Chanda, M., Dey, P., De, S., Sarkar, C.K.: Novel charge plasma based dielectric modulated impact ionization MOSFET as a biosensor for label-free detection. Superlattices Microstruct. **86**, 446–455 (2015)
50. Parihar, M.S., Kranti, A.: Enhanced sensitivity of double gate junctionless transistor architecture for biosensing applications. Nanotechnology **26**(14), 145201 (2015)
51. Singh, S., Kondekar, P.N., Jaiswal, N.K.: Label-free biosensor using nanogap embedded dielectric modulated schottky tunneling source impact ionization MOS. Microelectron. Eng. **149**, 129–134 (2016)

Detecting Signature of Virus Using Metamaterial-Based One-Dimensional Multi-layer Photonic Crystal Structure Under Polarized Incidence

Arpan Deyasi, Nabanita Pramanik, and Angsuman Sarkar

Abstract Photonic crystal (PhC) is the present-day building block in high-frequency communication systems due to its inherent capacity of noise reduction by offering filter characteristic around the choice of centre wavelength (Nagata in security, privacy, and applied cryptography engineering, SPACE 2019. Lecture notes in computer science, vol 11947. Springer, Cham, 2018 [1]; Huynh et al. in Asia-Pacific symposium on electromagnetic compatibility, 2013 [2]). This becomes possible due to the formation of photonic bandgap (Wang et al. in advances in neural networks, ISNN 2006. Lecture notes in computer science, vol 3973. Springer, Berlin, 2006 [3]), along the direction of electromagnetic wave propagation, where nature and shape of the bandgap critically depend on the characteristic of constituent material parameters (Loudon in J Phys A 3:233–245, 1970 [4]). Initial works are carried on the positive index materials (Blanco and López in annual review of nano research, pp 81–152, 2006 [5]); however, metamaterials become popular choice (Kuramochi in J Mater Chem C 4:11032–11049, 2016 [6]) for designing the components of all-optical circuits using PhC owing to improved response. However, tuning of filter is also dependent on dimensional parameters, which cannot be altered after fabrication. But slight alteration of these layer widths can be tailored so that the filter properties can be deviated before sending the design to fab-lab, where instead of both rise and fall bands, only a sharp peak can be found. Width of the peak can be made infinitesimally small, and with change of angle of polarized input wave, peak can be shifted. This can be set equivalent to the characteristic wavelength of a virus, as magnitude of the shift of peak is in the range nanometer. In the present chapter, shift of peak is analytically investigated in presence of both TE and TM mode incidences along with width of the peaks for both the cases. For ternary and quaternary structures, these results are graphically represented, and comparative study is made

A. Deyasi (✉) · N. Pramanik
Department of Electronics and Communication Engineering, RCC Institute of Information Technology, Kolkata 700015, India
e-mail: deyasi_arpan@yhoo.co.in

A. Sarkar
Department of Electronics and Communication Engineering, Kalyani Government Engineering College, Kalyani 741235, India

G. Dutta et al. (eds.), *Modern Techniques in Biosensors*, Studies in Systems, Decision and Control 327, https://doi.org/10.1007/978-981-15-9612-4_9

with existing results obtained for positive index materials. Result shows improvement for the present nano-fishnet structures, which can potentially speak in favour of biosensor applications.

Keywords Photonic biosensor · Photonic crystal · Nano-fishnet structure · Signature of virus · Polarized incidence · Shift of peak

1 Introduction

Evolution of electronic circuits in the twentieth century revolves around the VLSI chip where custom-specific design and reconfigurable design are getting importance due to the need of various applications [1–3] keeping the prime factors unchanged, i.e. lower power dissipation, less chip area, and higher speed. However, owing to the ever-increasing difference with Moore's guideline, researchers are forced to think alternatives, which lead to the concept of circuit design using photonic properties, instead of electronic counterparts. This novel concept leads to the foundation of a new branch of engineering, termed as photonics. Though initial foundation stone was laid much earlier, however, goes unnoticed as the electronic engineering puts its footstep in solid form in the last decade. But at the end of the twentieth century, physicists re-investigated the phenomenon of propagation of electromagnetic wave [4] inside bounded media, more precisely inside grating-like structure, and therefore, the term 'photonic crystal' (PhC) is coined. This is quite a revolution, realized as the twenty-first century is progressing, and is now considered as building block of next generation communication system [5, 6].

Photonic crystal, may be considered as one of the fruits of science and technological research, is already considered as one of the possible best replacements of its electronic counterparts. Propagation of electromagnetic wave inside the device creates waveguiding feature by allowing selected part of the total spectrum to flow [7]. This feature can only be exhibited if the propagation of electromagnetic wave is along the direction of refractive indices variation [8], i.e. electromagnetic wave has to be traversed through more than one medium, and the variation should be periodic in nature, according to Bragg's law. This periodic perturbation leads to the formulation of electromagnetic bandgap, more popularly known as photonic bandgap [9]. This is quite analogous to electronic bandgap, as observed in semiconductor devices, and has the similar property, but only the difference is the total analysis which has to be made in terms of propagating wave instead of carrier transport. Bandgap inside the first Brillouin zone for this type of structure is already investigated in detail through various numerical techniques, as reported elsewhere [10–13]. Nature and shape of the bandgap critically depend on the characteristic of constituent material parameters [14, 15], dimensions of different layers along the wave propagation direction [16, 17], atomic distributions [18], and incidence angle [19]. Depending on the formation of photonic bandgap (PBG), several novel optical devices have been proposed, and these show better performance compared to existing opto-electronic devices.

Among different devices, optical transmitters [20, 21] and optical receivers [22, 23] are researched a lot due to their enormous potential applications, precisely when incorporated in all-optical integrated circuit. Lasers made by PhC have better coherence and less loss factor and, therefore, more efficient when compared with electronic counterpart. Similar property has been observed for detectors also. Works are also extended further to apply the structure for quantum information processing [24]. It brings a revolution in the communication application due to the successful realization of photonic crystal fibre (PCF) [25, 26]. When coupled with waveguide, PCF can suppress the crosstalk [27] with a higher success rate. Also PhC is effectively utilized to make switches [28] which works at THz frequency range.

The most important application among all the other PhC made devices is probably sensors. Different types of sensors [29] are proposed for various optical applications. Among different sensors, it is used for colorimetric purpose [30], pressure sensor [31], imaging purpose [32], etc. Novel pressure sensor is recently proposed with defected structure [33] and having extended dynamic range. Role of material is critically important [34] to enhance sensitivity of the device. Fluid sensors are made to identify fluids from mixtures [35]. Spiral photonic crystal-based sensors are used for gas sensing purpose [36]. Cavities inside the structure are intentionally made [37] in some specific cases to utilize the device for biochemical application. pH of a chemical is also sensed very recently [38], whereas PCFs are applied to detect various parameters [39]. Hybrid PCFs are also used for chemical sensing.

Sensors are nowadays the integrated part of any electro-optical circuit, and its area is expanded in the biological applications. Very recently, researchers proposed it for early cancer detection [41, 42]. Later, PhC-based sensor is suggested for haemoglobin detection [43]. Novel experiments are carried out to design human blood sensor using PCF [44]. Amount of glucose in tear can be measured using 2D PhC-based sensor [45]. Transmission characteristics of photonic crystal are applied for biosensor design at optical spectrum [46]. Resonance peak of the characteristic curve is applied for disease detection when waveguide is constructed [47] using PCF. Real-time monitoring of bio-molecules can be performed using the inherent colour exhibited by photonic crystal, as reported in 2017 [48]. Resonant peak is basically applied in most of the sensor design, and its correlation with refractive index of the materials is the prime for biosensor functioning [49]. Nano-ring resonators are exercised for protein concentration detection [50]. All these works reveal the fact that the formation of photonic bandgap and corresponding resonance peak is the essence of designing various biosensors, as reported so far.

2 Objective of Present Work

All the sensors reported so far using photonic crystal have used the dependence of transmission spectra on optical properties of constituent materials. However, one point at this juncture may be considered from fabrication point-of-view that after production, it is not possible to change or tune the refractive index of the materials, and therefore, shift of resonance peak is impossible to obtain in this manner. The same argument is applicable for structural dimensions. Therefore, alternative ways have to find out, if the model has to be materialized, where tuning of structural widths or material properties should not be considered. Novelty of the present work lies here that it also theoretically characterizes the virus and similar nano-size devices, whose detection is possible by allowing incident of polarized wavelength on the device. With change of incidence angle, either in TE or in TM mode, peak of the reflection coefficient shifts. This shift, it can be in the range of nano-dimension, then becomes equivalent to the characteristic signature of virus or sub-atomic species. Once wavelengths of both the things match, then the corresponding living can easily be detected. In this point, it may be considered that the variation of incidence angle can be limited within a specified range and therefore provides a limitation of the proposal. However, the proposal is practically feasible as alteration of layer dimensions or material constituents are not proposed to match the wavelength of living object. Henceforth, the present proposal may be considered reliable and experimentally achievable, where choice of material plays a key factor.

3 Choice of Material

Material will play the pivotal role in designing photonic crystal-based applications, as the magnitude of shift is critical, and has to be in the nanometric range. Hence, different material systems are utilized, and multi-layer systems are also taken under numerical investigation, before fabrication. An optimization method is adopted, which gives better result when metamaterial-based systems are adopted for the said purpose. It has already been established by different researchers that metamaterial-based PhC works better than the conventional system [51, 52]. Also the workers established the supremacy of metamaterial-based bandpass filter compared to positive index-based material [53]. In the present chapter, authors, therefore, considered the negative index-based material system for the investigation. We have considered parallel nanorod ($n = -0.3$) with air-filled inter-space, and results are computed graphically under both the polarization conditions. The optimization result centred on the fact that magnitude of reflection coefficient should be very high so that the profile can easily be distinguished compared to that generated ripple or undue oscillations around the desired peak. In this context, another important function comes in terms of periodicity of the structure. After optimization, it has been found that ternary structure for the considered material system exhibits better result than binary or quaternary structures. Therefore, simulation is carried out for ternary parallel nanorod structure under TE and TM mode of polarizations, keeping an eye of reflection profile.

4 Results and Discussions

After choice of materials, work is progressed to compute reflectivity for the system under polarized conditions. We separately consider both TE and TM mode of incidences. With change of incidence angle, we have calculated the peak magnitude of reflection spectrum and corresponding wavelength. For each case, we have measured the shift of the peak w.r.t normal incidence. Magnitude of shift in the wavelength scale is measured, and it is found that the shift is in nanometer range. In this section, first we analyse the data for TE mode, and then we can proceed for TM mode. Comparative study for both the modes is discussed with relevant data thereafter.

4.1 Results for TE Mode

Figure 1 shows the shift of reflection peak with incidence angle for TE mode of operation. From the detailed analysis, it is found that for TE mode, maximum 17° incidence angle can be allowed for which the desired property can be obtained. Among them, 5 selected incidence angles are chosen for graphical representation, and all are plotted under Fig. 1. Figure 1a shows the shift for 6° angle, and higher angles are plotted for Fig. 1b (8°), c (9°), d (10°), e (11°). The shifts are measured for each case.

A closer look among the 5 graphs can reveal that though magnitude of peak remains within 60 dB for lower incidence angle, it is increased as angle of incidence becomes more. This is quite reflected if we compare with Fig. 1d, e with rest of the plots. Also the shift of peak point w.r.t reference peak increases. This is more evident if one can compare the next two plots w.r.t previous three plots.

4.2 Results for TM Mode

As per the previous subsection, we have also analysed the data for TM mode. All the results are plotted under Fig. 2. It may be mentioned in this context that we have plotted results for same angle of incidences as in Fig. 1. Therefore, Fig. 2a–e represents TM mode for 6°, 8°, 9°, 10°, 11° respectively. Similar increase in peak shift observed for TM mode as we have seen earlier for TE mode. However, one major fact may be noted out in this context that the magnitude of peak reflection coefficient decreases w.r.t normal incidence for TM mode of propagation. This rate enhances as angle of incidence increases.

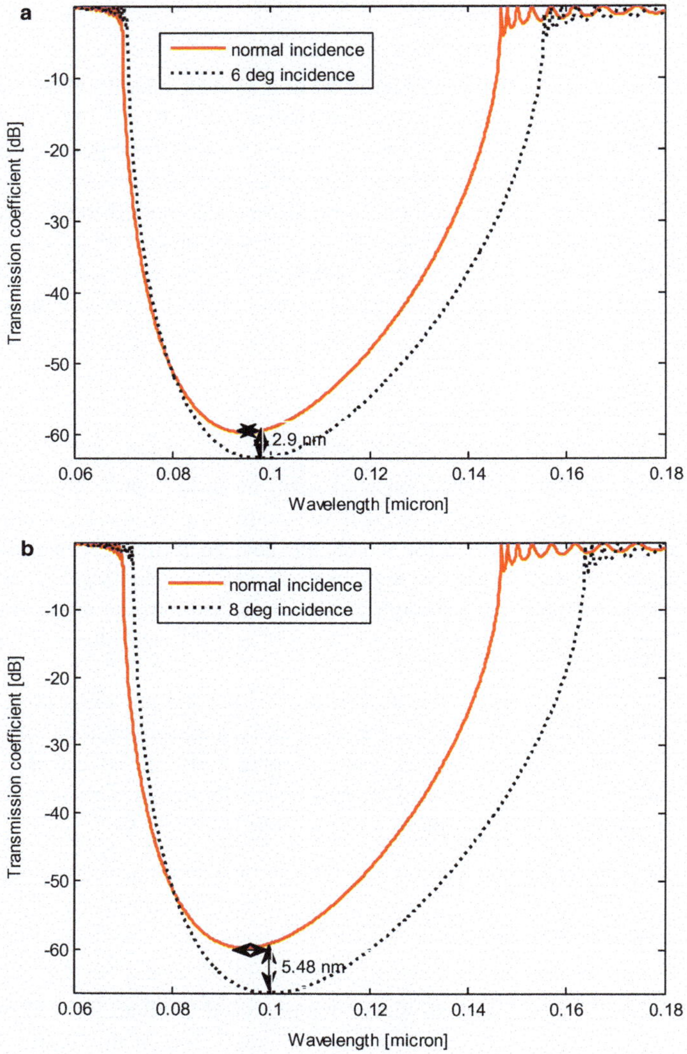

Fig. 1 **a** Shift of reflection peak w.r.t normal incidence for 6° angle of incidence under TE mode. **b** Shift of reflection peak w.r.t normal incidence for 8° angle of incidence under TE mode. **c** Shift of reflection peak w.r.t normal incidence for 9° angle of incidence under TE mode. **d** Shift of reflection peak w.r.t normal incidence for 10° angle of incidence under TE mode. **e** Shift of reflection peak w.r.t normal incidence for 11° angle of incidence under TE mode

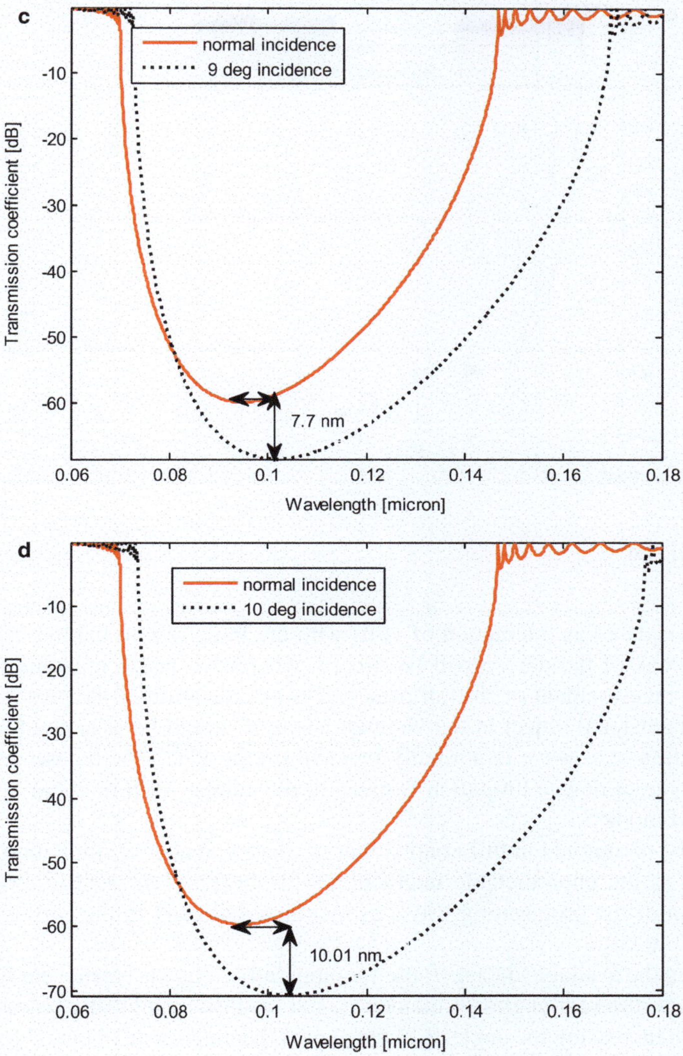

Fig. 1 (continued)

4.3 Significance of Results

The simulated findings show that with change of incidence angle, peak of reflection coefficient makes a redshift w.r.t the position of peak obtained for normal incidence. The change of incident angle may be caused due to change in electromagnetic wave interaction with the structure concerned. This is generally caused by the interaction of similar dimensional object/wave with the grating/PhC, and one clear example

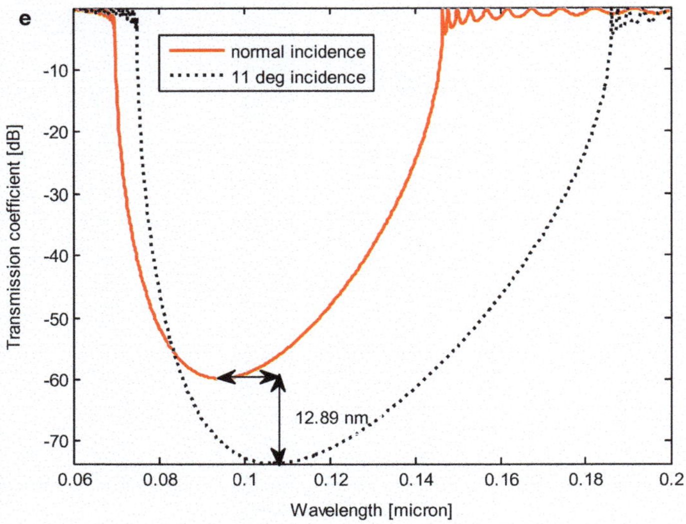

Fig. 1 (continued)

may be considered as interaction of virus with the PhC. Due to the interaction, the reflection peak of the device will be shifted, whatever smaller magnitude can be. Therefore, measurement of shift of peak is a clear indication of the presence of the smaller dimensional object in the vicinity, where it should be assumed that before the interaction, e.m wave is normally incident on the PhC. The dataset within the feasible range of device limitation is given in the tabular form in Table 1, for both TE and TM modes.

It may be mentioned in this context that we have considered simulation for both the modes as the unpredictable interaction with the PhC can lead to any type of polarization of the propagating wave, as mentioned already in the earlier reported literatures [46–50].

Based on the available data in Table 1, comparative study is carried out for reflection peaks at different angles of incidence, and result is graphically represented in Fig. 3. With increasing the angle, difference w.r.t the magnitude obtained for normal incidence increases. Corresponding to that, shift of the peak is also measured and represented in Fig. 4. Table 2 gives the datasheet. From Fig. 4, it may be concluded that for higher angle of incidence, TE mode provides greater shift than for TM mode.

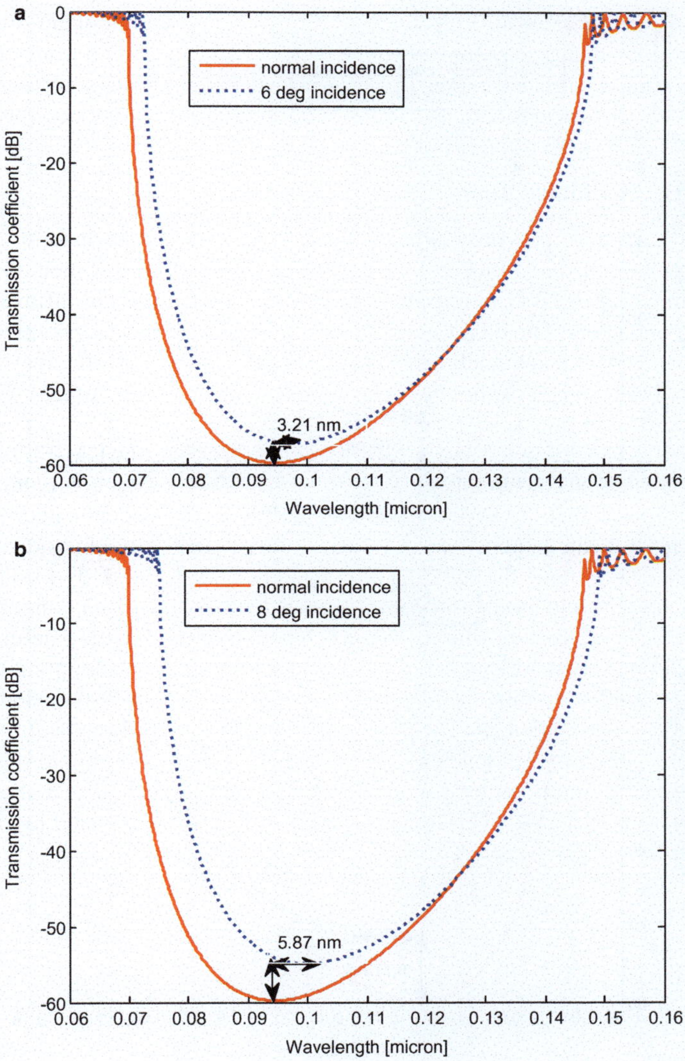

Fig. 2 **a** Shift of reflection peak w.r.t normal incidence for 6° angle of incidence under TM mode. **b** Shift of reflection peak w.r.t normal incidence for 8° angle of incidence under TM mode. **c** Shift of reflection peak w.r.t normal incidence for 9° angle of incidence under TM mode. **d** Shift of reflection peak w.r.t normal incidence for 10° angle of incidence under TM mode. **e** Shift of reflection peak w.r.t normal incidence for 11° angle of incidence under TM mode

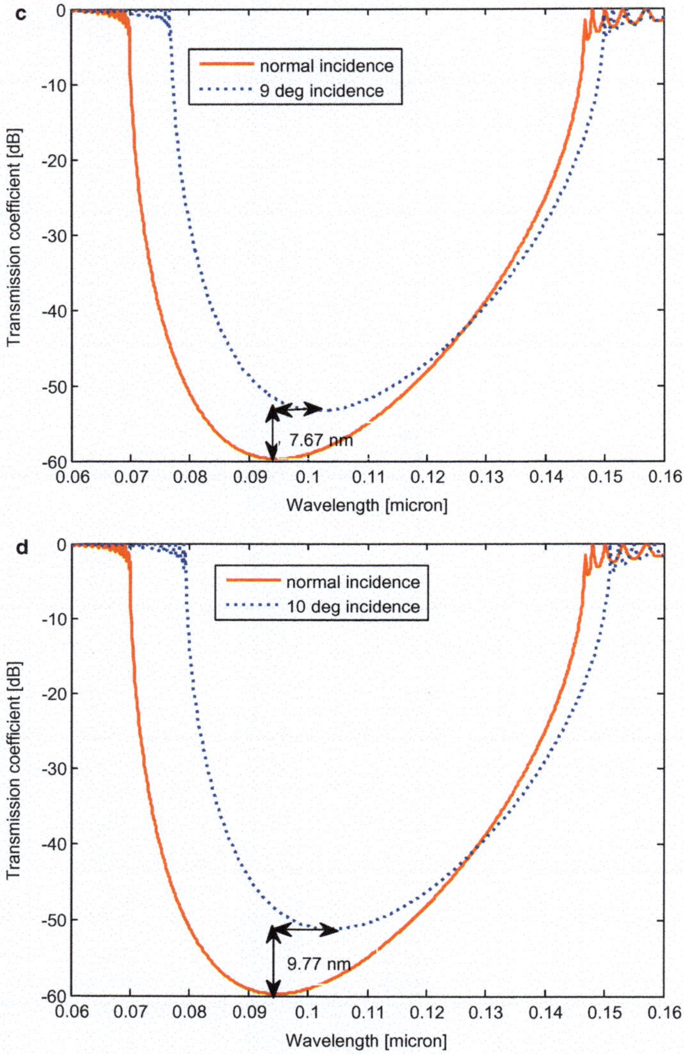

Fig. 2 (continued)

5 Conclusion

A closer inspection of the shift of peak reflection reveals that interaction of submicron objects with one-dimensional photonic crystal leads to the polarized wave propagation inside the structure which makes shift of peak reflection coefficient towards higher wavelength. Magnitude of the shift as well as the peak value of the profile is tailored with smaller change of angle of incidence of the interacting wave, which

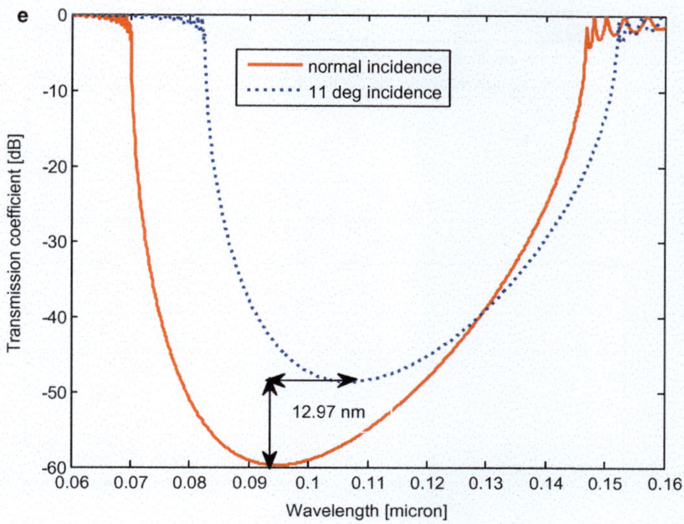

Fig. 2 (continued)

Table 1 Dataset for magnitude of reflection peak with angle of incidence

Angle of incidence (°)	TE mode		TM mode	
	Wavelength (μ)	Peak magnitude of reflection coefficient	Wavelength (μ)	Peak magnitude of reflection coefficient
0	0.09462	−59.74	0.09463	−59.74
1	0.09473	−59.83	0.0951	−59.67
2	0.09565	−60.1	0.0951	−59.48
3	0.09568	−60.59	0.09568	−59.15
4	0.09565	−61.27	0.09568	−58.67
5	0.09662	−62.17	0.09662	−58.04
6	0.09753	−63.31	0.09784	−57.22
7	0.09851	−64.7	0.09851	−56.18
8	0.1001	−66.39	0.1005	−54.88
9	0.1023	−68.41	0.1023	−53.26
10	0.1046	−70.82	0.1044	−51.23
11	0.1075	−73.69	0.1076	−48.65
12	0.1108	−77.12	0.1108	−45.35
13	0.1165	−81.22	0.1155	−40.99
14	0.125	−86.16	0.1241	−35.04
15	0.1371	−92.21	0.1354	−26.51
16	0.1644	−99.62	0.1108	−45.35

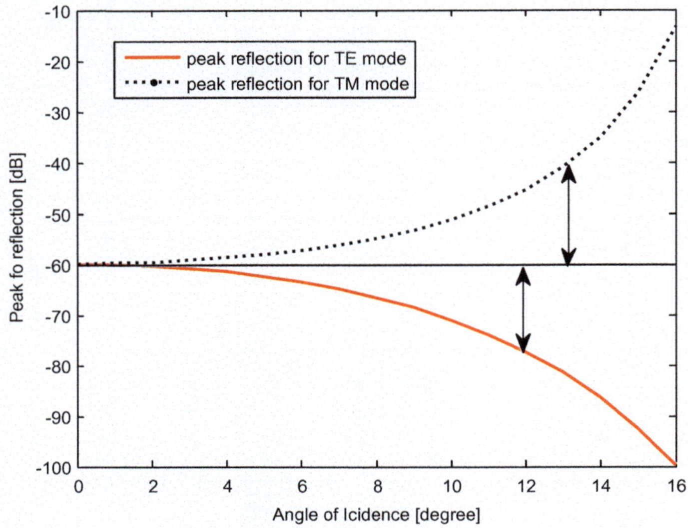

Fig. 3 Variation of peak reflection coefficient with angle of incidence for both TE and TM modes

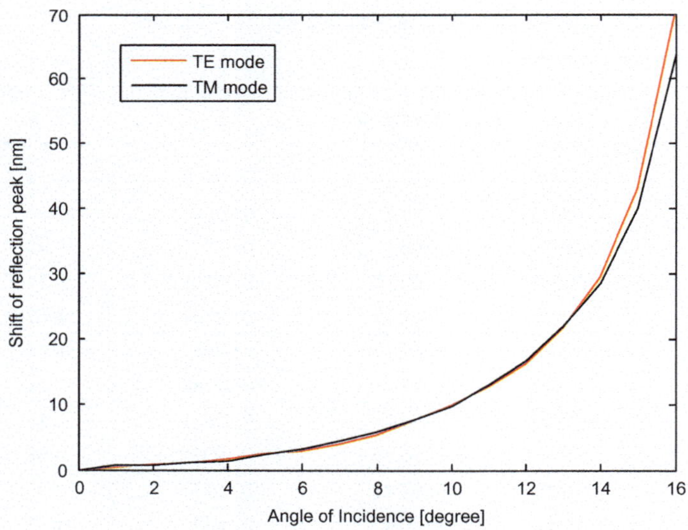

Fig. 4 Shift of peak for TE and TM modes with angle of incidence

Table 2 Dataset for shift of reflection peak with angle of incidence

Angle of incidence (°)	Shift of peak for TE mode (nm)	Shift of peak for TM mode (nm)
0	0	0
1	0.61	0.71
2	0.97	0.71
3	1.21	1.21
4	1.72	1.41
5	2.51	2.35
6	2.91	3.21
7	3.89	4.45
8	5.48	5.87
9	7.71	7.67
10	10.01	9.77
11	12.89	12.97
12	16.18	16.61
13	21.89	21.91
14	29.51	28.47
15	43.01	40
16	70.11	63.71

results in deviation w.r.t to the reflection profile obtained for normal incidence. If the interaction leads to TE mode of propagation, peak magnitude increases, whereas for TM mode, it monotonically decreases. Therefore, the interaction is predictable, and magnitude of reflection peak is also helpful to determine the nature of interaction. Therefore, in this context, the structure can be well behaved as a biosensor, which is sufficiently able to determine the presence of virus or similar submicron particle. Since each submicron particle has its own characteristic signature, therefore magnitude of shift also serves as a crucial parameter in this regard to predict the interaction. Here lies the novelty of the present work, and one-dimensional photonic crystal can primarily play the role of a biosensor device.

References

1. Nagata, M.: Deployment of EMC-compliant IC chip techniques in design for hardware security. In: Bhasin, S., Mendelson, A., Nandi, M. (eds.) Security, Privacy, and Applied Cryptography Engineering, SPACE 2019. Lecture Notes in Computer Science, vol. 11947. Springer, Cham (2019)
2. Huynh, H.A., Kim, K.S., Nah, W.S., Kim, S.Y.: EMC/EMI verification methodology for semi-custom design. In: Asia-Pacific Symposium on Electromagnetic Compatibility (2013)
3. Wang, Q., Li, A., Li, Z., Wan, Y.: A design and implementation of reconfigurable architecture for neural networks based on systolic arrays. In: Wang, J., Yi, Z., Zurada, J.M., Lu, B.L., Yin,

H. (eds.) Advances in Neural Networks, ISNN 2006. Lecture Notes in Computer Science, vol. 3973. Springer, Berlin (2006)
4. R Loudon 1970 The propagation of electromagnetic energy through an absorbing dielectric J. Phys. A 3 233 245
5. Blanco, A., López, C.: Photonic crystals: fundamentals and applications. In: Annual Review of Nano Research, pp. 81–152 (2006)
6. E Kuramochi 2016 Manipulating and trapping light with photonic crystals from fundamental studies to practical applications J. Mater. Chem. C 4 11032 11049
7. E Yablonovitch 1987 Inhibited spontaneous emission in solid-state physics and electronics Phys. Rev. Lett. 58 2059 2061
8. Dey, R., Banerjee, M., Das, A., Deyasi, A.: Effect of incidence angle on optical bandwidth in ternary photonic crystal for filter application. In: Springer Series of Lecture Notes in Networks and Systems: Industry Innovative Innovations in Science, Engineering and Technology, pp. 403–410 (2018)
9. IS Fogel JM Bendickson MD Tocci MJ Bloemer M Scalora CM Bowden JP Dowling 1998 Spontaneous emission and nonlinear effects in photonic bandgap materials, pure and applied optics J. Eur. Opt. Soc. Part A 7 393 408
10. Adhikary, M., Uppu, R., Harteveld, C.A.M., Vos, W.L.: An optical probe of a 3D photonic band gap. In: Conference on Lasers and Electro-Optics Europe and European Quantum Electronics Conference: OSA Technical Digest, ck_p_15 (2019)
11. Y Trabelsi N Ben Ali W Belhadj M Kanzari 2019 Photonic band gap properties of one-dimensional generalized Fibonacci photonic quasicrystal containing superconductor material J. Supercond. Novel Magn. 32 3541 3547
12. BK Singh A Bijalwan PC Pandey V Rastogi 2019 Photonic bandgaps engineering in double graded hyperbolic, exponential and linear index materials embedded one-dimensional photonic crystals Eng. Res. Express 1 2 025004
13. Chakraborty, P., Ghosh, R., Adhikary, A., Deyasi, A., Sarkar, A.: Electromagnetic bandgap formation in two-dimensional photonic crystal structure with DNG materials under TE mode. In: IEEE International Conference on Devices for Integrated Circuit (2019)
14. A Deyasi A Sarkar 2018 Variation of optical bandwidth in defected ternary photonic crystal under different polarization conditions Int. J. Nanopart. 10 1–2 27 34
15. Villa-Villa, F., Gaspar-Armenta, J.A., Mendoza-Suárez, A.: Surface modes in one dimensional photonic crystals that include left handed materials. J. Electromagn. Waves Appl.**21**, 485–499 (2007)
16. S Golmohammadi MK Moravvej-Farshi A Rostami A Zarifkar 2007 Spectral analysis of the Fibonacci-class one-dimensional quasi-periodic structures Prog. Electromagn. Res. 75 69 84
17. Y Kim YJ Lee S Hong K Moon SH Kwon 2018 Photonic crystal cavity with a thin low-index layer for silicon-compatible nanolight source Appl. Sci. 8 1552
18. Mukherjee, S., Roy, A., Deyasi, A., Ghosal, S.: Dependence of photonic bandgap on material composition for two-dimensional photonic crystal with triangular geometry. In: Acharyya, A. (ed.) Foundations and Frontiers in Computer, Communication and Electrical Engineering, chap. 52, pp. 259–263. CRC Press, Boca Raton (2016)
19. D Mao Z Ouyang JC Wang 2008 A photonic-crystal polarizer integrated with the functions of narrow bandpass and narrow transmission angle filtering Appl. Phys. B 90 127 131
20. T Tajiri S Takahashi Y Ota K Watanabe S Iwamoto Y Arakawa 2019 Three-dimensional photonic crystal simultaneously integrating a nanocavity laser and waveguides Optica 6 296 299
21. Yoshida, M., De Zoysa, M., Hatsuda, R., Tanaka, Y., Ishizaki, K., Noda, S.: Elliptical double-hole photonic-crystal surface-emitting lasers. In: IEEE Conference on Lasers and Electro-Optics Pacific Rim (2017)
22. K Nozaki S Matsuo T Fujii K Takeda M Ono A Shakoor E Kuramochi M Notomi 2016 Photonic-crystal nano-photodetector with ultrasmall capacitance for on-chip light-to-voltage conversion without an amplifier Optica 3 483 492

23. Wang, Y., Lee, J.H., Borsa, T., Wounjhang, P., Zeghbroeck, B.V.: Photo-response of integrated photonic crystal-photodiode micro-electro-optic filters. In: Proceeding of SPIE: Nanoengineering: Fabrication, Properties, Optics, and Devices II, vol. 5931, 59310K (2005)
24. RS Daveau KC Balram T Pregnolato J Liu EH Lee JD Song V Verma R Mirin SW Nam L Midolo S Stobbe K Srinivasan P Lodahl 2017 Efficient fiber-coupled single-photon source based on quantum dots in a photonic-crystal waveguide Optica 4 178 184
25. Shepard, S.R.: Photonic crystal fiber for the simultaneous transmission of information and power in optical smart grids. In: Advanced Photonics: JTu4A.29 (2016)
26. Kaur, V., Singh, S.: Hybrid design of photonic crystal fiber sensor for lower indexed chemicals. In: 13th International Conference on Fiber Optics and Photonics: Th3A.53 (2016)
27. V Liu S Fan 2013 Compact bends for multi-mode photonic crystal waveguides with high transmission and suppressed modal crosstalk Opt. Express 21 8069 8075
28. Reininger, P., Kalchmair, S., Gansch, R., Andrews, A.M., Detz, H., Zederbauer T., Ahn S.I., Schrenk, W., Strasser, G.: Optimized photonic crystal design for quantum well infrared photodetectors. In: Proceedings of SPIE, vol. 8425, 84250A (2012)
29. J Guo RA Norte S Gröblacher 2017 Integrated optical force sensors using focusing photonic crystal arrays Opt. Express 25 9196 9203
30. H Wang KQ Zhang 2013 Photonic crystal structures with tunable structure color as colorimetric sensors Sensors 13 4 4192 4213
31. Biallo, D., D'Orazio, De Sario, M., Marrocco, V., Petruzzelli, V., Prudenzano, F.: Photonic crystal sensors. In: IEEE International Conference on Transparent Optical Networks (2006)
32. G Pitruzzello TF Krauss 2018 Photonic crystal resonances for sensing and imaging J. Opt. 20 073004
33. KV Shanthi S Robinson 2014 Two-dimensional photonic crystal based sensor for pressure sensing Photonic Sens. 4 3 248 253
34. AM Ahmed A Mehaney 2019 Ultra-high sensitive 1D porous silicon photonic crystal sensor based on the coupling of Tamm/Fano resonances in the mid-infrared region Sci. Rep. 9 6973
35. S Amrehn X Wu C Schumacher T Wagner 2015 Photonic crystal-based fluid sensors: toward practical application Phys. Status Solidi A 212 6 1266 1272
36. MI Islam K Ahmed S Asaduzzaman BK Paul T Bhuiyan S Sen MS Islam S Chowdhury 2017 Design of single mode spiral photonic crystal fiber for gas sensing applications Sens. Bio-Sens. Res. 13 55 62
37. Y Zhang Y Zhao T Zhou Q Wu 2018 Applications and developments of on-chip biochemical sensors based on optofluidic photonic crystal cavities Lab Chip 18 57 74
38. AZ Mohammed 2018 Photonic crystal fibre Mach-Zehnder interferometer pH sensing AIP Conf. Proc. 2045 020010
39. Pinto, A.M.R., Lopez-Amo, M.: Photonic crystal fibers for sensing applications. Fiber Optic Sens. 2012:598178 (2012)
40. S Asaduzzaman K Ahmed T Bhuiyan T Farah 2016 Hybrid photonic crystal fiber in chemical sensing SpringerPlus 5 748
41. N Ayyanar GT Raja M Sharma DS Kumar 2018 Photonic crystal fiber-based refractive index sensor for early detection of cancer IEEE Sens. J. 18 17 7093 7099
42. Sharma, P., Sharan, P., Deshmukh, P.: A photonic crystal sensor for analysis and detection of cancer cells. In: IEEE International Conference on Pervasive Computing (2015)
43. Singh, B.K., Rastogi, V., Pandey, P.C.: Investigation of 1-D quasi-periodic photonic crystal based sensor for detection of hemoglobin. OSA Advanced Photonics Congress: JT4A.22 (2019)
44. CS Boopathi KV Kumar S Sheebarani K Selvakumar ANZ Rashed P Yupapin 2018 Design of human blood sensor using symmetric dual core photonic crystal fiber Results Phys. 11 964 965
45. C Chen ZQ Dong JH Shen HW Chen YH Zhu ZG Zhu 2018 2D Photonic crystal hydrogel sensor for tear glucose monitoring ACS Omega 3 3211 3217
46. J Divya S Selvendran AS Raja 2018 Photonic crystal-based optical biosensor: a brief investigation Laser Phys. 28 066206

47. H Chopra RS Kaler B Painam 2016 Photonic crystal waveguide-based biosensor for detection of diseases J. Nanophotonics 10 3 036011
48. H Chen R Lou Y Chen L Chen J Lu Q Dong 2017 Photonic crystal materials and their application in biomedicine Drug Deliv. 24 1 775 780
49. K Tong J Guo P Dang M Wang F Wang Y Zhang M Wang 2018 Surface plasmon resonance fiber optic biosensor-based graphene and photonic crystal Mod. Phys. Lett. B 32 6 1850072
50. AM Bahabady S Olyaee H Arman 2017 Optical biochemical sensor using photonic crystal nano-ring resonators for the detection of protein concentration Curr. Nanosci. 13 4 421 425
51. Čtyroký, J., Wangüemert-Pérez, J.G., Kwiecien, P., Richter, I., Litvik, J., Schmid, J.H., Molina-Fernández, I., Ortega-Moñux, A., Dado, M., Cheben, P.: Design of narrowband Bragg spectral filters in subwavelength grating metamaterial waveguides. Opt. Express **26**(1), 179–194 (2018)
52. Z Jiao R Ning Y Xu J Bao 2016 Tunable angle absorption of hyperbolic metamaterials based on plasma photonic crystals Phys. Plasmas 23 063301
53. A Deyasi A Sarkar 2019 Computing optical bandwidth of bandpass filter using metamaterial-based defected 1D PhC AIP Conf. Proc. 2072 020003

Shared Medical Decision-Making and Patient-Centered Collaboration

Arunima Ghosh and Sajjad Ahmed

Abstract Artificial intelligence is benefiting modern society in forecasting weather, detecting fraud, recognizing faces, deciphering genomics, and breast cancer but broadly speaking this role in medical practice, especially in SMDM is still unanswered. Shared Medical Decision-Making (SMDM) is an interaction model between patient and healthcare professionals. It is a mutual agreement between a patient, provider, and other healthcare professionals, based on which further treatment decisions are taken. SMDM helps in determining the probabilities about the outcome of the severity of the disease. Through various research, it has been identified that SMDM can be used in the early detection of some chronic diseases, which in the future may lead to a severe condition if not treated at the early stage. According to international guidelines in nephrology, it is suggested that patients suffering from Advanced Kidney Disease (AKD) can take preventive measures well in advance before AKD heads toward End-Stage Kidney Disease (ESKD). Research on diabetic medication adherence through patient-centered collaboration has proved to be one of the effective measures in avoiding hospitalizations. We make a critical review before we go for the implementation of proposed SMDM model in healthcare with the integration of AI. We have underlined the present concerns and components that limit the adoption of shared decision-making from the patient and provider level. It has been noticed that decision and communication aids are one of the effective measures to expedite the instrumentation of SDM in clinical practice. Additionally, health literacy and lack of time during clinical consultation also have a direct impact on SDM. On the other side, AI plays a prominent role to understand the pattern in certain areas like image analysis in radiology, pathology, and dermatology. Due to the sensitiveness of medical decision-making, it is important to go into depth before decision-making. Based on the steps of SDM, we proposed shared medical decision-making with the integration of artificial intelligence techniques that can be helpful for

A. Ghosh (✉)
Cognizant Technology Solutions India Private Limited ("Cognizant"), MBP, Bengaluru 560045, India
e-mail: g.arunima@yahoo.co.in

S. Ahmed
FHNW University of Applied Sciences and Arts Northwestern, 4600 Basel, Switzerland
e-mail: sajjadarid@gmail.com

G. Dutta et al. (eds.), *Modern Techniques in Biosensors*, Studies in Systems, Decision and Control 327, https://doi.org/10.1007/978-981-15-9612-4_10

future decision-making. A combination of machine plus physician reliably enhances the overall process of SMDM.

Keywords Shared medical decision making · Chronic diseases · Evidence-based medicine · Health literacy · Patient–physician communication · Decision aids

1 Introduction

Health decisions usually occur without an adequate collaboration between patients and physicians. This is a legitimate threat to: patients receiving care. Patient care is the combination of the patient's preferences about the treatment, informed decision-making, patient's satisfaction, the quality of physician–patient communication, and health system sustainability. Shared Decision-Making (SDM) can avert these undesirable outcomes; it is a mutual agreement about the selection of the therapeutic goals to the treatment options between patient, physicians, and other healthcare professionals. Thus, it is defined as a process during which patient's care givers and healthcare professionals collaborate and influence each other while making healthcare decisions [1]. Along with patient's involvement, it needs physicians having experience in evidence-based practice and communication [2]. The collaboration between patient and physician to seek for better treatment has been prevalent in patients with chronic health conditions. According to the research on mental health service delivery, the two strategies that were focused are SDM and self-directed care. Enforcement of shared medical decision-making in the healthcare industry can be boosted by including AI techniques in medical data. As health data deal with the huge amount of datasets, which is out of the scope of the human capacity to analyze, interpret, and take decisions on the same so the integration of AI techniques with healthcare data can result in the enhancement of personalized medication, reduction in the length of hospital stay in certain cases, thereby making it cost-effective from patients' level and time and decision management from physicians' level. In some cases, as time becomes constrain so the physicians usually treat the illness instead of the patient.[1] This affects the personal values of the patient. In the present era, where the shift from paternalistic treatment toward a shared decision is heading slowly there using different algorithms of AI, and techniques of machine learning can be advantageous if the decision about treatment options is shared between the patient, physician, and medical AI.[2]

---

[1] https://pubmed.ncbi.nlm.nih.gov/27284799/.

[2] https://ceur-ws.org/Vol-2542/MOD-KI8.pdf.

2 Literature Review

The vigilant approach to healthcare delivery is embedded in the Hippocratic ideal of a clinician, a person who is knowledgeable enough to make decisions taking medical regulations into consideration, based on patients' interest. Since ages, this practice is predominant. It was assumed that there was a single "right way" of treating an illness. The patients no longer could be driven under paternalistic umbrella of medical practitioners; rather, a physician should provide all information to patients in a comprehensible manner so that the patients can take decisions. For promoting best outcomes, the policymakers is also putting stress on the voice of the patient and family members at the time of medical decision-making [3]. In this process, the health decisions are contemplated along with the joint effort of the patient and one or more than one health professionals and referring to the best available evidence, the patients' values and preferences [4, 5]. The SDM conceptual models are usually limited to the clinical encounter between a patient and a single doctor.

An inter professional SDM conceptual model was designed to demonstrate that SDM models should consider the involvement of multiple players including inter professional teams [6, 7]. Starting from initial preferences to informed preferences highlighted all sections in Fig. 1 till to reach final decision.

2.1 *SDM Application in AKD*

According to international guidelines in nephrology, it is suggested that patients suffering from Advanced Kidney Disease (AKD) can take well-timed treatment modality decisions that match with their values and preferences. Inadequate research contribution in the areas of SDM, challenges are faced during the adoption of SDM in clinical practice. AKD is defined as an estimated Glomerular Filtration Rate (eGFR) of less than 30 mL/min/1.73 m^2, and from here, it heads toward End-Stage Kidney

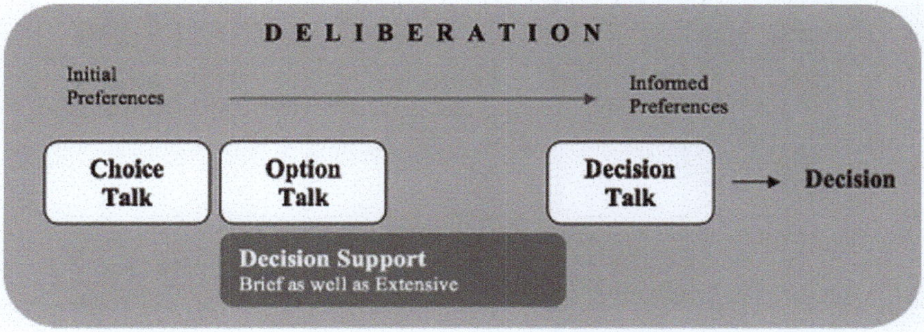

Fig. 1 Shared decision making: a model for clinical practice. *Source* Elwyn et al. [7]

Disease (ESKD). It may take months or years for change of AKD to ESKD. At this point of time, it demands treatment modality decision from patient's perspective.

In support to SDM, Patients Decision Aids (PtDAs) have already been developed [8], recently emphasizing toward Patient-Reported Outcome Measures (PROMs), which is known for bench marking, organization of care, and as novel tools to foothold the decision-making. A functioning vascular access is a lifeline for patients requiring Hemodialysis (HD); then again, it is also vulnerable to complications [9, 10]. In certain cases, vascular access complications account for hospitalizations in patients on HD which results in substantial healthcare costs. Initiative was taken by the International Standardized Outcomes in Nephrology (SONG) to stress on a core outcome measure for vascular access function in case of hemodialysis trials [11]. On the same page, an action was initiated by the Standardized Outcomes in Nephrology—Kidney Transplantation (SONG-Tx) to forecast a root cause for graft loss for all trials in kidney transplantation [12]. Based on the shared priorities of all the stakeholders, the Standardized Outcomes in Nephrology-Peritoneal Dialysis (SONG-PD) aimed to identify the core outcomes for trials in patients on peritoneal dialysis which further was used to evaluate the results of most relevance for decision-making, and that the interventions can likewise be analyzed dependably [13]. Conceptual schema can be seen in Fig. 2 for better understanding.

SDM can be categorized from patient's level and healthcare professional's level. As per systemic review from patient's level.

According to the PtDAs, the elementary issue that restricts the adoption of SDM are not detailed and uninvestigated [14]. From healthcare professional's level, according to Cochrane review, it was uncertain if any interventions to ramp up use of SDM were effective. The evidence to this was not clear [1].

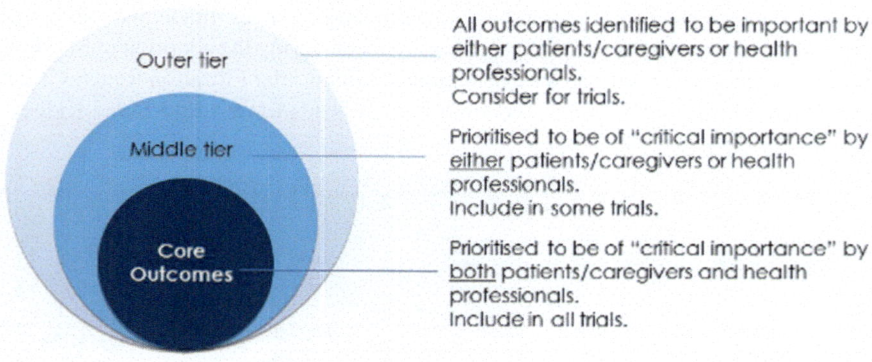

Fig. 2 Conceptual schema of core outcome domains [13]

2.2 SDM and Patient Centered Collaboration in Diabetic Patient

For various chronic health conditions, adherence to suboptimal medication is considered to be the primary preventive measure. Diabetic medication adherence through patient-centered collaboration has been proved to be one of the effective measures, where hospitalizations can be avoided. No adherence is considered to be a multifactorial problem according to a report by World Health Organization [15], which reflects the synergism between patients, Health Care Providers (HCPs), and the health care system; the insights of the condition and its treatment; and broader socioeconomic factors. Another development is the focus of uplifting only one factor (e.g., patient understanding) which is likely to have less effect on population adherence. To make this effective in a broader sense, the adherence promotion should be multifactorial [16] involving synchronous changes in the patient and provider behavior, health care system, etc. The factors contributing toward non-adherence at different levels are:

Patient level—Understanding of the patient's familiarity about taking of diabetic medications, grasp about the challenges of the diabetic management will help the providers to identify the adherence barriers which in turn will help to find out the pathway for stronger collaboration. Patient—provider interaction level—This level determines the relationship between patient and the healthcare team [17]. Providers fail to promote patients engagement in the areas of self-management and therapeutic alliance, until they win their patients' confidence and make recommendations based on the awareness of how diabetes could be fixed into their patients' lives.

Supported by the organized approach, the healthcare system should be restricted to boost patient centered collaboration. Many times it was observed that dearth of patients' knowhow skills or the urge to sustain necessary behaviors, usually intensifies provider's focus on disease management. However, the time invested in building the therapeutic relationship might have better enduring payoffs for patient engagement. While providers spend more time to know their patients, and patients listen to the people they know and believe this on contrary builds a bond of trust. A change in the role of healthcare providers has been perceived from telling patients what to do to hear out the patients, but most of providers are still lacking in this shift due to dearth of time and fee for the service rendered that rewards quantity over quality.

Steps toward the growth of adherence will additionally benefit from focusing on patients at highest risk for poor health outputs. High-risk patients account for a majority of healthcare spending, avoidable hospitalizations, morbidity, and mortality thus mediations successfully targeting adherence in this group are likely to have huge impact on population health. Additionally, patient from racial and ethnic minority groups usually face resistance in communications with HCP. With the promotion of medication adherence in diabetes, providers can get the evidence and feasible tools which additionally also help them to promote to precise medication-taking behavior among patients and healthcare teams. In future, a collaborative work between patients and providers, with proper utilization of time, and the applications can enhance medication adherence.

2.3 Role of Decision Aid in Implementation of SDM

Decision aids are the type of interventions with the help of which patients can explicitly make their decisions. Not only this, but also it provides various alternatives with clarity on the pros and cons of the same. It also helps to clarify the coherence between decisions and personal values. Through various researches, it has been identified that decision aids in comparison to usual care can enhance people's knowledge more about the options, thereby reducing the decisional conflict stemming from the feeling of being uninformed and unclear about their personal values. In addition, decision aids lowered the proportion of people remaining undecided. However, for affluent implementation of decision aid, several conditions may be required including—amiable quality decision aids that meet the needs of the population; clinicians who are willing to use decision aids in their routine; effective systems for delivering decision support; and clinicians and healthcare consumers who are skilled in shared decision-making [5]. Six key elements of SDM have been identified—situation diagnosis, choice awareness, option clarification, discussion of harms and benefits, patient preferences deliberation, and making the decision. According to The International Patient Decision Aid Standards (IPDAS), a Decision Aid (DA) should support these key elements [18]. One of few productive measures in promoting adoption of SDM in clinical practice is decision and communication aids. A systematic review in case of elective surgical procedures like—joint replacement for osteoarthritis or treatment for early stage breast cancer or localized prostate cancer demonstrated that well-developed educational information provided through interactive multimedia, computer, or DVD based which may amplify the decision-making process [19–21] (Fig. 3).

Methodical review determined empirical proof linking patient outputs (affective cognitive, behavioral, and health) and SDM. Advanced and affective cognitive outcomes can be attained when SDM is perceived by patients. However, there is still lack of evidence about the collaboration between empirical measures of SDM and patient behavioral and health end results [22]. It was seen through system review and meta = analysis that not only the children were not involved in decision-making, but also interventions to engage patients, and parents are often not studied. Thus, apart from SDM intervention in adults, physicians who care for children may also be interested in implementing SDM in their practice but due to limited availability of comprehensive review in this particular area, they fail to implement the same [23]. A comprehensive review through theoretical and empirical work revealed that more the information received by the patients matching their preferences, the better is their adjustment toward treatment, and subsequently, there are less chances of emotional dysphoria [24]. Physicians can support the patients in understanding the importance of their values and keenness through shared decision making. Various studies have shown that if patients are nurtured with options for best treatment, screening test, or diagnostic procedures, then most of them actively engage with the clinicians in making the choice. Thus, a well-developed patient-centered healthcare environment can be established when an informed woman or a man can succeed in taking decisions

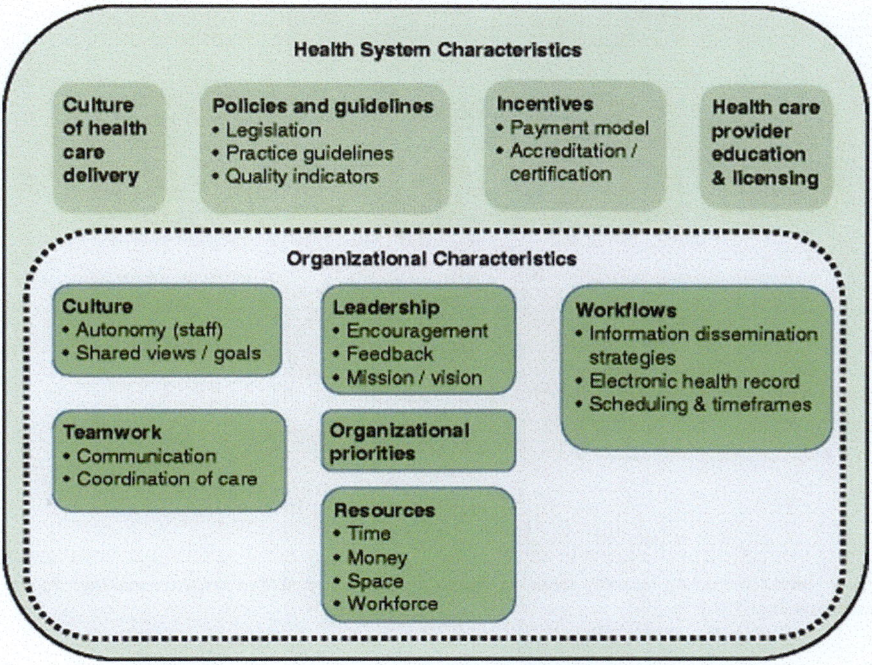

Fig. 3 Organizational- and system-level characteristics that influence implementation of shared decision-making and strategies to address them—a scoping review. *Source* Scholl et al. [20]

on whether to have a screening mammogram or to have a prostate-specific antigen screening test, respectively, without their physicians calling the decision "wrong" based on several beliefs and preferences [25].

2.4 *Interdependence of Evidence Based on Medicine and Shared Decision-Making*

SDM is a low-risk intervention as it helps patient to become alert well in advance before heading to high risk condition. This can be considered as low cost from patient's perspective as it reduces the chances of hospitalization in certain cases. Also it is a high yield evidence-based health intervention that helps in improving health and decision outcomes [26] (Fig. 4).

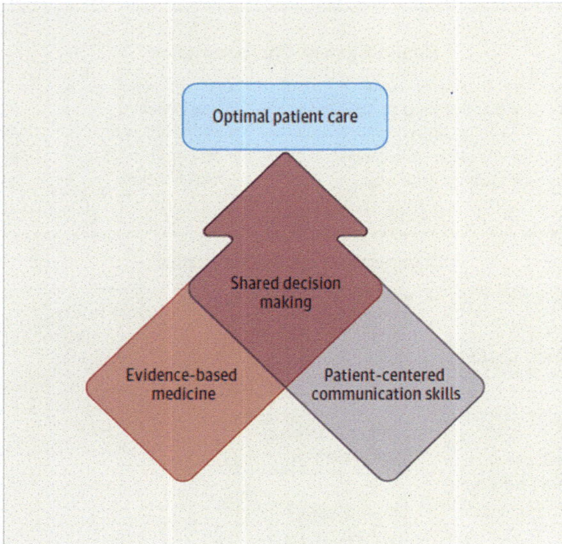

Fig. 4 Interdependence between evidence-based medicine and shared decision-making. *Source* Hoffmann et al. [26]

2.5 Role of Artificial Intelligence in the Field of Medical Decision-Making

Computer programs while assisting with medical decision-making do not underline models of clinical reasoning; instead, they have used statistical techniques such as Bayes' theorem or formal decision analysis for assigning structure to a domain. Aggregation of patient data and the generation of advice have mainly been emphasized by computer-based decision aids. Conversely, computer-based decision aids may be integrated with an ability to explain decisions. This can be achieved if the system is given an acceptable internal model of the logic that it uses along with conveying capability and intelligibility to the physician user. The incorporation of explanation capabilities may be a crucial step toward effectively favoring a system's use. This will provide fair benefits, thereby developing an environment in which the physician can feel comfortable and efficient. Artificial intelligence is the vital area of success for the researchers in the sector of computer-based clinical decision-making.

Big data science and artificial intelligence, which are considered to be the fourth industrial revolution [27], will not only bring enhancement in the field of radiology and pathology rather will also have their significant impact on other medical facilities in the long run. Enlitic, a technology company in Silicon Valley was successful in highlighting the radiographs with fractures and was able to highlight the fractures using deep learning [28]. Another study proposed by **Watson** and his team and they developed IBM's DeepQA [29]. The system was able to recognize pulmonary embolism on the basis of Computed Tomography (CT). It can also identify abnormal wall motion on echocardiography [30]. Radiology has advanced from subjective perception to objective science due to the availability of abundant and complex dataset [31]. The core purpose of radiologists is to provide medical information; the image is only to mean information. Thus, some of the complex data could be detected and analyzed by artificial intelligence, such as screening for lung cancer on CT. This involves identifying, measuring, and characterizing a lung nodule, the management of which is standardized [32]. Likewise, pathologists are also information specialists because both radiologists and pathologists extract medical information from images. Certain areas in pathology like identifying grade and stage of lung cancer can be better identified by computer study as compared to pathologists [33]. To identify the similarities between radiology and pathology, researchers are utilizing operant conditioning trained pigeons to detect abnormal calcifications on mammograms and locate breast cancer on histology [34]. Thus, the information specialists and artificial intelligence as a single unit can better manage individuals and populations, thereby guiding medical information and rendering better patient care at low cost and less time.

2.6 Role of Biosensor in Shared Medical Decision-Making

Biosensors play an intrinsic role in the enhancement of clinical decision-making in patients with Parkinson's disease. Wearable biosensor can aid in scrutinizing of motor symptoms in patients with PD via ankle and wrist sensors. The usual monitoring of motor symptoms in patients with PD is processed using Rotigotine Transdermal Patch which is applied to the skin once/day. However, there are certain drawbacks in the use of Rotigotine as the motor symptoms keep on varying by time in a day. Based on various research outcomes, it has been demonstrated that wearable devices provide more accurate monitoring of the motor symptoms during the day to day activities [35–37]. This kind of biosensors is an advantage in improving personalized medication titration which in turn can also enhance shared clinical decision-making [38]. Currently, this technology is still limited to the monitoring and assessment of certain diseases only; therefore, further research needs to be undertaken for the application of biosensors in SMDM.

3 Method

Shared Medical Decision-Making is the intersection of patient centered communication skills and Evidence based Medicine. The general process of shared medical decision-making is described below with an example of a patient suffering from migraine pain.

3.1 Case Study

A 34-year-old female patient is encountered with the onset of severe headache accompanied by nausea and vomiting at early morning. The patient is taken to the emergency department by her family members. Due to severe headache, the patient requested for some strong pain relief medicine. The ED physician decides to treat the patient with Zofer 4 MG injection which is an antiemetic agent used to treat nausea and vomiting and hydromorphone, an Opioid medicine to reduce headache. While rendering the complete evaluation and management service to the patient, the physician comes to know that the patient frequently encounters migraine attack.

The simple steps that can be outlined related to shared medical decision-making in this scenario are as follows:

1. If the patient is having little knowledge about her medical conditions, her past treatment about the similar symptoms, then she can inform the physician about her frequent attacks of migraine headache.
2. Instead of treating the patient with hydromorphone, the patient could have given an option of non-opioid alternatives such as metoclopramide and triptans, which have already been proven to be effective in treatment of migraine.
3. The provider could have informed the patient about the side effects of Opioid, and how the patient can avoid hospitalization, because the use of Opioid in ED to treat patients suffering from headaches or migraines may have several negative ramifications which include an increased risk of revisit, hospital admission, and increased length of stay in ED.

3.2 Proposed Diagram

Post literature review, we suggested shared medical decision-making model which portrays the close relationship between evidence-based medicine and patient-centered communication skills with shared medical decision-making that aids to render an appropriate care to the patient.

When a patient visits a provider due to some health problem, then as per the chronological description of the development of a patient's present illness from the first sign and symptom or from the past history of encounter the History of Present Illness (HPI) is determined. The inventory of body systems (ROS) is obtained by series of questions to identify signs and symptoms; the patient may be experiencing currently or has experienced in the past. The PFSH is the patient's past history about illness, operations, injuries, and treatments along with patient's family and social history. After hearing the patient's version about the problem, the provider examines several organ systems or single organ system (Physical Exam). Based on the HPI, ROS, physical exam the provider takes the decision about the treatment. Here is the role of shared medical decision-making, where physician can provide all information about the various treatment options to the patient in a comprehensible manner so that

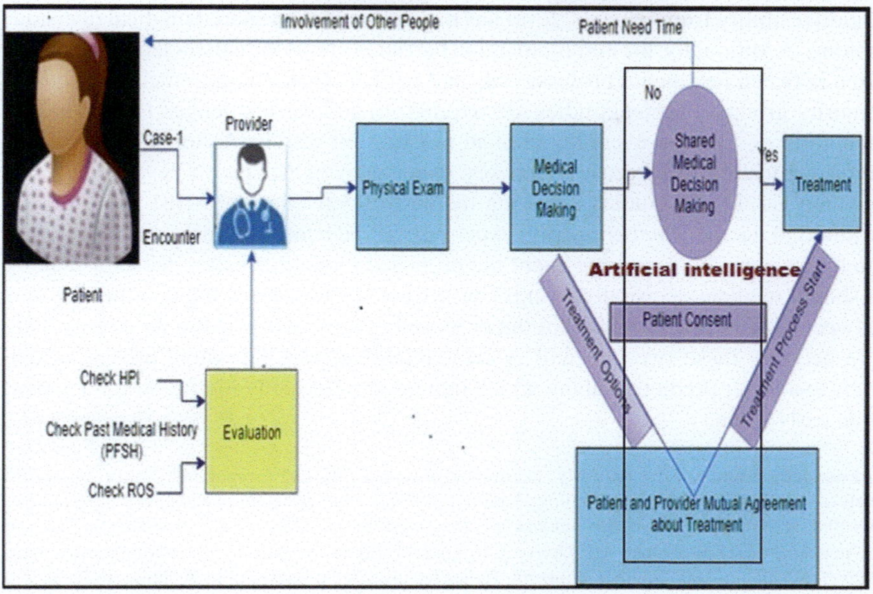

Fig. 5 Proposed diagram for shared medical decision-making. *Source* Own design

the patient can also take part in decision making. Based on the mutual agreement between patient and provider, the final treatment option starts. If the patient is not agreeing to the treatment options, then the patient may take further time to decide on the same. Here, AI will come that will start from treatment options to patient provider mutual agreement until the final treatment phase. The integration of AI techniques and SMDM can help the provider to overcome the problem of early diagnosis detection in patient. The proposed model in Fig. 5 could be the starting point of shared decision-making and AI for better identification and early decision-making, especially in those cases, where an urgent decision required.

4 Conclusion

Most of the papers published on shared decision-making either refers to particular areas of decision-making or their methodological context which restricts their validity because of their incredulity in the reproducibility of the findings. This limits healthcare professionals' adoption of SDM and also limits further investigations in this field. There are certain roadblocks to the implementation of SDM in the healthcare sector from both provider and patients' perspective. Apparent lack of time during clinical consultation and clinical applicability from providers' and patients' cognitive ability from patients' level has a direct impact on shared medical decision-making. Additionally, the communication barrier is one of the salient features noticed often between healthcare professionals and patients' belonging to racial and ethnic minority groups. Further, studies are required to address challenges toward implementation of SDM in everyday clinical practice. Researchers should also conduct empirical testing to forecast which key element of SDM should be introduced in decision aid. Future studies assessing the cost-effectiveness of SDM measures are required to decide whether upfront expenses will be beneficial monetarily or not. In the future, our idea is to implement the proposed model for better decision-making in shared medical decision-making. Our goal is not to replace the communications between patient and provider but the proposed system will perform an intermediate role between these two and will systematically support the whole process which leads to a better prediction about which treatment most likely important for a patient at an early stage.

Acknowledgements We gratefully acknowledges to Prof. Dr. Gorachand Dutta (Assistant Professor, Indian Institute of Technology Kharagpur) for his encouragement and guidance to publish this chapter. Arunima Ghosh also renders her sincere gratitude to Michelle, Yearwood (Sr. Medical Review Analyst, Digital Ops—HC Delivery, Cognizant, USA) and Karthikeyan Manickam (Service Delivery Manager, Healthcare, Cognizant, India) for always encouraging in bringing up ideas, out of the box.

References

1. Légaré, F., Adekpedjou, R., Stacey, D., et al.: Interventions for increasing the use of shared decision making by healthcare professionals. Cochrane Database Syst. Rev. **7**, CD006732 (2018)
2. Hoffmann, T.C., Legare, F., Simmons, M.B., McNamara, K., McCaffery, K., Trevena, L.J., et al.: Shared decision making: what do clinicians need to know and why should they bother? Med. J. Aust. **201**(1), 35–39 (2014)
3. Harrison, C.: Treatment decisions regarding infants, children and adolescents. Paediatr. Child Health 2004; 9. Mercurio, M.R., Adam, M.B., Forman, E.N., Ladd, R.E., Ross, L.F., Silber, T.J.: American Academy of Pediatrics policy statements on bioethics: summaries and commentaries: Part 1. Pediatr. Rev. **29**, e1–e8 (2008)
4. Charles, C., Gafni, A., Whelan, T.: Shared decision-making in the medical encounter: what does it mean? (Or it takes at least two to tango). Soc. Sci. Med. **44**, 681–692 (1997)
5. Stacey, D., Légaré, F., Lewis, K., et al.: Decision AIDS for people facing health treatment or screening decisions. Cochrane Database Syst. Rev. **4**, CD001431 (2017)
6. Légaré, F., Stacey, D., Gagnon, S., et al.: Validating a conceptual model for an inter-professional approach to shared decision making: a mixed methods study. J. Eval. Clin. Pract. **17**, 554–564 (2011)
7. Elwyn, G., Frosch, D., Thomson, R., et al.: Shared decision making: a model for clinical practice. J. Gen. Int. Med. **27**, 1361–1367 (2012)
8. Stacey, D., Légaré, F., Col, N.F., et al.: Decision AIDS for people facing health treatment or screening decisions. Cochrane Database Syst. Rev. **28** (2014)
9. Verberne, W.R., Das-Gupta, Z., Allegretti, A.S., et al.: Development of an international standard set of value-based outcome measures for patients with chronic kidney disease: a report of the

International Consortium for health outcomes measurement (ICHOM) CKD Working group. Am. J. Kidney Dis. **73**, 372–384 (2019)
10. Breckenridge, K., Bekker, H.L., Gibbons, E., et al.: How to routinely collect data on patient reported outcome and experience measures in renal registries in Europe: an expert consensus meeting. Nephrol. Dial. Transplant. **30**, 1605–1614 (2015)
11. Viecelli, A.K., Tong, A., O'Lone, E., et al.: Report of the standardize outcomes in nephrology-hemodialysis (SONG-HD) consensus workshop on establishing a core outcome measure for hemodialysis vascular access. Am. J. Kidney Dis. **71**, 690–700 (2018)
12. Tong, A., Sautenet, B., Poggio, E.D., et al.: Establishing a core outcome measure for graft health: a standardized outcomes in Nephrology-Kidney transplantation (SONG-Tx) consensus workshop report. Transplantation **102**, 1358–1366 (2018)
13. Manera, K.E., Tong, A., Craig, J.C., et al.: Standardized outcomes in Nephrology-Peritoneal dialysis (SONG-PD): study protocol for establishing a core outcome set in PD. Perit. Dial. Int. **37**, 639–647 (2017)
14. Elwyn, G., Scholl, I., Tietbohl, C., et al.: "Many miles to go …": a systematic review of the implementation of patient decision support interventions into routine clinical practice. BMC Med. Inf. Decis. Mak. **13** (2013)
15. Sabaté, E.: Adherence to Long-term Therapies: Evidence for Action. Switzerland, World Health Organization, Geneva (2003)
16. Sapkota, S., Brien, J., Greenfield, J., Aslani, P.: A systematic review of interventions addressing adherence to anti-diabetic medications in patients with type 2 diabetes: components of interventions. PLoS ONE **10**, e0128581 (2015)
17. Wylie, J.L., Wagenfeld-Heintz, E.: Development of relationship-centered care. J. Healthc. Qual. **26**(1), 14–45 (2004)
18. Wieringa, T.H., Rodriguez-Gutierrez, R., Spencer-Bonilla, G., De Wit, M., Ponce, O.J., Sanchez-Herrera, M.F., Espinoza, N.R., Zisman-Ilani, Y., Kunneman, M., Schoonmade, L.J., Montori, V.M. (2019). Decision aids that facilitate elements of shared decision making in chronic illnesses: a systematic review. Syst. Rev. **8**(1), 121
19. Pham, C., Lizarondo, L., Karnon, J., Aromataris, E., Munn, Z., Gibb, C., Fitridge, R., Maddern, G.: Strategies for implementing shared decision making in elective surgery by health care practitioners: a systematic review. J. Eval. Clin. Pract. **26**(2), 582–601 (2020)
20. Scholl, I., LaRussa, A., Hahlweg, P., et al.: Organizational- and system-level characteristics that influence implementation of shared decision-making and strategies to address them—a scoping review. Implement Sci. **13**, 40 (2018)
21. Légaré, F., Ratté, S., Gravel, K., et al.: Barriers and facilitators to implementing shared decision-making in clinical practice: update of a systematic review of health professionals' perceptions. Patient Educ. Couns **73**, 526–535 (2008)
22. Shay, L.A., Lafata, J.E.: Where is the evidence? A systematic review of shared decision-making and patient outcomes. Med. Decis. Making **35**, 114–131 (2015)
23. Wyatt, K.D., List, B., Brinkman, W.B., et al.: Shared decision-making in paediatrics: a systematic review and meta-analysis. Acad. Pediatr. **15**, 573–583 (2015)
24. Kiesler, D.J., Auerbach, S.M.: Optimal matches of patient preferences for information, decision-making and interpersonal behaviour: evidence, models and interventions. Patient Educ. Couns. **61**, 319–341 (2006)
25. Barry, M.J., Edgman-Levitan, S.: Shared decision-making—the pinnacle of patient-centered care. N. Engl. J. Med. **366**, 780–781 (2012)
26. Hoffmann, T.C., Montori, V.M., Del Mar, C.: The connection between evidence-based medicine and shared decision making. JAMA **312**(13), 1295–1296 (2014)
27. The return of the machinery question. The Economist. June 25, 2016. https://www.economist.com/news/special-report/21700761-after-many-falsestarts-artificial-intelligence-has-taken-will-it-cause-mass. Accessed 15 Nov 2016
28. Jha, S.: Will computers replace radiologists? Medscape, 12 May 2016. https://www.medscape.com/viewarticle/863127. Accessed 15 Nov 2016

29. Ferrucci, D., Levas, A., Bagchi, S., Gondek, D., Mueller, E.T.: Watson: beyond jeopardy! Artif. Intell. **199**, 93–105 (2013)
30. McMillan, R., Dwoskin, E.: IBM crafts a role for artificial intelligence in medicine. Wall Street J. 11 Aug 2015. https://www.wsj.com/article_email/ibm-crafts-a-role-for-artificial-intelligence-in-medicine-1439265840-lMyQjAxMTI2NjA3NTAwMDUxWj. Accessed 4 Oct 2016
31. Gillies, R.J., Kinahan, P.E., Hricak, H.: Radiomics: images are more than pictures, they are data. Radiology **278**(2), 563–577 (2016)
32. Lung-RADS Version 1.0 Assessment Categories. April 28, 2014. https://www.acr.org/~/media/ACR/Documents/PDF/QualitySafety/Resources/LungRADS/AssessmentCategories.pdf. Accessed 4 Oct 2016
33. Yu, K.H., Zhang, C., Berry, G.J., et al.: Predicting non-small cell lung cancer prognosis by fully automated microscopic pathology image features. Nat. Commun. **7**(7), 12474 (2016)
34. Levenson, R.M., Krupinski, E.A., Navarro, V.M., Wasserman, E.A.: Pigeons (Columba livia) as trainable observers of pathology and radiology breast cancer images. PLoS ONE **10**(11), e0141357 (2015)
35. Rovini, E., Maremmani, C., Cavallo, F.: How wearable sensors can support Parkinson's disease diagnosis and treatment: a systematic review. Front. Neurosci. **11**, 555 (2017)
36. Godinho, C., Domingos, J., Cunha, G., Santos, A.T., Fernandes, R.M., Abreu, D., Al-Jawad, A.: A systematic review of the characteristics and validity of monitoring technologies to assess Parkinson's disease. J. Neuroeng. Rehabil. **13**(1), 1–10 (2016)
37. Boroojerdi, B., Ghaffari, R., Mahadevan, N., Markowitz, M., Melton, K., Morey, B., Stumpp, O.: Clinical feasibility of a wearable, conformable sensor patch to monitor motor symptoms in Parkinson's disease. Parkinsonism Related Disorders **61**, 70–76 (2019)
38. Isaacson, S.H., Boroojerdi, B., Waln, O., McGraw, M., Kreitzman, D.L., Klos, K., Markowitz, M.: Effect of using a wearable device on clinical decision-making and motor symptoms in patients with Parkinson's disease starting transdermal rotigotine patch: a pilot study. Parkinsonism Related Disorders **64**, 132–137 (2019)

Chemosensors Development for Selective Detection of Biologically Relevant Small Molecules and Biomolecules

Avijit Kumar Das and Nathan D. McClenaghan

Abstract The detection of different biologically relevant species like cations, anions, reactive species and biomolecules is an important challenge. Various chemosensors have been developed for the selective detection of cations like metal ions which play a vital role in our daily physiological life and some toxic metal ions, which cause serious health and environmental problems. Simultaneously, the detection of anions and reactive species is crucial as they also play a vital role in a range of physiological and pathological processes. On the other hand, the study of interactions of nucleic acids with various ligands is an active research area as the binding ligands represent paradigms for drug design in anticancer therapy. In this chapter, various fluorescent chemosensors are reported for the detection of biologically and/or environmentally important species as well as the basic principles involved in the design of the chemosensors for specific analytes associated with problems as well as challenges.

Keywords Chemosensors · Cations · Anions · Biomolecules · Fluorescence · Cell imaging

1 Introduction

Fluorescent chemosensors are typically designed with the luminescent structures (hosts) integrating binding sites for selective detection of cations, anions and biomolecules (guest). The chemosensor has been developed by the reversible and noncovalent interaction between host and guest [1], on the contrary the binding interaction of host and guest in the chemodosimeter is based on irreversible chemical reaction [2] (Fig. 1). The detection of various analytes by various fluorescent chemosensors is usually achieved through one or more common photophysical mechanisms, including chelation induced enhanced fluorescence (CHEF) [3], intramolecular charge transfer (ICT) [4], photoinduced electron transfer (PET) [5]

A. K. Das (✉) · N. D. McClenaghan
Institut des Sciences Moléculaires, UMR 5255 CNRS/Université Bordeaux, 351, Cours de la Libération, 33405 Talence Cedex, France
e-mail: sanjuavi.das@gmail.com

G. Dutta et al. (eds.), *Modern Techniques in Biosensors*, Studies in Systems, Decision and Control 327, https://doi.org/10.1007/978-981-15-9612-4_11

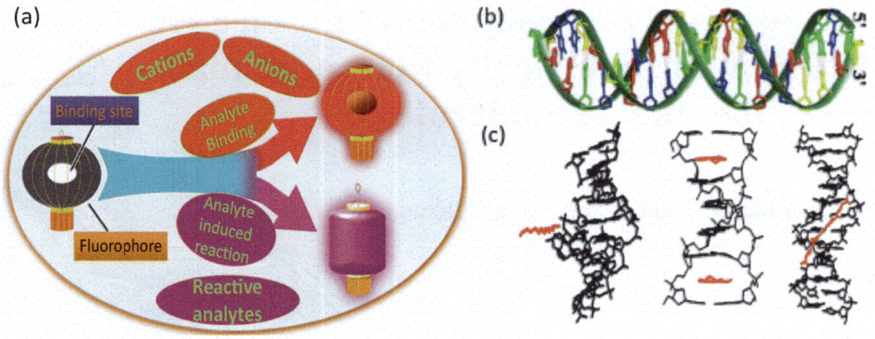

Fig. 1 **a** Schematic representation of chemosensor development for the different pathway of analytes sensing (Fig. 1a is reproduced with the permission from Ref. [1] of American Chemical Society). **b** Structure of duplex DNA. **c** Different binding mode of DNA with ligands. Left (outside-edge binding), middle (intercalation binding), right (groove binding)

and aggregation-induced emission (AIE) [6], and the number of approaches is still expanding. At very low doses, sometimes less than 1 mg/L, metal ions may have significant biological effects in animals and the environment. Therefore, detecting metal ions at low concentrations is very crucial for determining health risks and monitoring the environment. Metal ions regulate vital biological functions in living cells, and many biological reactions do not occur without their catalytic intervention.

Among metal ions, Na^{+}, K^{+}, Mg^{2+} and Ca^{2+} ions are essential neurotransmitters that interact with negatively charged phosphate groups by electrostatic interactions. On the other hand, as a consequence of metal-DNA interaction, many non-essential metal ions like Cd^{2+}, Hg^{2+}, Co^{2+}, Ni^{2+}, etc. are carcinogenic and even mutagenic [7]. F. Goppelsröder developed the first fluorescent chemosensor in 1867 for detection of Al^{3+} by forming a strongly fluorescent morin chelate [8]. On the other hand, anions and reactive species also play a crucial role in various physiological and pathological processes, and some of the environmental pollutants analytes are also a danger to public health and safety [9]. Thus, the development of various fluorescent chemosensors for selective detection of such analytes is necessary.

The study of interactions between nucleic acids with the fluorescent ligands is an active research area, and the nucleic acid binding ligands represent paradigms for drug design in anticancer therapy [10]. Thus, the detection of nucleic acids may be performed with appropriate dye molecules (e.g., annelated quinolizinium derivatives, polycyclic azoniahetarenes, chromene derivatives, heptamethinecyanine dyes, benzothiazoloquinolinium derivatives, metal-crown complexes, etc.) that bind to DNA [11]. In most cases, the binding interactions are easily detected by a change of the absorption and emission spectra of the dye upon DNA addition. And the fluorimetric detection of biomacromolecules has developed as a key method in this research area mainly because due to the high sensitivity of emission spectroscopy. In this regard, the significant enhancement of emission intensity of various fluorescent probes upon complex formation with DNA or proteins is very useful, and these probes are called "light-up probes" [12]. This significant fluorescence enhancement of various fluorescent probes upon DNA binding is often applied in analytical biochemistry, biology and also in medicine [13].

2 Fluorescent Chemosensors for Alkali and Alkaline Earth Metal Ions

Goppelsröder reported that morin forms a strongly fluorescent chelate with Al^{3+} in 1867. Various fluorescent probes have been developed for the selective detection of Al^{3+} mainly based on the Schiff base type of ligand (**1**). All these Schiff base ligands (**1a–1f**) [14–19] selectively bind Al^{3+} and result in changing the absorption and emission color. All the ligands show strong light-up effect by binding with Al^{3+}, and the light-up effect is achieved by ICT, CHEF and inhibition of "C=N" rotation. Notably, the detection limit (DL) for all the ligands with Al^{3+} in micromolar level with significant binding constant values (K_{ass}) is reported accordingly.

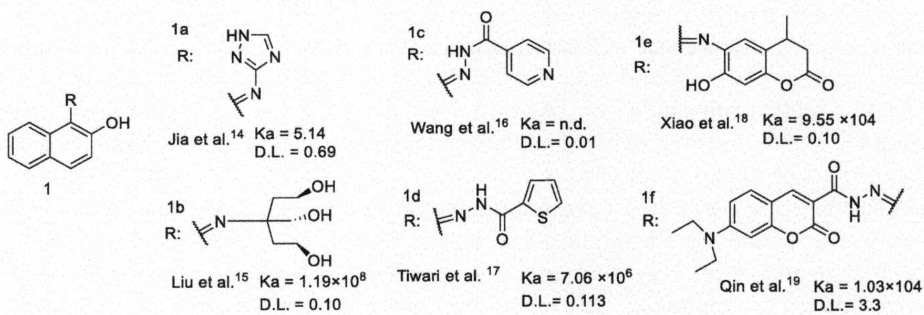

Based on N-(o-methoxyphenyl)aza-15-crown-5, He and co-workers reported a receptor **2** which shows a large fluorescence enhancement at 540 nm by an ion-binding blocking of a PET process with a desired binding constant ($K_{ass} \approx 80$ mM) [20]. For practical applications, probe **2** can be immobilized in a hydrophilic polymer layer for the analysis of sodium ion in serum as well as in blood samples.

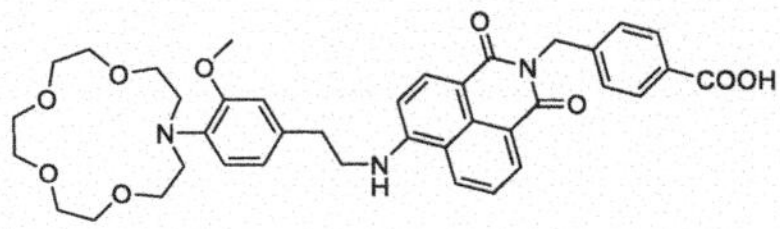

2

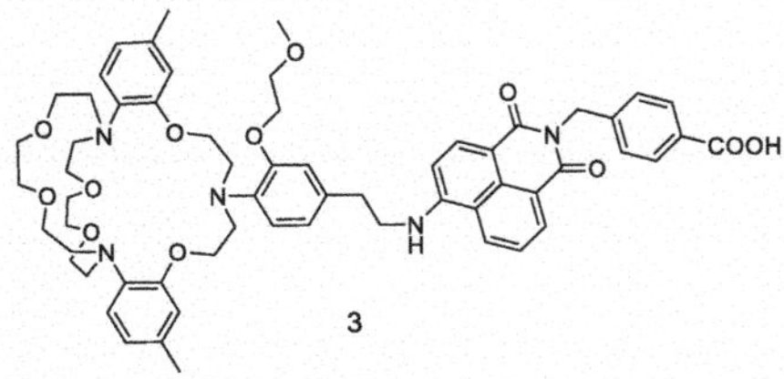

3

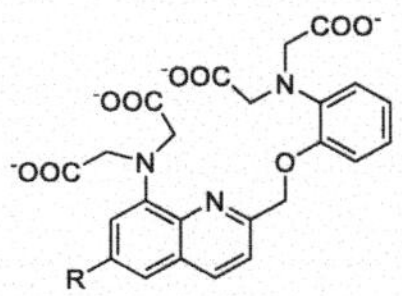

4 R = H
5 R = OCH_3

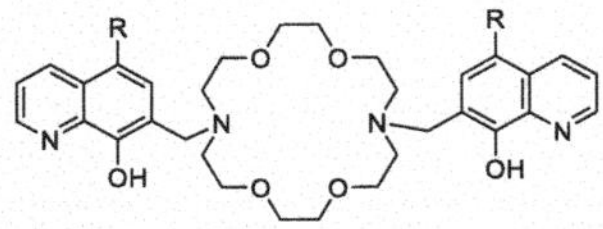

6 R = H
7 R = Cl

By using a comparatively larger crown ether, Minta and Tsien et al. reported a recognition unit based on 1,10-diaza-4,7,13,16-tetraoxacyclooctadecane and benzofuran skeleton for the detection of K^+, but the sensor was not ideal on the sensitivity issue toward K^+ [21]. He et al. developed a PET probe **3** with triazacryptand as a recognition motif to increase the selectivity and sensitivity for K^+ over Na^+ by increasing the fluorescence at 540 nm [21]. In 1980, R. Y. Tsien described two fluorescent chemosensors **4** and **5** for Ca^{2+} based on 1,2-bis(2-aminophenoxy) ethane-N,N,N′,N′-tetraacetic acid (BAPTA), which exhibit good selectivity and sensitivity toward Ca^{2+} as compared to Mg^{2+} and their in vivo application developed the understanding of biochemical processes within cells [22]. For the detection of cellular magnesium ions, two 8-hydroxyquinoline-based fluorescent chemosensors **6** and **7** have been presented by Farruggia and co-workers for the selective detection of Mg^{2+}, and both ligands show very weak emission in absence of Mg^{2+} due to intermolecular photoinduced proton transfer (PPT) and PET processes [22]. But, on Mg^{2+} binding with **6** and **7**, the PPT and PET process are blocked generating remarkable fluorescence enhancement with significant binding constants (K_d) of 44 and 73 μM, respectively.

Beside the development of various fluorescent sensors for alkali and alkaline earth metals, various fluorescent chemosensors have been developed for transition metal ions. Among the transition metal ions, copper (Cu^{2+}) is the third most abundant transition metal in the human body. A rhodamine-B-hydrazine-based derivative **8** was developed in 1997 as the pioneered work for the development of a fluorescent sensor for selective detection of Cu^{2+} by the opening of the spirolactam ring followed by the copper catalyzed hydrolysis to generate the fluorescent rhodamine-B [23]. Simultaneously, various chemosensors have been developed for the selective detection of Cu^{2+} [24]. As the second most abundant d-block metal in the human body, zinc (Zn^{2+}) acts as the pools of mobile ions in specific tissues, and various zinc selective moieties like polyamines, di-2-picolylamine (DPA), iminodiacetic acid, quinoline, bipyridine and Schiff bases have been utilized as Zn^{2+}-binding motifs [25].

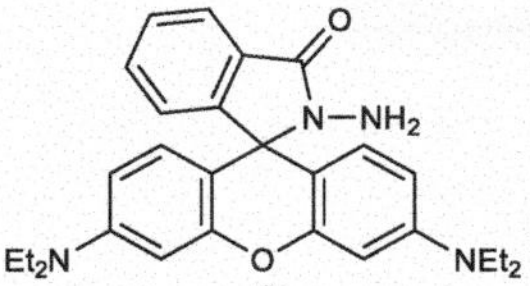

8

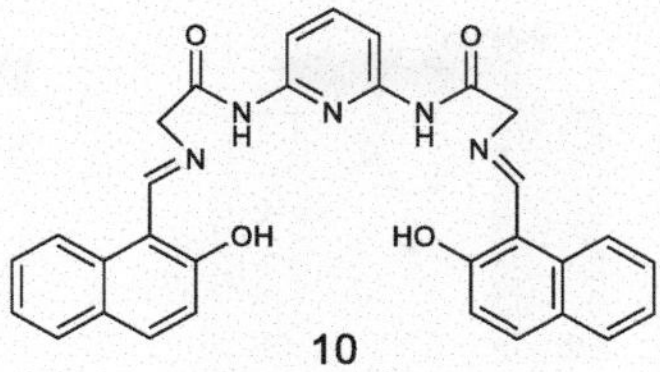

10

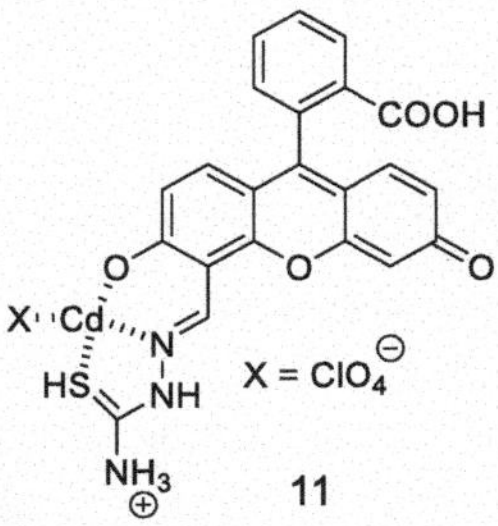

11

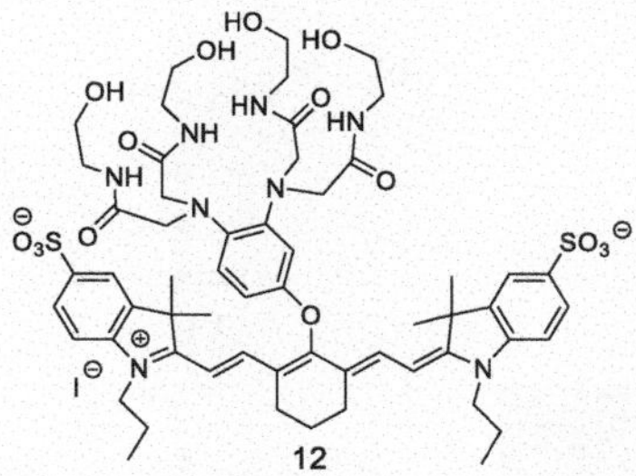

12

Das et al. developed a ratiometric water-soluble fluorescent zinc sensor **9** of carboxamidoquinoline about a 15-fold increase in fluorescence quantum yield with 100 nm red-shift of the increasing fluorescence band on binding with Zn^{2+} [26] (Fig. 2a). This fluorescence also has been detected inside a human lung cancer cell line (A 549). A condensed hydroxynaphthyl pyridine Schiff base type ligand **10** was synthesized by Das and co-workers for selective detection of Zn^{2+} by increasing the fluorescence by tenfold ($K_{ass} = 1 \times 10^4\ M^{-1}$) which was further quenched by addition of pyrophosphate (PPi) [27]. However, most Zn^{2+} selective ligands can bind with Cd^{2+} as these two metals exist in the same group having similar coordination properties [28]. Liu and co-workers reported a water-soluble fluorescein-based fluorescent probe **11** for selective detection of Cd^{2+} over other metal ions by the metal-induced complexation having elevated binding constant ($K_d = 0.123\ \mu M$) through the inhibition of C=N isomerization [29]. Polyamide-based ligands have also been utilized for the selective detection of Cd^{2+}, and Yang et al. demonstrated a cyanine-based polyamide ligand **12** which can selectively bind ($K_{ass} = 6.0 \times 10^{-10}\ M^2$) Cd^{2+} through the enhancement of fluorescence by inhibition of PET [30].

Mercury (Hg) is considered as one of the most dangerous metals, which are produced from many sources like gold production, coal plants, thermometers, barometers and mercury lamps. By utilizing thiophilic character of Hg^{2+}, Czarnik and co-workers developed PET sensor **13** which shows fluorescence by a Hg^{2+}-induced desulfurisation reaction [31] (Fig. 2b). In 2013, a novel colorimetric and fluorescent chemosensor based on 7-(diethylamino)-3-(pyrimidin-4-yl)-2H-chromen-2-one (**14**) was synthesized by Das et al. for the selective detection of Hg^{2+} forming 1:1 complex ($K_{ass} = 4.0 \times 10^4\ M^{-1}$) displaying notable absorption color change from green to red ($\Delta\lambda = 80$ nm) and the fluorescence changes from yellowish green to light orange [32] (Fig. 2c). Iron (Fe^{3+}) is another transition metal ion whose detection in the physiological environment is very important due to its prominent role in human and animal health. A rhodamine-B-based chemosensor **15** has been reported, comprising an

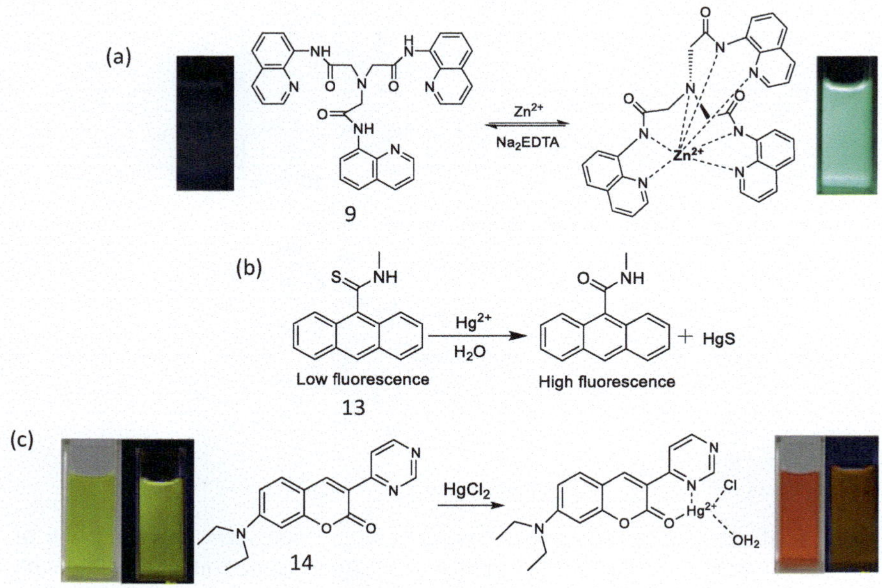

Fig. 2 **a** Probable binding mode and fluorescence changes of ligand **9** on binding with Zn^{2+} (Fig. 2a is reproduced with the permission from Ref. [26] of Royal Society of Chemistry). **b** Probable chemodosimetric approach of ligand **13** with Hg^{2+} (Fig. 2b is reproduced with the permission from Ref. [31] of American Chemical Society). **c** Probable binding mode and absorption and fluorescence changes of ligand **14** with Hg^{2+} (Fig. 2c is reproduced with the permission from Ref. [32] of Royal Society of Chemistry)

electron-withdrawing carbonyl group adjacent to the thiourea moiety. The decrease of electron density on the thiourea sulfur atom lowered the affinity of **15** to other metal ions, leading to a higher binding affinity toward Fe^{3+} [33]. An Fe^{3+} induced spiro-lactam ring-opening process is the mechanism for the metal ion selectivity. Detection of another trivalent transition metal, namely Cr^{3+}, is crucial as its deficiency in the human body leads to a variety of diseases, including diabetes and cardiovascular disease [34]. In 2008, Zhou et al. [35] and Goswami et al. [36] in 2014 reported two chemosensors **16** (rhodamine-based) and **17** (spirobenzopyran-based) respectively for selective detection of Cr^{3+} based on metal-induced ring-opening mechanism. Significantly Cr^{3+} induced spirolactam ring-opening of **16** resulted in the appearance of emission at 594 nm which corresponds to the rhodamine moiety even upon excitation of the 1,8-naphthalimide chromophore as a result on an intramolecular FRET process. But in case of **17**, Cr^{3+} binding with spirobenzopyran ring results to the opening of benzopyran ring leading to the formation of Cr^{3+} bounded merocyanine form showing a naked eye yellow-visible color and red fluorescence at 675 nm. These two chemosensors **16** and **17** could also be used to monitor intracellular Cr^{3+} corresponding to an in vitro fluorescence response (Fig. 3).

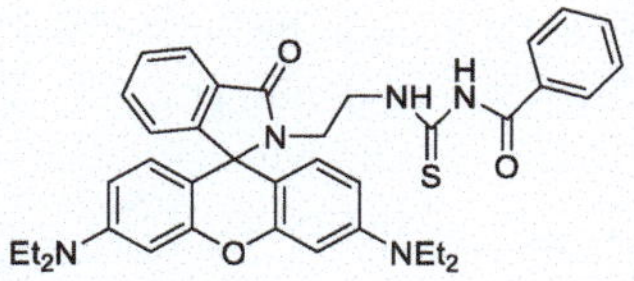

15

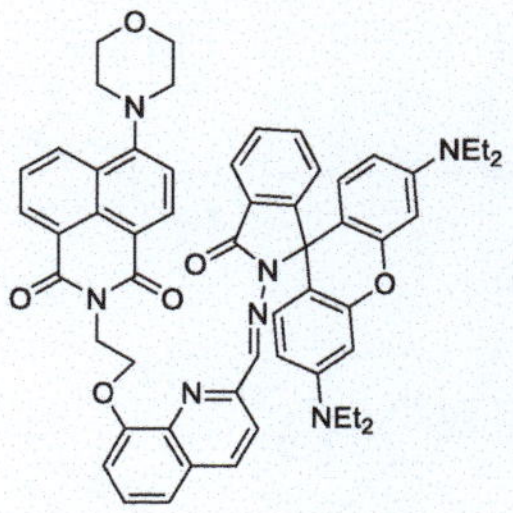

16

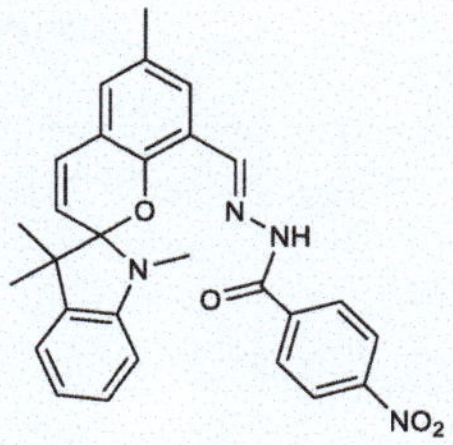

17

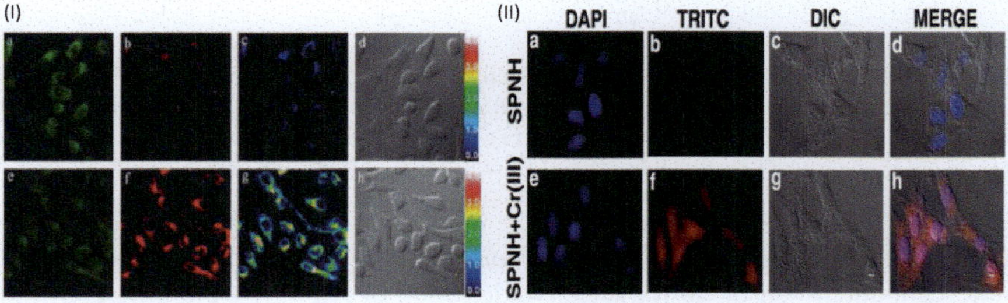

Fig. 3 **I** FRET images of **16** for Cr^{3+} in HeLa cells. Top, **a**–**d** Cells incubated with 5 μM 16 for 30 min. Bottom, **e**–**f** Cells incubated with both 50 mM Cr^{3+} and 5 μM 16 for 30 min. Emission was collected by the green channel at 530 ± 20 nm (**a**, **e**) and the red channel at 610 ± 40 nm (**b**, **f**). **c**, **g** FRET images with ratio function with red and green channel (λ_{ex} = 405 nm) (Fig. 3I is reproduced with the permission from Ref. [35] of Royal Society of Chemistry). **II** HeLa cells showed intense red fluorescence in the presence of both 50 μM each of the receptor **17** and Cr^{3+} (**f**) and did not show any fluorescence in the absence (**b**) of 50 μM of Cr^{3+}. Corresponding DAPI images (**e**, **a**), differential interference contrast (DIC) images (**g**, **c**) and merge images (**h**, **d**) of the cells were shown (Fig. 3II is reproduced with the permission from Ref. [36] of Royal Society of Chemistry)

Pb^{2+} is a highly neurotoxic pollutant metal ion, whose presence leads to behavioral disturbances, learning disability, hearing loss and reduced cognitive functioning in humans [37]. To aid its detection, Yoon and co-workers developed a dipicolylamine coupled rhodamine derivative **18** as a fluorescent chemodosimeter for selective detection of Pb^{2+} [38]. Pb^{2+} induced spirolactam ring-opening of the rhodamine analogue resulted in a significant fluorescence red-shift approximately ~30 nm and a 100-fold of fluorescence enrichment providing an association constant of 1.95×10^5 M^{-1}. The increasing industrial use of toxic silver (Ag^+) presents a potential risk to human health associated with organ failure and reduction in mitochondrial function through the promotion of oxidative stress [39]. Hence to develop the fluorescent Ag^+ chemodosimeter, Ahn et al. devised a rhodamine-B derivative **19** which selectively detects Ag^+ based on rhodamine-based oxazoline **19′** formation upon binding of Ag^+ to the iodide, which induces spirolactam ring-opening of the rhodamine derivative [40] (Fig. 4a). While gold is considered a stable and biocompatible metal in many instances, its reactive ionic forms are very potentially toxic to humans because they can strongly interact with DNA and the nervous systems [41]. So, it is beneficial

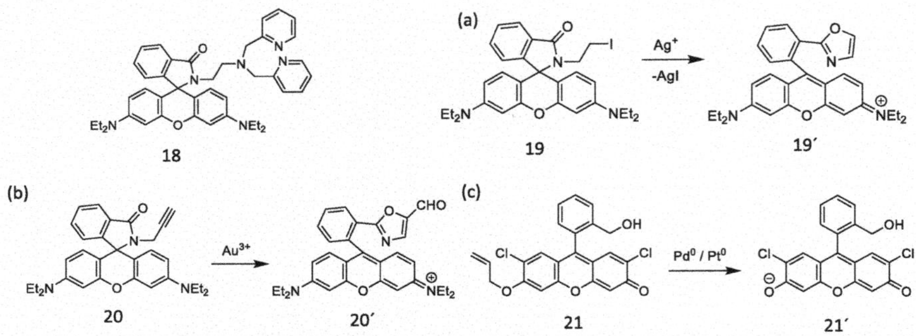

Fig. 4 Chemodosimetric approaches for the selective detection of **a** Ag^+ with ligand **19** (Fig. 4a is reproduced with the permission from Ref. [40] of American Chemical Society). **b** Au^{3+} and **c** Pd^0/Pt^0 by the ligands **20** and **21**, respectively (Fig. 4b, c are reproduced with the permission from Refs. [42, 43] respectively of Royal Society of Chemistry)

to find gold ions in biological systems. Yoon et al. first reported a rhodamine-based fluorescent chemosensor **20** for selective detection of Au^{3+} by exhibiting 250-fold fluorescence enhancement via Au^{3+}-induced spirolactam ring-opening through the formation of oxazolecarbaldehyde **20′** in the rhodamine backbone by intramolecular cyclization mechanism [42] (Fig. 4b).

Among the precious metal ions, palladium and platinum are commonly used for dental crowns, catalytic converters, fuel cells and jewelry, but in light of these advantages features, certain forms of these metals are responsible for various health hazards. To facilitate their detection, Garner et al. developed a fluorescein-based chemodosimeter **21** for the detection of Pd/Pt based on Pd/Pt-catalyzed deallylation reaction [43]. The cleavage of the allylic ether of nonfluorescent ligand **21** by Pd^0/Pt^0 produces fluorescent **21′** (Fig. 4c). Generally, a π-complex was formed by the insertion of M^0L_{4-n} ($L = PR_3$ or solvent) generated from the reduction of stable metal species $M^{II/IV}$ ($M = Pd^{II}/Pt^{II/IV}$) by phosphine into the C–O bond of the nonfluorescent allylic ether **21**. Finally, π-complex reacts with the nucleophile forming of the fluorescent product **21′**.

3 Fluorescent Chemosensors for Anions

Due to the strong hydration of anions, the development of anion selective chemosensors lagged in comparison with the progress in cation chemosensor development. Nevertheless, due to the important roles of anions in biological and industrial processes, the challenging field of anion sensing continues to grow. Among the different anions, fluoride ion has a vital role in the broad range of biological, chemical, medical treatment processes, mostly in dental care, osteoporosis treatment, water supply treatment and chemical warfare agents [44]. Therefore, chemosensor development for selective detection of fluoride is important and is often based on interactions with various fluorophores containing thioureas, ureas, amides, pyrrole, imidazole and imidazolium moieties, via an intramolecular hydrogen transfer process or interaction between active N–H group and F^- [45]. However, this type of interaction is most effective in aprotic/organic solvents, and the presence of water inhibits the sensing pathway, making this type of sensor not applicable for bioimaging. To allow fluoride detection inside the living system, chemodosimeter type of sensors is preferable. In this way, fluoride-triggered Si–O and Si–C bond cleavage pathways are very well known for the detection of fluoride in aqueous media.

Resorufin-based chemodosimeter **22** was first reported by Kim and co-workers based on F^- induced Si–O bond cleavage in the presence of H_2O releasing of the fluorescent resorufin [46]. Similarly, Das et al. also described a ratiometric fluorescent turn-on probe (decreasing emission at 407 nm and concomitant increase at 477 nm) **23** for fluoride ion, based on the mechanism of the excited-state intramolecular proton transfer (ESIPT) pathway by chemodosimetric fluoride prompted desilylation reaction [47]. Furthermore, the fluoride detection of **23** in aqueous media has been utilized for the fluorescence imaging of living cells (Fig. 5). On the other hand, a similar type of phenomenon has been achieved by fluoride-triggered cleavage reaction of the Si–C bond in –C≡C–$SiMe_3$, which is easier as compared to cleavage of Si–O bond. Fu et al. developed a chemosensor **24** which shows blue-shifted fluorescence from the decrease of emission maximum at 571 nm to an increase of the intense emission peak at 554 nm by the cleavage of a Si–C bond [48].

Acetate is another significant anion that plays a prominent role in pharmaceutical science, biochemistry as well as in the environment having a significant contribution in many metabolic processes, and acetate has been utilized as an indicator of organic decomposition of marine sediments [49].

Based on a hydrogen bonding interaction, Das et al. reported two ligands (**25** and **26**) which can bind selectively with acetate by hydrogen bonding interactions.

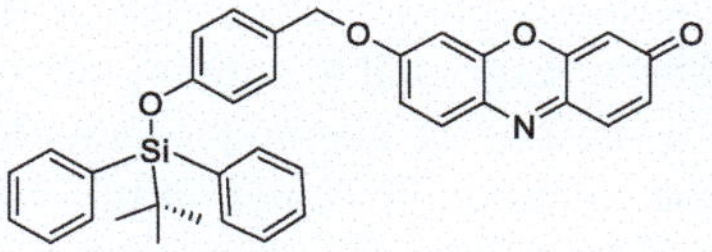

22

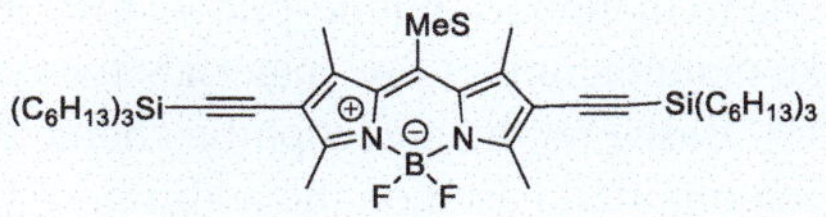

24

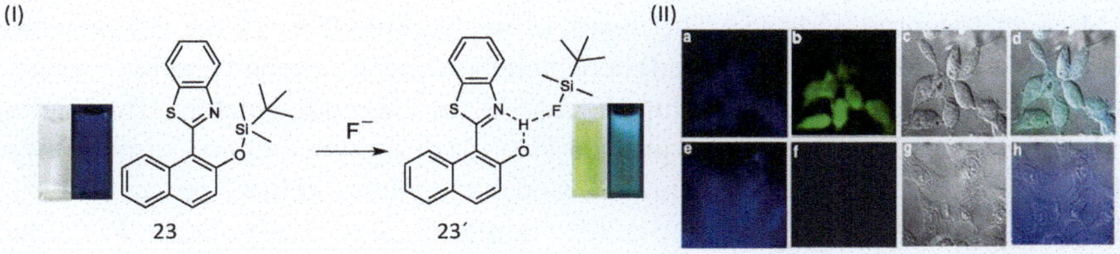

Fig. 5 **I** Possible fluoride-induced desilylation mechanism in aqueous phase. **II** Fluorescence microscope images in HeLa cells incubated with 50 μM of **23** in presence (**a, b**) and in absence (**e, f**) of fluoride ion. Corresponding bright field images (**c, g**) and merge images (**d, h**) of the cells were reported (Fig. 5I, II are reproduced with the permission from Ref. [47] of Elsevier)

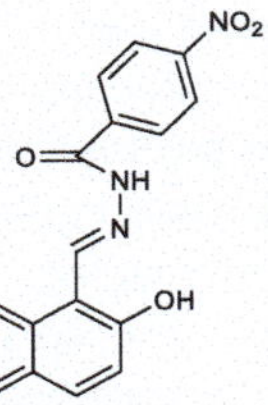

25

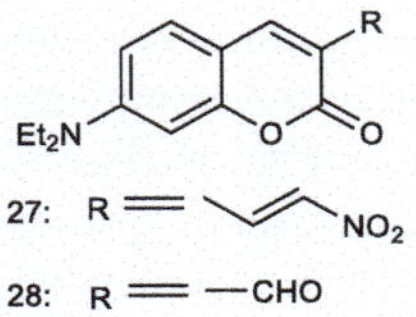

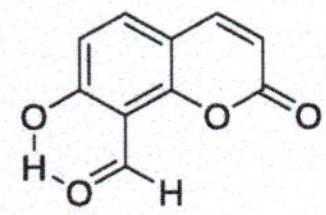

29

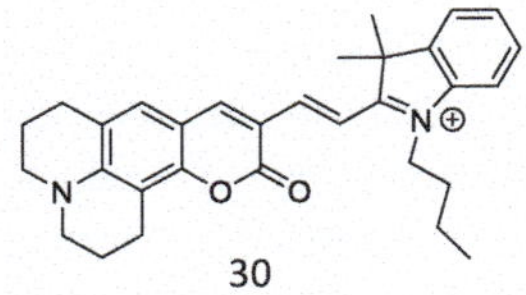

30

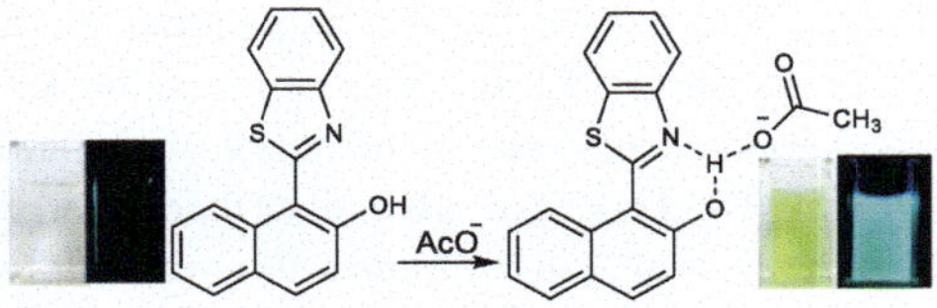

26 26′

Fig. 6 Probable binding of mode of **26** in solution phase with AcO$^-$ (Fig. 6 is reproduced with the permission from Ref. [51] of Elsevier)

Ligand **25** selectively binds with acetate with the detection limit of 5.83 μM with an accompanying absorption color change from colorless to pink [50]. On the contrary, receptor **26** shows acetate selectivity with the detection limit of 4.4 μM by a changing ratiometric absorption (colorless to yellowish green color) as well as emission spectra (colorless to greenish fluorescence) by employing ESIPT through the formation of hydrogen bonding complex **26′** [51] (Fig. 6). In both cases, a strong hydrogen bonding interaction between these two ligands and AcO$^-$ leads to the deprotonation of the naphthyl hydroxyl group which could enhance electron delocalization reducing the energy of the π–π* transition and consequently spectral changes [52].

The development of an efficient chemodosimeter for cyanide (CN$^-$) is significant in environmental and biological samples as this ion is considered as extremely toxic to mammals, but despite its toxicity, CN$^-$ is widely used in different methods like manufacture of raw materials for synthetic fibers, resins and the mining of gold [53]. Based on the nucleophilic character of CN$^-$, most of the chemodosimeter has been reported based on the addition of CN$^-$ to the electrophilic center. In this way, Kim and co-workers reported two chemosensors **27** and **28** bearing α,β-unsaturated moiety connected to electron-withdrawing groups nitro and aldehyde respectively as the Michael acceptors [54, 55]. The attack of nucleophile CN$^-$ to the β/δ-position of the Michael acceptor ('C=C' bond) activated by nearest nitro and aldehyde group leads to the change of significant fluorescence of the chemodosimeters. The nucleophilic character of CN$^-$ has been utilized also for designing chemodosimeter like **29** based on intramolecular H-bonding-assisted nucleophilic addition processes [56]. The activation of the aldehyde group of the coumarin-based chemosensor **29** by the cyanide attack leads to the proton transfer by the intramolecular hydrogen transfer from the nearest ortho hydroxy group to the cyanide attacked aldehyde (cyanohydrin formation) showing high selectivity with the significant fluorescence changes. Another chemodosimetric approach of cyanide based on its nucleophilic character is the nucleophilic addition of CN^- with indolium group. An indolium-coupled coumarin **30** was used as a chemodosimeter for selective detection of CN^- based on operating ICT from donor amine group of coumarin moiety to the acceptor positively charged indolium group through a π-conjugated spacer. However, by the nucleophilic addition of CN^- onto the $-C{=}N^{\oplus}-$ functionality of **30** interrupts the strong ICT via breaking of the indolium conjugation displaying a distinct fluorescence change.

Similar type of sensing method has been employed for another anion HS^- which remains at equilibrium with the weak acid H_2S at physiological pH window and H_2S is generally well known for its traditional characteristic of rotten egg smell, highly diffusible, reactive and poisonous [57]. Furthermore, H_2S is considered for controlling of various physiological responses, and the changes of the levels of cellular H_2S lead to the cause of many diseases like Alzheimer's diseases, Down's syndrome to diabetes and also the liver cirrhosis [58]. Therefore, the detection of H_2S is very highly important. Das et al. developed two chemosensors **31** and **32** which show H_2S/HS^- selectivity via "addition-NGP (neighbouring group participation)-elimination" and addition-elimination techniques, respectively. The presence of aldehyde group adjacent to naphthyl hydroxyl group for **31** makes faster H_2S/HS^- sensing by NGP of thiol through the formation of probable six-membered transition state that undergoes cleavage of the dinitrophenyl ether moiety, but for **32**, the cleavage of dinitrophenyl ether moiety is comparable slower by direct addition of H_2S/HS^- forming a Meisenheimer complex [59] (Fig. 7a). The rate of the reactions reflects also to the

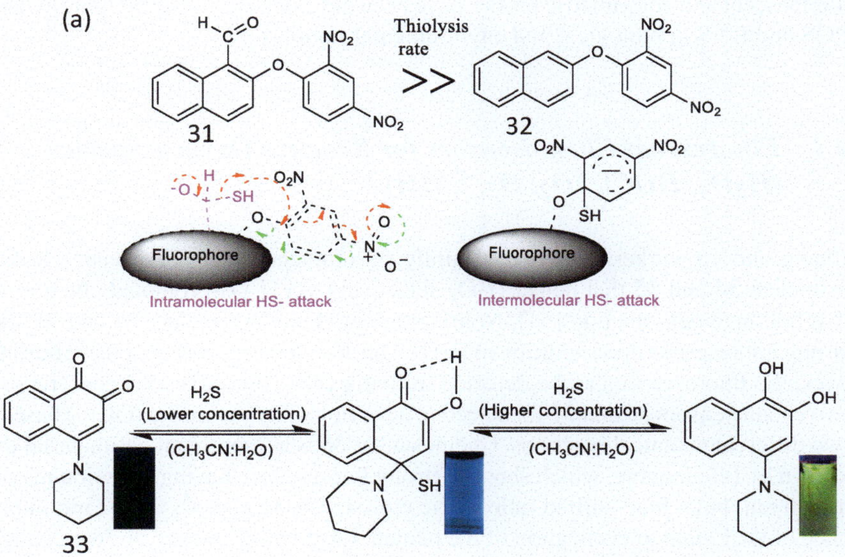

Fig. 7 Probable chemodesometric approach of H_2S **a** cleavage of dinitrophenylether linkage for the ligands **31** and **32** (Fig. 7a is reproduced with the permission from Ref. [59] of Royal Society of Chemistry). **b** Michael addition followed by reduction for the ligand **33** (Fig. 7b is reproduced with the permission from Ref. [60] of Royal Society of Chemistry)

fluorescence response of the two chemosensors **31** and **32** where the fluorescence light-up effect by addition of H_2S/HS^- for **31** is higher (14-fold) as compared to **32** (twofold). Das et al. have also developed a single chemosensor **33** for selective detection of H_2S/HS^- where H_2S/HS^- shows dual role-playing nucleophilic attack via Michael addition (lower concentration, blue fluorescence) followed by the reduction of 1,2-diketo functionality (higher concentration, green fluorescence) [60] (Fig. 7b). Two different fluorescence responses of **33** at various concentrations of H_2S/HS^- makes it a chemosensor with sensing a single analyte in relay pathway via a fluorescence "off–on-on" mechanism, and the fluorescence also can be detected inside the cell.

4 Fluorescent Chemosensors for Reactive Species

As described above the host–guest interaction allows for the detection of cations and anions. Equally, a variety of fluorescent chemosensors have been developed also to detect reactive oxygen (ROS) and nitrogen species (RNS). During vital processes in living organisms, various ROSs derive from molecular oxygen [61], and on the other hand, RNS modulates cardiovascular, nervous and immune systems [62]. Due to this pathophysiological significance, ROS and RNS have been the subject of extensive investigations aimed at determining their biological roles, and therefore, various luminescent and colorimetric probes have been developed as powerful tools to detect ROS and RNS in biological and environmental samples [63].

4.1 Fluorescent Chemosensors for Reactive Oxygen Species (H_2O_2, HOCl, $^1O_2$, $O_2^{\cdot-}$, $^{\cdot}OH$)

Chang and co-workers reported a family of xanthene-based diboronate chemodosimeters **34** and **35** that detect H_2O_2 selectively via H_2O_2-promoted chemoselective boronate deprotection [64]. Nonfluorescent chemodosimeters **34** and **35** show strong fluorescence upon addition of H_2O_2. Due to cell permeability of these boronate dyes, the fluorescence can be detected in living cells (Fig. 8II). Das et al. reported two chemosensors **36** and **37** for selective detection of hypochlorite (OCl^-) based on two different chemical reactions. Chemosensor **36** is a fluorescent antimalarial drug known as Tafenoquine, which shows a strong fluorescence having emission maxima at 476 nm, but a blue-shifted ratiometric emission band appears at 361 nm in presence of OCl^- due to cyclization of the pentan-1,4-diamine moiety of **36** creating an azine type compound **36** (Fig. 8I) [65].

On the other hand, chemosensor **37** is a rhodamine-based chiral hydrazide derivative which detects OCl^- in aqueous medium at nanomolar level [66]. Hypochlorite triggers the oxidation of the sensor **37** followed by hydrolysis followed by spirolactam

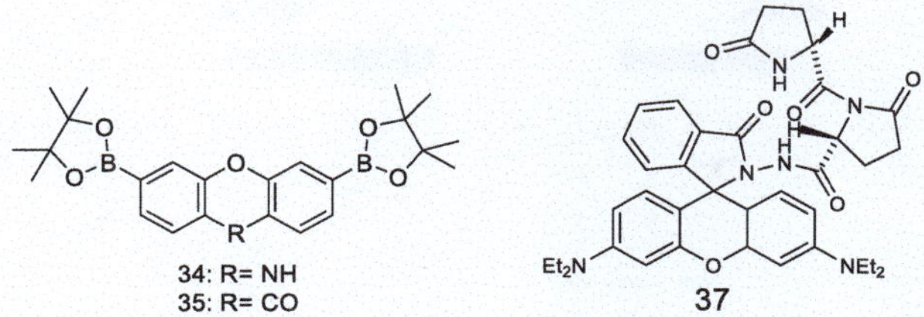

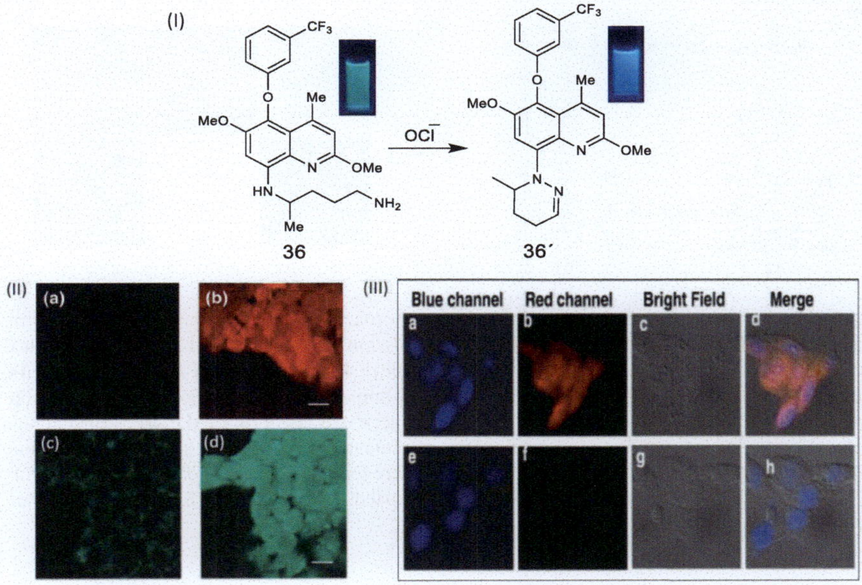

Fig. 8 **I** Proposed mechanism for OCl^--induced fluorescence response of **36**. (Fig. 8I is reproduced with the permission from Ref. [65] of Royal Society of Chemistry). **II** Fluorescence image of HEK cells incubated with 10 μM of **34** (**a**) and **35** (**c**) for 10 min at 25 °C. Fluorescence image of **34** (**b**) and **35** (**d**)-stained HEK cells treated with 100 μM H_2O_2 for 30 min at 25 °C. Excitation and two-photon excitation were provided at 543 nm and 704 nm for **34** and **35**, respectively, and emission was collected in a window from 548 to 644 nm. Scale bar = 18 μm (Fig. 8II is reproduced with the permission from Ref. [64] of American Chemical Society). **III** Fluorescence images of HeLa cells incubated with 50 μM of the **37** in presence (**a**, **b**) and in absence (**e**, **f**) of 50 μM of OCl^-. Corresponding bright field images (**c**, **g**) and merged images (**d**, **h**) of the cells (Fig. 8III is reproduced with the permission from Ref. [66] of American Chemical Society)

ring-opening which leads to releases of the rhod-B generating strong reddish-yellow fluorescence and pink-colored absorption. The chemosensor **37** has been utilized for bio-activity and also for detection of OCl^- in aqueous media (millipore deionized water, distilled water, pond water, canal water, rainwater and tap water). For selective detection of $^1O_2$, Tanaka et al. reported fluorescein derivatives containing anthracene **38** and **39**, which are almost nonfluorescent due to the favorable PET process [67]. However, upon reaction with $^1O_2$, highly fluorescence endoperoxide derivatives of

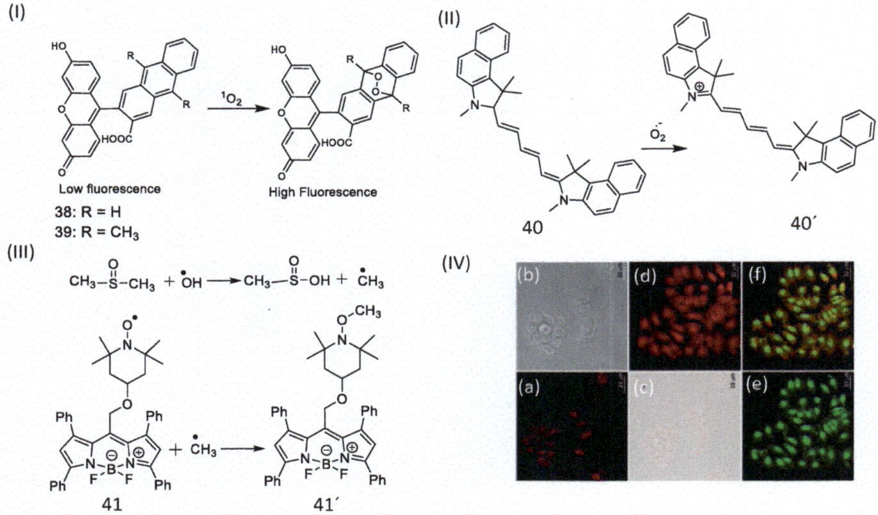

Fig. 9 Proposed mechanism for chemodosimetric approach for selective detection of **I** $^1O_2$ by ligands **38** and **39** (Fig. 9I is reproduced with the permission from Ref. [67] of American Chemical Society). **II** $O_2^{\cdot -}$ by ligand 40 (adapted from Ref. [68]). **III** ·OH (adapted from Ref. [69]). **IV** Imaging of normal human liver (HL-7702) cells and human hepatoma (HepG2) cells. **a** Probe-loaded HL-7702 fluorescence image; **b** brightfield image; **c** overlay of (**a**, **b**). **d** Fluorescence image of HepG2 cells loaded with the probe; **e** fluorescence image of HepG2 cells loaded with Acridine orange; **f** overlay of (**d**, **e**). HL-7702 and HepG2 cells were separately incubated in 1.0 mm TEMPO–BDP and 0.10 m DMSO for 15 min and then were rinsed 3 times with PBS (adapted from Ref. [69])

38 and **39** are formed (Fig. 9I). Another oxidative reactive species $O_2^{\cdot -}$ is very effective for the oxidation of hydrocyanine-like **40** [68]. Normally, the hydrocyanine-like **40** is weakly fluorescent because of its disrupted π-conjugation, but $O_2^{\cdot -}$ triggers the oxidation of the cyanine dyes forming **40′** by regenerating an extended π-conjugation, accompanied by significant enhancement of the fluorescence intensity (Fig. 9II).

Another important reactive oxygen species the hydroxyl radical (·OH) is formed non-enzymatically from H_2O_2 via a metal-based reaction. Its high reactivity triggers cellular disorders and cytotoxic effects which damage DNA, proteins and lipids. Detection of this radical is therefore very challenging, but seldom probes have been developed for its detection so far. Normally, the fluorophore which is covalently linked to a nitroxide moiety has been utilized as chemodosimeter for ·OH and this type of chemosensor exhibits lower fluorescence due to an electron exchange mechanism. Li et al. reported chemosensor **41** containing a nitroxide moiety [69]. Normally, a quantitative reaction between ·OH and DMSO forms $\cdot CH_3$, which reduces the nitroxide moiety of **41** to its corresponding diamagnetic product **41′**, and this results in the elimination of the intramolecular quenching pathway a significant fluorescence enhancement (Fig. 9III). In particular, **41** can respond rapidly to ·OH with a detection limit of 18 pM, which has been also used for selective detection of intracellular ·OH in living mice macrophages, normal human liver cells and human hepatoma cells, without causing cellular damage (Fig. 9IV).

4.2 *Fluorescent Chemosensors for Reactive Nitrogen Species (NO, HNO and ONOO$^-$)*

Nagano et al. developed two NIR tricarbocyanine probes **42** and **43** as chemodosimeters for selective detection of NO. Typically, the probes attached with a phenyldiamine are weakly fluorescent due to the PET process operating from *o*-phenylenediamine to the excited cyanine fluorophore [70] (Fig. 10a). However, during detection of NO with these fluorophores, phenyldiamine reacts with NO forming a less electron donating group triazole moieties (**42′** and **43′**) inhibiting a PET process which results in the increase of the fluorescence in the near IR region. Furthermore, the fluorescence of probes **42** and **43** fluorescence has been studied in vivo via the detection of the fluorescence from outside the kidney.

Nitroxyl (HNO), the one-electron reduced and protonated form of NO under physiological conditions, displays important roles in pharmacological processes distinct from those of nitric oxide, and therefore, significant effort has been focused

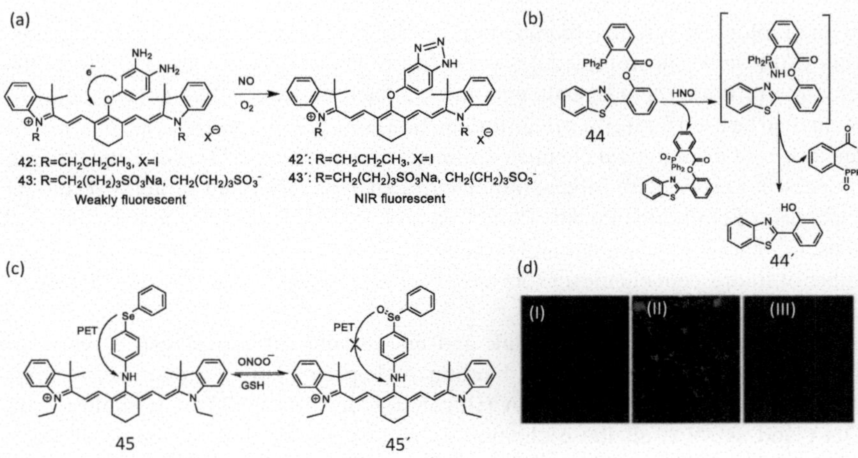

Fig. 10 Proposed mechanism for chemodosimetric approach for selective detection of **a** NO by ligand **42** and **43** (Fig. 10a is reproduced with the permission from Ref. [70] of American Chemical Society). **b** HNO by ligand **44** (Fig. 10b is reproduced with the permission from Ref. [71] of Elsevier) and **c** ONOO$^-$ by ligand **45** (Fig. 10c is reproduced with the permission from Ref. [72] of American Chemical Society). **d** Confocal fluorescence microscopy images of RAW264.7 cells treated with **I** **45** alone, (middle) LPS, IFN-γ, PMA + **45** and **III** H_2O_2 + **45** (Fig. 10d is reproduced with the permission from Ref. [72] of American Chemical Society)

on the synthesis of probes for HNO detection. Zhang et al. developed the 2-(2-hydroxyphenyl)-benzothiazole (HBT)-based fluorescent probe **44** containing a triphenylphosphine moiety for HNO sensing [71]. Initially, the probe **44** is non-fluorescent, but in the presence of HNO, the probe shows strong fluorescence at 459 nm due to the release of a highly fluorescent benzothiazole moiety **44′** (Fig. 10b). Significantly HNO promotes oxidation of a triphenylphosphine moiety in **44** leading to the formation of phosphine oxide and aza-ylide, which then undergoes Staudinger ligation to produce an amide derivate and fluorescent benzothiazole **44′**.

Peroxynitrite ($ONOO^-$), a highly reactive oxidant, can be formed in vivo by the reaction of nitric oxide with the superoxide radical. Han et al. reported an organoselenium functionalized NIR reversible fluorescent probe **45** for selective detection of peroxynitrite oxidation based on oxidation reaction under physiological conditions (Fig. 10c) [72]. The chemosensor **45** is weakly fluorescence due to quenching by PET between the selenide ether and the cyanine dye but the addition of $ONOO^-$ oxidizes selenide preventing the operated PET, which results in a strong fluorescence at 775 nm. Moreover, the probe **45** was successfully employed to real-time imaging of peroxynitrite in living RAW264.7 cells (Fig. 10d).

5 Fluorescent Chemosensors for Biomacromolecules

In living biological systems, biomacromolecules play important role, and the abnormality of these biomacromolecules often has severe consequences. Therefore, the interaction of fluorescent ligands with the biomacromolecules like nucleic acids or proteins can be used for its fluorimetric detection as emission spectroscopy has proved a versatile tool in analytical chemistry, biology and medicine [73]. Since the binding of fluorescent probe with biomacromolecules such as DNA and proteins results in the increase of emission intensity ("light-up probes"), this type of probes can be used as a useful marker in genomics and proteomics [74]. Over the past several decades, a number of fluorescent chemosensors have been developed for fluorometric detection of DNA [75] and protein [76].

As the pioneering work, Czarnik and co-workers first developed a series of anthrylpolyamines **46a–d** for the binding of heparin, poly-L-glutamate, dsDNA (double-stranded DNA) and ssDNA (single-stranded DNA) [77]. On binding with dsDNA and ssDNA, all these chemosensors display a redshifted emission band with decrease of the emission intensity but **46b** and **46c** effectively bind with polyglutamate and heparin, respectively.

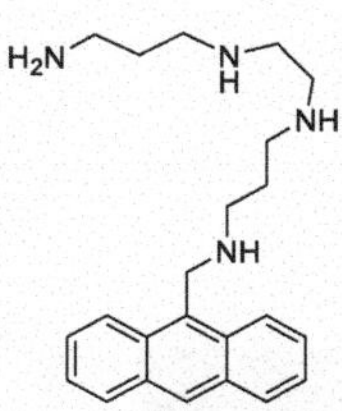
46a

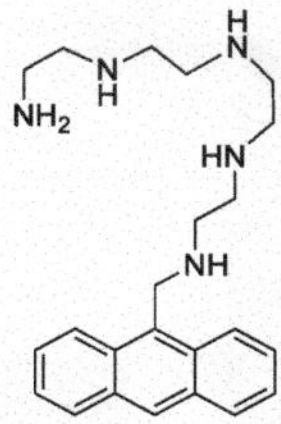
46b

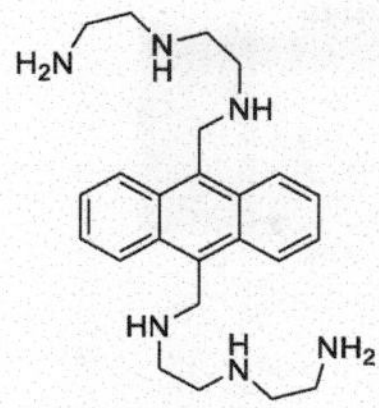
46c

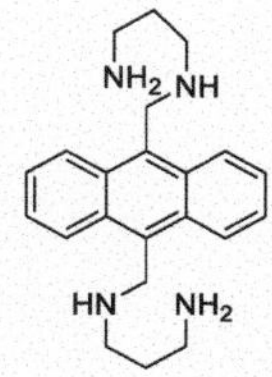
46d

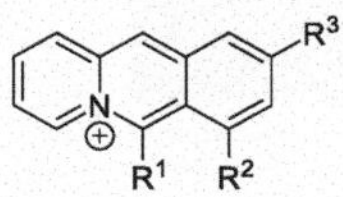

47a: $R^1 = R^2 = R^3 = H$
47b: $R^1 = NH_2$; $R^2 = R^3 = H$
47c: $R^1 = NH_2$; $R^3 = Br$; $R^2 = H$
47d: $R^2 = NH_2$; $R^1 = R^3 = H$
47e: $R^3 = NH_2$; $R^1 = R^2 = H$

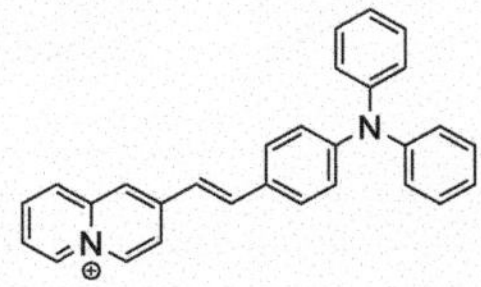
48

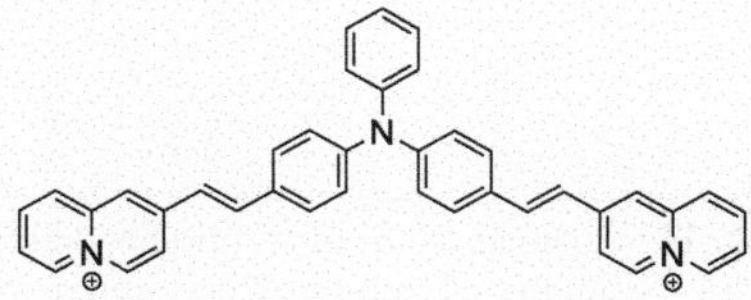
49

For the binding of DNA and proteins, Ihmels and co-workers reported different amino-substituted benzo[b]quinolizinium derivatives **47a–e** which bind with DNA via intercalation with binding constants $K_{ass} = 10^4$–10^5 M^{-1} showing significant changes in spectrophotometric titrations, DNA denaturation studies and viscometric titrations [78]. The radiationless deactivation of the excited state and the restricted conformational flexibility about the exocyclic C_{ar}–NH_2 bond of the intercalated ligands results in the increase of the emission intensity. Moreover, the derivatives **47b** and **47c** show enhancement of the emission intensity upon association with various proteins like bovine serum albumin (BSA), human serum albumin (HSA) and chicken egg white albumin (CEA). On contrary, based on the fluorescence response, the proteins have no essentially effect to the derivatives **47d** and **47e**. Another triphenyl amine-based two ligands **48** and **49** have been developed by Das et al. for binding duplex DNA (ct DNA) [79]. Due to the significant structural variation, less sterically demanding ligand **48** intercalates into duplex DNA whereas the more encumbered ligand **49** favors groove binding. Both **48** and **49** show very weak fluorescence in absence of DNA due to radiationless deactivation by conformational relaxation of the excited state, but upon association with duplex DNA, the emission intensity of the ligands increases significantly by a factor of 11 (**48**) and >100 (**49**) with emission bands between 600 and 750 nm due to the restricted conformational freedom of movement of the ligand within the binding site.

Das et al. also developed a hydroxystyryl–quinolizinium photoacid **50** for colorimetric and fluorimetric detection of duplex DNA (ct DNA) [80]. The ligand **50** shows two absorption bands at 376 and 520 nm giving pink absorption color, but on addition of ct DNA, two absorption bands decrease with the increase of the new absorption bands at 383 and 420 nm giving a yellow color (Fig. 11). Simultaneously, the emission intensity increases at 515 nm on addition of ct DNA to **50** by a factor of 200. Here, the ligand 50 binds with DNA via intercalation as well as groove binding based on ligand to DNA ratio (LDR). Indeed at lower LDR, the ligand favors groove binding and at higher LDR preferably binds via intercalation, which is proved by

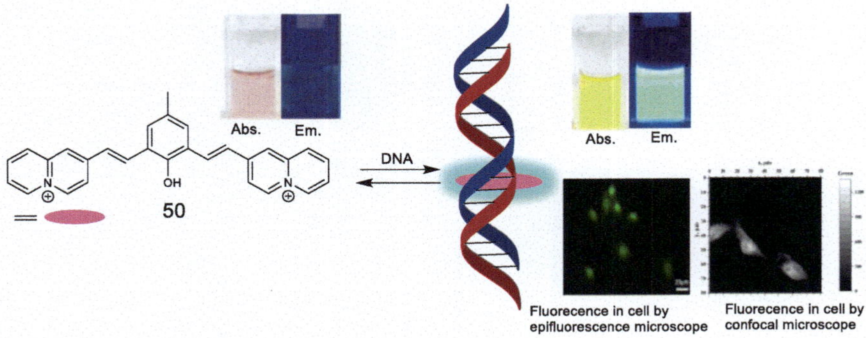

Fig. 11 Changes of the absorption and emission color of the ligand **50** on binding with duplex ct DNA. Examples of epifluorescence microscopy images of NIH 3T3 mouse fibroblasts after fixation and staining with **50** (2.5 mM in PBS) for 1 h and counterstained with Hoechst 33258. λ_{ex} = 450–490 nm, λ_{em} > 515 nm (green channel). Confocal fluorescence microscopy images of NIH 3T3 mouse fibroblasts after fixation and incubation with **50** (2.5 mM) for 1 h; λ_{ex} = 485 nm, λ_{em} = 515–652 nm (adapted from Ref. [80])

CD and LD spectra. The fluorimetric detection of DNA in cells has been studied also with epifluorescence and confocal fluorescence microscopy (Fig. 11).

6 Conclusions and Outlook

After discovery of the first fluorescent chemosensor for Al^{3+} by Goppelsröder, the field of fluorescent chemosensors has been greatly broadened over the years based on the sensing of various analytes according to their environmental and biological importance. This book chapter highlights some key representative examples of fluorescent chemosensors for various biologically important analytes including cations, anions, reactive species and biomacromolecules. Various well-known fluorophores, such as rhodamine, fluorescein, cyanine, coumarin, quinoline, xanthene, naphthalene, 1,8-naphthalimide, anthracene and BODIPY, have also been integrated into chemosensors for detection of various analytes harnessing a variety of mechanisms including PET, ICT, ESIPT and FRET. Fluorophores like anthracene, benzo[b]quinolizinium, quinolizinium and triphenylamine have shown an inherent propensity for binding with proteins and DNA with significant changes of absorption and emission properties. The fluorescent turn-on behavior or emission wavelength shifts of the various chemosensors on binding with analytes and biomacromolecules makes them ideal for application in bioimaging. The main aim of this book chapter is to provide an overview of how various chemosensors have been utilized for selective sensing of various analytes as well as for future aspects to develop a general design leading to more chemosensors based on the reported idea. Chemosensor research will undoubtedly continue to expand and develop for new applications or approaches by integrating existing fluorophores into chemosensors and harnessing appropriate fluorescence switching mechanisms. Indeed the continued elucidation of biologically relevant analytes, many of which are transient in nature, offers significant challenges for detection, particularly in situ.

References

1. (a) Desvergne, J.P., Czarnik, A.W.: Chemosensors for Ion and Molecule Recognition. Kluwer Academic Publishers, London (1997). (b) Suksai, C., Tuntulani, T.: Chromogenic anion sensors. Chem. Soc. Rev. **32**, 192–202 (2003). (c) Kim, H.N., Guo, Z.Q., Zhu, W.H., Yoon, J., Tian, H.: Recent progress on polymer-based fluorescent and colorimetric chemosensors. Chem. Soc. Rev. **40**, 79–93 (2011). (d) Yang, Y., Zhao, Q., Feng, W., Li, F.: Luminescent chemodosimeters for bioimaging. Chem. Rev. **113**, 192−270 (2013)
2. (a) Martínez-Máñez, R., Sancenón, F.: Fluorogenic and chromogenic chemosensors and reagents for anions. Chem. Rev. **103**, 4419–4476 (2003). (b) Cho, D.G., Sessler, J.L.: Modern reaction-based indicator systems. Chem. Soc. Rev. **38**, 1647–1662 (2009). (c) Duong, T.Q., Kim, J.S.: Fluoro- and chromogenic chemodosimeters for heavy metal ion detection in solution and biospecimens. Chem. Rev. **110**, 6280–6301 (2010)
3. Thompson, R.B., Bozym, R.A., Cramer, M.L., Stoddard, A.K., Westerberg, N.M., Fierke, C.A.: Fluoresc. Sens. Biosens. **107** (2006)
4. De Silva, A. P., Gunaratne, H. Q. N., Gunnlaugsson, T., Huxley, A. J. M., McCoy, C. P., Rademacher, J. T., Rice, T. E.: Signaling Recognition Events with Fluorescent Sensors and Switches, Chem. Rev. **97**, 1515–1566 (1997).
5. Daly, B., Ling, J., and de Silva, A. P.: Current developments in fluorescent PET (photoinduced electron transfer) sensors and switches. Chem. Soc. Rev. **44**, 4203–4211 (2015).
6. Kwok, R. T. K., Leung, C. W. T., Lam, J. W. Y., Tang, B. Z.: Biosensing by luminogens with aggregation-induced emission characteristics. Chem. Soc. Rev. **44**, 4228–4238 (2015).
7. (a) Hare, D.J., New, E.J., de Jonge, M.D., McColl, G.: Imaging metals in biology: balancing sensitivity, selectivity and spatial resolution. Chem. Soc. Rev. **44**, 5941−5958 (2015). (b) Zhou, W., Saran, R., Liu, J.: Metal sensing by DNA. Chem. Rev. **117**, 8272–8325 (2017). (c) Yannone, S.M., Hartung, S., Menon, A.L., Adams, M.W.W., Tainer, J.A.: Metals in biology: defining metalloproteomes. Curr. Opin. Biotechnol. **23**, 89−95 (2012)
8. Czarnik, A. W.: Fluorescent Chemosensors for Ion and Molecule Recognition, American Chemical Society, Washington, DC, 1993.
9. (a) Gale, P.A., Caltagirone, C.: Anion sensing by small molecules and molecular ensembles. Chem. Soc. Rev. **44**, 4212–4227 (2015). (b) Chen, X., Wang, F., Hyun, J.Y., Wei, T., Qiang, J., Ren, X., Shin, I., Yoon, J.: Recent progress in the development of fluorescent, luminescent and colorimetric probes for detection of reactive oxygen and nitrogen species. Chem. Soc. Rev. **45**, 2976–3016 (2016)
10. Neidle, S., Waring, M., Eds. (1993) Molecular Aspects of Anticancer Drug-DNA Interactions, CRC Press, Boca Raton, FL.
11. (a) Bohne, C., Faulhaber, K., Giese, B., Häfner, A., Hofmann, A., Ihmels, H., Köhler, A.K., Pera, S., Schneider, F., Sheepwash, M.A.L.: Studies on the mechanism of the photo-induced DNA damage in the presence of acridizinium salts involvement of singlet oxygen and an unusual source for hydroxyl radicals. J. Am. Chem. Soc. **127**, 76–85 (2005). (b) Granzhan, A., Ihmels, H.: Playing around with the size and shape of quinoliziniumderivatives: versatile ligands for duplex, triplex, quadruplex and abasic site-containing DNA. Synlett **27**, 1775–1793 (2016)

12. Morozkin, E.S., PLaktionov, P., Rykova, E.Y., Vlassov, V.V.: Fluorometric quantification of RNA and DNA in solutions containing both nucleic acids. Anal. Biochem. **322**, 48 (2003)
13. Gowan, S.M., Harrison, J.R., Patterson, L., Valenti, M., Read, M.A., Neidle, S., Kelland, L.R.: A G-quadruplex-interactive potent small-molecule inhibitor of telomerase exhibiting in vitro and in vivo antitumor activity. Mol. Pharmacol. **61**, 1154–1162 (2002)
14. Jia, T. J., Cao, W., Zheng, X. J., Jin, L. P.: A turn-on chemosensor based on naphthol – triazole for Al (III) and its application in bioimaging. Tetrahedron Lett. **54**, 3471-3474 (2013)
15. Liu, Z., Li, Y., Ding, Y., Yang, Z., Wang, B., Li, Y., Li, T., Luo, W., Zhu, W., Xie, J., Wang, C.: Water-soluble and highly selective fluorescent sensor from naphthol aldehyde-tris derivate for aluminum ion detection. Sensors and Actuators B. **197**, 200-205 (2014) .
16. Wang, D. F., Ke, Y. C., Guo, H. X., Chen, J., Weng, W.: A novel highly selective colorimetric sensor for aluminum (III) ion using Schiff base derivative. Spectrochim. Acta Part A Mol. Biomol. Spectrosc. **122**, 268-272 (2014).
17. Tiwari, K., Mishra, M., Singh, V. P.: A highly sensitive and selective fluorescent sensor for Al^{3+} ions based on thiophene-2-carboxylic acid hydrazide Schiff base. RSC Adv. **3**, 12124-12132 (2013).
18. Xiao, H., Chen, K., Jiang, N., Cui, D., Yin, G., Wanga, J., Wang, R.: A highly selective turn-on fluorescent probe for Al (III) based on coumarin and its application in vivo. Analyst **139**, 1980-1986 (2014).
19. Qin, J., Yang, Z.: Ratiometric fluorescent probe for Al^{3+} based on coumarin derivative in aqueous media. Anal. Methods. **7**, 2036-2040 (2015).
20. He, H., Mortellaro, M. A., Leiner, M. J. P., Young, S. T., Fraatz, R. J., Tusa, J. K., A Fluorescent Chemosensor for Sodium Based on Photoinduced Electron Transfer. Anal. Chem. **75**, 549-555 (2003).
21. (a) Minta, A., Tsien, R.: Fluorescent indicators for cytosolic sodium. J. Biol. Chem. **264**, 19449–19457 (1989). (b) He, H., Mortellaro, M.A., Leiner, M.J.P., Fraatz, R.J., Tusa, J.K.: A fluorescent sensor with high selectivity and sensitivity for potassium in water. J. Am. Chem. Soc. **125**, 1468–1469 (2003)
22. (a) Tsien, R.Y.: New calcium indicators and buffers with high selectivity against magnesium and protons: design, synthesis, and properties of prototype structures. Biochemistry. **19**, 2396–2404 (1980). (b) Farruggia, G., Iotti, S., Prodi, L., Montalti, M., Zaccheroni, N., Savage, P.B., Trapani, V., Sale, P., Wolf, F.I.: 8-Hydroxyquinoline derivatives as fluorescent sensors for magnesium in living cells. J. Am. Chem. Soc. **128**, 344–350 (2006)
23. Dujols, V., Ford, F., Czarnik, A. W.: A Long-Wavelength Fluorescent Chemodosimeter Selective for Cu(II) Ion in Water. J. Am. Chem. Soc. **119**, 7386–7387 (1997) .
24. Udhayakumari, D., Naha, S., Velmathi, S.: Colorimetric and fluorescent chemosensors for Cu^{2+}. A comprehensive review from the years 2013–15. Anal. Methods. **9**, 552-578 (2017).
25. Xu, Z., Yoon, J., Spring, D. R.: Fluorescent chemosensors for Zn^{2+}. Chem. Soc. Rev. **39**, 1996–2006 (2010).
26. Goswami, S., Das, A. K., Aich, K., Manna, A., Maity, S., Khanra, K., Bhattacharyya, N.: Ratiometric and absolute water-soluble fluorescent tripodal zinc sensor and its application in killing human lung cancer cells. Analyst. **138**, 4593-4598 (2013).
27. Goswami, S., Das, A. K., Pakhira, B., Roy, S. B., Maity, A. K., Saha, P., Sarkar S.: Pyrophosphate selective fluorescent chemosensors: cascade recognition of nuclear stain mimicking DAPI. Dalton Trans. **43**, 12689-12697 (2014).
28. Nolan, E. M., Ryu, J. W., Jaworski, J., Feazell, R. P., Sheng, M., Lippard, S. J.: Zinspy Sensors with Enhanced Dynamic Range for Imaging Neuronal Cell Zinc Uptake and Mobilization. J. Am. Chem. Soc. **128**, 15517-15528 (2006) .
29. Liu, W., Xu, L., Sheng, R., Wang, P., Li, H., Wu, S.: A Water-Soluble "Switching On" Fluorescent Chemosensor of Selectivity to Cd^{2+}. Org. Lett. **9**, 3829-3832 (2007).
30. Yang, Y., Cheng, T., Zhu, W., Xu, Y., Qian, X.: Highly Selective and Sensitive Near-Infrared Fluorescent Sensors for Cadmium in Aqueous Solution. Org. Lett. **13**, 264-267 (2011).
31. Chae, M.Y., Czarnik, A.W.: Fluorometric chemodosimetry. Mercury(II) and silver(I) indication in water via enhanced fluorescence signaling. J. Am. Chem. Soc. **114**, 9704–9705 (1992)

32. Goswami, S., Das, A.K., Maity, S.: 'PET' vs. 'push–pull' induced ICT: a remarkable coumarinyl-appended pyrimidine based naked eye colorimetric and fluorimetric sensor for the detection of Hg^{2+} ions in aqueous media with test trips. Dalton Trans. **42**, 16259–16263 (2013)
33. Tang, L. J., Li, Y., Nandhakumar, R., Qian, J. H.: An unprecedented rhodamine-based fluorescent and colorimetric chemosensor for Fe^{3+} in aqueous media. Monatsh. Chem. **141**, 615-620 (2010).
34. Vincent, J. B.: Quest for the molecular mechanism of chromium action and its relationship to diabetes. Nutr. Rev. **58**, 67-72 (2000).
35. Zhou, Z. G., Yu, M. X., Yang, H., Huang, K. W., Li, F. Y., Yi, T., Huang, C. H.: FRET-based sensor for imaging chromium(iii) in living cells. Chem. Commun. **44**, 3387-3389 (2008).
36. Goswami, S., Das, A. K., Maity, A. K., Manna, A., Aich, K., Maity, S., Saha, P., Mandal, T. K.: Visual and Near IR (NIR) Fluorescence detection of Cr^{3+} in Aqueous Media via Spirobenzopyran ring opening with Application in Logic gate and Bio-imaging. Dalton Trans. **43**, 231-239 (2014).
37. Chen, L. J., Yang, X. Q., Jiao, H. L., Zhao, B. L.: Effect of Tea Catechins on the Change of Glutathione Levels Caused by Pb^{2+} in PC12 Cells. Chem. Res. Toxicol. **17** 922-928 (2004).
38. Kwon, J. Y., Jang, Y. J., Lee, Y. J., Kim, K. M., Seo, M. S., Nam, W., Yoon, J. Y.: A Highly Selective Fluorescent Chemosensor for Pb^{2+}. J. Am. Chem. Soc. **127**, 10107-10111 (2005).
39. Ratte, H. T.: Bioaccumulation and toxicity of silver compounds: a review. Environ. Toxicol. Chem. **18**, 89-108 (1999).
40. Chatterjee, A., Santra, M., Won, N., Kim, S., Kim, J. K., Kim, S. B., Ahn, K. H. Selective fluorogenic and chromogenic probe for detection of silver ions and silver nanoparticles in aqueous media. J. Am. Chem. Soc. **131**, 2040-2041 (2009) .
41. Habib, A., Tabata, M.: Oxidative DNA damage induced by HEPES (2-[4-(2-hydroxyethyl)-1-piperazinyl] ethanesulfonic acid) buffer in the presence of Au (III). J. Inorg. Biochem. **98**, 1696 -1702 (2004) .
42. Jou, M. J., Chen, X. Q., Swamy, K. M. K., Kim, H. N., Kim, H. J., Lee, S., Yoon, J.: Highly selective fluorescent probe for Au^{3+} based on cyclization of propargylamide. Chem. Commun. **45**, 7218-7220 (2009).
43. Garner, A. L., Koide, K.: Studies of a fluorogenic probe for palladium and platinum leading to a palladium-specific detection method. Chem. Commun. **45**, 86-88 (2009).
44. (a) Connett, P.: Professionals mobilize to end water fluoridation worldwide. Fluoride **40**, 155–158 (2007). (b) Ayoob, S., Gupta, A.K.: Fluoride in drinking water: a review on the status and stress effects. Crit. Rev. Environ. Sci. Technol. **36**, 433–487 (2006)
45. (a) Sessler, J.L., Camiolo, S., Gale, P.A., Pyrrolic and polypyrrolic anion binding agents. Coord. Chem. Rev. **240**, 17–55 (2003). (b) Gale, P.A.: From anion receptors to transporters. Acc. Chem. Res. **44**, 216–226 (2011)
46. Kim, S. Y., Hong, J. I.: Chromogenic and fluorescent chemodosimeter for detection of fluoride in aqueous solution. Org. Lett.**9**, 3109-3112 (2007).
47. Goswami, S., Das, A. K., Manna, A., Maity, A. K., Fun, H. K., Quah, C. K., Saha, P.: A colorimetric and ratiometric fluorescent turn-on fluoride chemodosimeter and application in live cell imaging: high selectivity via specific SiO cleavage in semi aqueous media and prompt recovery of ESIPT along with the X-ray structures. Tetrahedron Lett. **16**, 2633-2638 (2014).
48. Fu, L., Jiang, F. L., Fortin, D., Harvey, P. D., Liu, Y.: A reaction-based chromogenic and fluorescent chemodosimeter for fluoride anions. Chem. Commun. **47**, 5503-5505 (2011).
49. (a) Mànez, R.M., Sacenón, F.: Fluorogenic and chromogenic chemosensors and reagents for anions. Chem. Rev. **103**, 4419–4476 (2003). (b) Gunnlaugsson, T., Glynn, M., Tocci, G.M., Kruger, P.E., Pfeffer, F.M.: Anion recognition and sensing in organic and aqueous media using luminescent and colorimetric sensors. Coord. Chem. Rev. **250**, 3094–3117 (2006). (c) Kral, V., Andrievsky, A., Sessler, J.L.: A covalently linked sapphyrin dimer. A new receptor for dicarboxylate anions. J. Am. Chem. Soc. **117**, 2953–2954 (1995)
50. Goswami, S., Das, A. K., Sen, D., Aich, K., Fun, H. K., Quah, C. K.: A simple naphthalene-based colorimetric sensor selective for acetate. Tetrahedron Lett., **53** 4819–4823 (2012).

51. Goswami, S., Das, A. K., Aich, K., Manna, A.: Competitive intra- and inter-molecular proton transfer in hydroxynaphthyl benzothiazole: selective ratiometric sensing of acetate. Tetrahedron Lett. **54**, 4215-4220 (2013).
52. Quin, H.J., He, Y.B., Hu, Ch.G., Chen, Zh.H., Hu, L.: Enantioselective fluorescent sensor for dibenzoyl tartrate anion based on chiral binaphthyl derivatives bearing an amino acid unit. Tetrahedron Asymmetry **18**, 1769–1774 (2007)
53. Holland, M. A., Kozlowski, L. M.: Clinical features and management of cyanide poisoning. Clin. Pharm.**5**, 737-741(1986).
54. Kim, G.-J., Kim, H.-J.: Doubly activated coumarin as a colorimetric and fluorescent chemodosimeter for cyanide. Tetrahedron Lett. **51**, 185-187 (2010).
55. Kim, G.-J., Kim, H.-J.: Coumarinyl aldehyde as a Michael acceptor type of colorimetric and fluorescent probe for cyanide in water. Tetrahedron Lett. **51**, 2914-2916 (2010).
56. Lee, K. S., Kim, H. J., Kim, G. H., Shin, I., Hong, J. I.: Fluorescent Chemodosimeter for Selective Detection of Cyanide in Water. Org. Lett. **10**, 49-51 (2008).
57. Beauchamp Jr. R. O., Bus, J. S., Popp, J. A., Boreiko, C. J., Andjelkovich, D. A., Leber, P.: A Critical Review of the Literature on Hydrogen Sulfide Toxicity. Crit. Rev. Toxicol. **13**, 25-97 (1984).
58. Kamoun, P., Belardinelli, M.C., Chabli, A., Lallouchi, K., Chadefaux-Vekemans, B.: Endogenous hydrogen sulfide overproduction in down syndrome. Am. J. Med. Genet. **116A**, 310–311 (2003)
59. Das, A. K., Goswami, S., Quah, C. K., Fun, H. K.: Neighbouring group participation of thiol through aldehyde group assisted thiolysis of active ether: ratiometric and vapor phase fast detection of hydrogen sulfide in mixed aqueous media. New J. Chem. **39**, 5669-5675 (2015).
60. Das, A. K., Goswami, S., Dutta, G., Maity, S., Mandal, T. K., Khanra, K., Bhattacharyya N.: A concentration dependent auto-relay-recognition by the same analyte: a dual fluorescence switch-on by hydrogen sulfide via Michael addition followed by reduction and staining for bio-activity. Org. Biomol.Chem. **14**, 570-576 (2016).
61. Dickinson, B. C., Chang, C. J.: Chemistry and biology of reactive oxygen species in signaling or stress responses. Nat. Chem. Biol. **7**, 504-511 (2011).
62. Fukumura, D., Kashiwagi, S., Jain, R. K.: The role of nitric oxide in tumour progression. Nat. Rev. Cancer. **6**, 521–534 (2006).
63. Chen, X., Tian, X., Shin, I., Yoon, J.: Fluorescent and luminescent probes for detection of reactive oxygen and nitrogen species. Chem. Soc. Rev. **40**, 4783–4804 (2011) .
64. Miller, E. W., Albers, A. E., Pralle, A., Isacoff, E. Y., Chang, C. J.: Boronate-Based Fluorescent Probes for Imaging Cellular Hydrogen Peroxide. J. Am. Chem. Soc. **127**, 16652–16659 (2005).
65. Das, A. K., Hayashi, N., Shiraishi,Y., Hirai, T.: An antimalarial drug, tafenoquine, as a fluorescent receptor for ratiometric detection of hypochlorite. RSC Adv. **7**, 30453-30458 (2017).
66. Goswami, S., Das, A. K., Manna, A., Maity, A. K., Saha, P., Quah, C. K., Fun, H. K., Abdel Aziz, H. A.: Nanomolar Detection of Hypochlorite by a Rhodamine-Based Chiral Hydrazide in Absolute Aqueous Media: Application in Tap Water Analysis with Live-Cell Imaging. Anal. Chem. **86**, 6315–6322 (2014).
67. Tanaka, K., Miura, T., Umezawa, N., Urano, Y., Kikuchi, K., Higuchi, T., Nagano, T.: Rational design of fluorescein-based fluorescence probes. mechanism-based design of a maximum fluorescence probe for singlet oxygen. J. Am. Chem. Soc. **123**, 2530–2536 (2001)
68. Kundu, K., Knight, S.F., Willett, N., Lee, S., Taylor, W.R., Murthy, N.: Hydrocyanines: a class of fluorescent sensors that can image reactive oxygen species in cell culture, tissue, and in vivo. Angew. Chem. Int. Ed. **48**, 299–303 (2009)
69. Li, P., Xie, T., Duan, X., Yu, F.B., Wang, X., Tang, B.: A new highly selective and sensitive assay for fluorescence imaging of ${}^{\cdot}OH$ in living cells: effectively avoiding the interference of peroxynitrite. Chem. Eur. J. **16**, 1834–1840 (2010)
70. Sasaki, E., Kojima, H., Nishimatsu, H., Urano, Y., Kikuchi, K., Hirata, Y., Nagano, T.: Highly Sensitive Near-Infrared Fluorescent Probes for Nitric Oxide and Their Application to Isolated Organs. J. Am. Chem. Soc. **127**, 3684-3685 (2005) .

71. Liu, C.Y., Cao, Z.M., Wang, Z.H., Jia, P., Liu, J., Wang, Z.K., Han, B.J., Huang, X., Li, X., Zhu, B.C., Zhang, X.L.: A highly sensitive and reductant-resistant fluorescent chemodosimeter for the rapid detection of nitroxyl. Sens. Actuators B **220**, 727–733 (2015)
72. Yu, F. B., Li, P., Li, G. Y., Zhao, G. J., Chu, T. S., Han, K. L.: A Near-IR Reversible Fluorescent Probe Modulated by Selenium for Monitoring Peroxynitrite and Imaging in Living Cells. J. Am. Chem. Soc. **133**, 11030–11033 (2011).
73. McDonagh, C., Burke, C. S., MacCraith, B. D., Optical Chemical Sensors, Chem. Rev. **108**, 400-422 (2008).
74. Prentø, P. A.: contribution to the theory of biological staining based on the principles for structural organization of biological macromolecules. Biotech. Histochem.**76**, 137-161(2001).
75. Neto, B. A. D., Lapis, A. A. M.: Recent Developments in the Chemistry of Deoxyribonucleic Acid (DNA) Intercalators: Principles, Design, Synthesis, Applications and Trends, Molecules.**14**, 1725-1746 (2009).
76. (a) Wang, B., Yu, C.: Fluorescence turn-on detection of a protein through the reduced aggregation of a perylene probe. Angew. Chem. Int. Ed. **49**, 1485–1488 (2010). (b) Srivastava, A., Singh, P.K., Kumbhakar, M., Mukherjee, T., Chattopadyay, S., Pal, H., Nath, S.: Identifying the bond responsible for the fluorescence modulation in an amyloid fibril sensor, Chem. Eur. J. **16**, 9257–9263 (2010)
77. Van Arman, S.A., Czarnik, A.W.: A general fluorescence assay for enzyme catalyzed polyanion hydrolysis based on template directed excimer formation. Application to heparin and polyglutamate. J. Am. Chem. Soc. **112**, 5376–5377 (1990)
78. Faulhaber, K., Granzhan, A., Ihmels, H., Otto, D., Thomas L., and Wells, S.: Studies of the fluorescence light-up effect of amino-substituted benzo[b]quinolizinium derivatives in the presence of biomacromolecules. Photochem. Photobiol. Sci. **10**, 1535-1545 (2011).
79. Das, A. K., Ihmels, H., Kölsch, S.: Diphenylaminostyryl-substituted quinolizinium derivatives as fluorescent light-up probes for duplex and quadruplex DNA. Photochem. Photobiol. Sci. **18**, 1373-1381 (2019).
80. Das, A. K., Druzhinin, S. I., Ihmels, H., Müller, M., Schönherr, H.: Colorimetric and fluorimetric DNA detection with a hydroxystyrylquinolizinium photoacid and its application for cell imaging. Chem. Eur. J. **25**, 12703–12707 (2019) .

Biosensor and Healthcare Vis-à-Vis Cloud Computing and IoT: Towards Sophisticated Healthcare Development—An Overview

P. K. Paul

Abstract Biosensor is a kind of analytical device which is required for the purpose of analysis and discovery of a chemical substance which integrates both the biological equipments and a physicochemical detector. Biosensor refers to powerful device responsible for the biological sensing element in wide range of utilizations, viz. drug discovery, biomedicine, food processing, food safety, environmental management, including monitoring, defense and homeland security, information security and so on. In 1962, Clark and Lyons invented biosensors to get the status and analysis of glucose in biological entities utilized the way of electrochemical detection of oxygen. And thereafter, huge development is made in the field of biosensor including its wide range of applications. The subjects, viz. nanotechnology, electrochemistry, bioelectronics, play an important role in development and progress of biosensor. Biosensors are increasing rapidly as it is the easiest to perform, cost-effective and very rapid and as thus it becomes an important name in the field of bio-sciences and medicine. Hence, biosensors are analytical devices depended, and it converts a biological response into a quantifiable signal. The core components of a biosensor may include the *bioreceptor* and transducer. As far as healthcare is concerned, the core aim and agenda of biosensor include better position, analysis and result in diagnosing, monitoring and even in preventive health. Different tools, techniques and technologies are emerging which are directly and indirectly associated with the biosensors and for its better result. As far as information technology field is concerned, it has great importance in this field, including the emerging IT components, viz. cloud computing, big data, Internet of Things, etc. This paper is conceptual and case study-based and mentions the applications of cloud computing and IoT in biosensors and healthcare for improved medical and healthcare systems. Paper also describes the market and commercial aspects of different types of biosensors, briefly.

Keywords Biosensor · Computational techniques · Cloud computing · Internet of Things (IoT) · Healthcare · Medical informatics

P. K. Paul (✉)
Department of CIS, Raiganj University (RGU), Raiganj, West Bengal, India
e-mail: pkpaul.infotech@gmail.com

G. Dutta et al. (eds.), *Modern Techniques in Biosensors*, Studies in Systems, Decision and Control 327, https://doi.org/10.1007/978-981-15-9612-4_12

1 Introduction

Biosensor is a type of analytical device and combines the biological equipment's integrated with physicochemical detector. It is needed in detection of a chemical substance. Various biologically connected or derived contents are as follows:

- Tissue,
- Cell receptors,
- Antibodies,
- Nucleic acids,
- Enzymes, etc.

Here, the field biomedical engineering can be considered as important for designing and development of above subjects and contents; moreover, the biosensors are an important part of this interdisciplinary branch. Nanotechnology and nanofabrication play an important role for healthy biosensor development; thus, interdisciplinary works and efforts are highly desirable for the advancement and innovative biosensors designing and activities [1, 5]. The biosensor substrate contains basically following three electrodes, viz.

- Reference electrode,
- Working electrode,
- Counter electrode.

The biosensing becomes itself a technology or field now and helps in the fields of healthcare, agriculture, food processing and nutrition, environmental monitoring, etc. And as a result, biosensor-related researches are also increasing rapidly. The integration of biosensors as well as technologies considered as many growths in other areas, and thus, its applications in biosensors are being integrated with cloud computing, big data, smartphone and telecommunications, machine learning, Internet of Things (IoT) and so on [2, 24, 30]. The emergence of biosensors in different fields results in other sub-fields, and branches of biosensors refer Table 1 for brief on each.

However, apart from these, biosensors are applicable for other important areas of medical, clinical and environmental sciences, and among these, a few important include

- Biosensors applications in drug discovery and drug analysis,
- Biosensor applications in stretchable electronics,
- Recognition receptors in biosensors,
- Biosensor applications in biodetection,
- Biosensors with fiberoptics,
- Biosensor applications in security-related areas,
- Biosensor applications in bioterrorism applications,
- Non-invasive biosensors in clinical analysis and clinical research,
- Biosensors for clinical diagnostics,

- Graphene-based biosensors and bioelectronics,
- Graphene-based tattoo-like skin biosensors,
- Biosensor applications in biophotonics, etc.

Table 1 Different types of biosensors and basic features

Biosensors	Aspects
Amperometric biosensors	Amperometric biosensors are applicable when a potential is applied between two electrodes having response times including dynamic ranges and sensitivities
Electrochemical biosensors	Electrochemical transducer basically uses in electrochemical biosensors
Potentiometric biosensors	It is kind of chemical sensor and useful in determination of the analytical concentration regarding a few components of analyte gas, or it may be a solution
Microbial biosensors	It is a kind of analytical device and biologically connected with a transducer, which is responsible for generating measurable signal indicating, and it is ideally suited for the analysis of extracellular chemicals, etc.
Optical biosensor	Optical biosensors offer the benefit of high specificity including the sensitivity and due to its cost-effectiveness. Further, it is small and best than traditional analytical techniques due to its feature of real-timeness and label-free detection systems in the chemical substances and biological element
Enzymatic biosensors	The enzymatic biosensors are used for oxidative stress biomarkers, and it comprises of an enzyme and ultimately producing a chemical signal, a transducer
Fluorescent biosensors	Fluorescent biosensors are molecules and help in concentrations, locations and other dynamics aspects in biomolecules and bioactivities, etc.
Microelectrode biosensors	It is provided real-time analyses, and here, two types of microelectrodes, cylindrically shaped wire electrodes and microfabricated multi-electrode needles
Impedimetric biosensors	It is constructed by immobilizing biological recognition elements
Organism- and whole cell-based biosensors	This is compute useful information and which affects analyte on the physiological function of living cells
Voltammetric biosensors	It is a kind of electro-analytical methods and also an example of amperometric technique

Apart from the traditional areas of biosensor, viz. clinical diagnosing, health and clinical monitoring and maintaining health, biosensor is also valuable in lashing healthy behaviors such as preventive health and physiologic functions [3, 4, 17]. Application and enhancement of biosensors in medicine and healthcare are noticeable and having wide research and scope. It is needed in healthcare due to its important features of rapidness, simplicity, effectively to hand-held testing devices in pharmacy and medicine.

2 Objective

The present paper entitled, "Biosensor Vis-à-Vis Cloud Computing and IoT: Toward Sophisticated Healthcare Development—*A Study*", is theoretical and conceptual in nature and deals with the following aim and agenda (but not limited to)

- To learn about the basic of biosensors including its feature, characteristics and core components.
- To get the knowledge on evolution and general applications of biosensors in different fields, briefly.
- To gather knowledge on biosensing technologies with its applications in environment and allied activities.
- To find out basic and general aspects of information technology specially cloud computing, with its foundation, types and applications to get the knowledge on biosensors.
- To learn about the Internet of Things including its origin, feature and functions with reference to its medical applications, etc.
- To find out the potential applications of cloud computing and IoT in biosensing and allied facets.
- To learn about the market and commercial aspects of different types of biosensors, in brief manner.

3 Methods

As the current title of the research is theoretical in nature as depicted its title on "Biosensor Vis-à-Vis Cloud Computing and IoT: Toward Sophisticated Healthcare Development—*A Study*", so it is a kind of theoretical work and more clearly review-based. Hence, various secondary and primary sources are played an important role and in this regard for preparing this work. Web review including reviewing contemporary biotechnology-based company sites also used to get the knowledge on biosensors specially the emerging technologies, viz. cloud computing, IoT, etc. All such knowledge gathered, analyzed and reported here in different sections in this work.

3.1 Biosensors: Basics and Technologies

Based on functions of biosensor, it may consider as three major components. The first part of the biosensor is treated as the biological element, that is, caused for the detection of the analyte and also responsible for generating signal, forms of the biosensor [6, 10, 23]. The second part is considered as transducer which ID engages in the activities of the generated signal toward transformed into a detectable response in the biological element, which is the most critical component in any biosensing device. Detector is the third part that amplifies as well as processes the generated signals before displaying the same by the electronic display system.

Biosensors are required for various kinds of *sensitive biological element* such as tissue, microorganisms, enzymes, antibodies and nucleic acids which are a kind of biomimetic components. Different type of biosensors is responsible for various activities. The components are depicted in Fig. 1.

Amperometric biosensors work when a potential is applied between 02 electrodes, and normally, it holds the response times as well as dynamic ranges similar to the potentiometric biosensors. Potentiometric biosensor gives a logarithmic response with higher and dynamic ranges, and here, potential is created at 0 current, whereas microbial biosensor is another kind of analytical device which helps in microorganism(s) to generate a measurable signal proportional [7, 13, 26].

The applications and role of biosensor in medicine, healthcare, food processing are increasing; as a result, it is continuing growth of international markets. It is applicable in traditional engineering and sciences for healthy and better analysis of biological systems. Here, role of materials sciences can also be noted and included in allied biological systems. However, healthy biosensing practice requires reliable methods for the stand-off detection of various components (CBRNE), viz.

- Chemical,
- Biological,

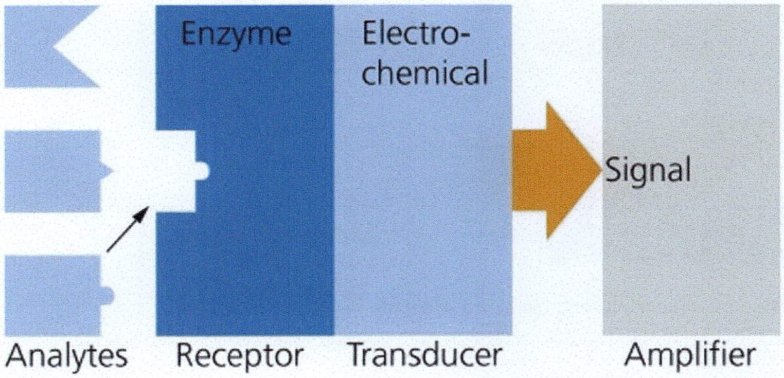

Fig. 1 Basic components of biosensor

- Radiological,
- Special nuclear,
- Explosive materials.

The technological needs are not always same and dynamic in biosensor space [8, 16, 25]. Biosensors are actively required for the following areas

- Sensing for the clinical sciences, healthcare and medicine,
- Biomechanics and allied subjects,
- CBRNE sensing,
- Ubiquitous devices which are required for the biodetection,
- Biosensing technologies for the agriculture and production including genetic engineering,
- Biosensing for the food quality and safety,
- Optical biosensing tools and technologies,
- Biosensing in security [9, 14, 22].

The applications of biosensing are rising, and biosensing is emerging rapidly in public healthcare as well due to the emergence of information technological applications, viz. cloud computing, IoT, etc. However, it is important to note that the component of biosensors is more clearly depicted in Fig. 2 herewith.

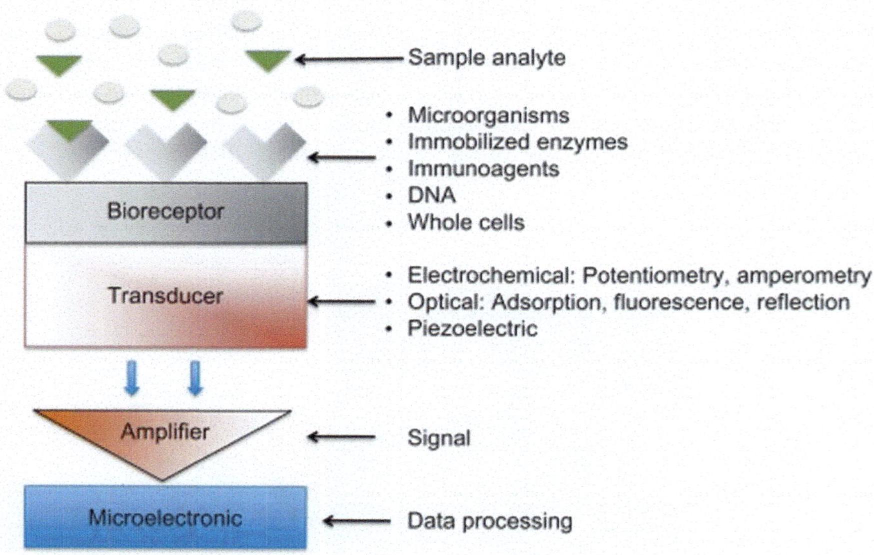

Fig. 2 Biosensors elements and facets are in details

3.2 Biosensors: Applications with Reference to Medicine

Biosensor gives us measurable signal. And among the emerging areas of biosensor applications, few popular and important are home-based pregnancy test kits, glucose detectors food analysis tools, deoxyribonucleic acid testing and drug detection, etc. [11, 15].

Many potential applications can be noted with biosensors; however, the requirements for a biosensor approach depend on the growth of the research as well as commercial application availability of the suitable biological element, and healthy disposable portable detection systems are required in some situations. There are many applications of biosensors, viz.

- Glucose monitoring in diabetes patients is the most common example of biosensor applications and we can understand and noted in medical systems.
- In environmental applications, many examples can be noted, viz. detection of the pesticides, remotely sensing of the airborne bacteria, remote sensing of water quality in coastal waters, etc.
- Counter and find out the bioterrorist activities can be detected with the help of biosensors [7, 18].
- Ethnology of some group, i.e., in anthropological studies including the biological rhythms, growth rates, spawning and similar activities.
- In analysis and routine analytical measurement of the Vitamin B12, folic acid, find out and analyze the drag in a food, analysis of antibiotics, etc. biosensors are useful in some context.
- Finding the compounds in a product, doing the activities of the protein engineering in biosensors, etc.

There is a common concept that biosensors are useful in healthcare and clinical settings; however, there are certain areas of environmental sciences where biosensors are widely using including bio-environmental surveillance, environmental monitoring, ecological management, etc. As it is a biophysical devices, thus it is responsible for the specific substances of environment such as sugars, proteins, hormones, pollutants as well as other kinds of toxins in the ecological components and environment [12, 28].

Ultimately following can be considered as important reasons for the uses of biosensors in environments, viz.

- Environmental monitoring and management,
- Anthropological applications,
- Plants and trees with environmental biosensors,
- Biological detection for the metal ions in water,
- Biosensors for oceanographical studies including marine monitoring of its components.

In most of the cases, the optical biosensors depend on SPR techniques, and it uses the property and gradients of some other materials, specifically a thin layer. Other optical biosensors are working on absorbance or fluorescence, viz. a fully operational prototype device detecting casein in milk has been fabricated [15, 20]. It is worthy to note that biosensor often also works in a genetically modified form of a native protein as well as in enzyme. The protein is configured to detect a specific analyte, and here, biosensors are working rapidly. Nanobiosensors normally use an immobilized bioreceptor which is selective regarding the target analyte molecules. Nanoscale materials demonstrate unique properties. And here, biosensors can be nicely used for. However, as far as its common medical uses are concerned, important include

- Common healthcare checking,
- Metabolites measurement,
- Insulin treatment and further processing,
- Clinical and medicinal psychotherapy,
- Diagnosis of disease and further treatment,
- Agricultural and veterinary healthcare and applications,
- In pharmacy including in drug improvement, with offense detection, etc.,
- Ecological pollution control that is related to the healthcare as well,
- Biotechnology biosensors,
- Chemosensors for chemical activities,
- Biomedicine as well as biomarkers, etc.

As far as industrial applications and marketing and commercialization are concerned, following can be considered as important requirements (but not limited to):

- For better and healthy result of the biosensor, bioreceptor is needed to be very specific regarding the analysis and should be stable within the normal stage; further should show the low variations as well between assays [12, 27].
- Moreover, the biosensor should be more firstly accurate, precise and also reproducible, linear in analytical range. Further, it needs minimal sample pretreatment as well.
- In biosensors, the detected is better to be independent of physical parameters. This is the pH and temperature and also with minimal pretreatment of samples.
- The biosensor must be cheap and effective, small or portable. Additionally must be simple for the semiskilled users.

4 Technologies and Biosensors

Information technology is the applied science and engineering branch responsible for information activities ranging from collection, selection, organization, processing, management and dissemination of the information. Further, for doing the task, IT uses its major sub-fields, viz.

- Software technology/engineering,
- Networking technology/engineering,

- Multimedia technology/engineering,
- Web technology/engineering,
- Database technology/engineering, etc.

As far as biosensor is concerned, information technology plays a leading role, specially the computational techniques, networking technologies, communications, etc. Optical biosensor is prime example in this regard.

As we know that, optical sensor field is growing rapidly as a research area over the years, and there are good advantages of optical sensing than the other transduction methods. Recently, one of the common approaches regarding the optical biosensor is the integration of the high sensitivity of fluorescence detection along with high selectivity offered by ligand-binding proteins. The field image processing (here, i.e., medical image analysis) helps in offering/enhancing the medical field emphasizing efforts and applications of the computer vision, virtual reality and robotics and so on. Among the possible areas, IT and computational sciences are applicable in the following

- 3D imaging interaction of various components,
- Live cell fluorescent biosensors,
- Theranostics as well as implantable biosensors,
- Novel biosensors for the purpose of live cell imaging,
- Image analysis (biomedical).

Information technology is also composed of more emerging fields and technologies of the IT such as

- Cloud computing and virtualization,
- Big data and analytics,
- Internet of Things (IoT),
- Wireless and converged network,
- Usability engineering and interface,
- Human-computer interaction, etc.

And all these emerging fields and technologies are directly and indirectly associated with the biosensor and medical systems and technologies. Cloud computing and Internet of Things are also leading and increasing in biosensing activities, but before getting the applications and integration of cloud/IoT with biosensors, here in the next section, basics of cloud computing and IoT provide first for more clarification on applications of these technologies into biosensor and healthcare market. Further, the marketing and business-related aspects are also discussed in this paper based on international market scenario.

5 Cloud Computing, IoT: Biosensor Perspective

Cloud computing is considered as one of the emerging technologies and emerging areas as a component of the information technology field [9, 19]. In other words, it is also called as virtualization technology as it is responsible for the creation of the virtual presence and helps in virtual reality. Cloud computing helps in making availability of different components of the computer and information technology with the help of Internet as well as similar infrastructure, with following core features and functionalities:

- Public Cloud Computing—*This kind of cloud computing services is basically offered by the third party by the Internet, and users can access their required information and tools, infrastructure, remotely.*
- Private Cloud Computing—*Cloud computing when designed, developed as well as built in the concerned institutions or organization based on need; then, it is called private cloud computing.*
- Hybrid Cloud Computing—*The third and most popular type of cloud computing is hybrid cloud computing, and this is the combination of both the computing and applicable in such cases where privacy and ownerships are very much important and valuable.*

It is important to note that whether public cloud, private cloud or hybrid cloud computing, all these are responsible for the remote-based IT services such as software and programs, storage, infrastructure, platform, security and privacy, etc., and these are offered with specified nomenclature, viz.

- Software as a service (to offer software and applications remotely to long distance),
- Infrastructure as a service (to offer infrastructure and hardware, etc. remotely to long distance),
- Storage as a service (to offer storage and databases remotely to long distance),
- Platform as a service (to offer platform and operating systems remotely to long distance),
- Security as a service (to offer security services remotely to long distance), etc.

There is an important role and linkage with the cloud computing with big data and other emerging technologies, specially IoT. Cloud computing is the modern and valuable computing service in the computing market with services of hardware, software, applications, platforms and complete IT by the Internet as well as similar services [3, 17]. The service models such as SaaS, IaaS and PaaS are also increasing day by day, rapidly in the emerging service models, viz. security, storage services, etc. The public cloud computing is valuable in many contexts. IT companies were initially started to get the cloud-based services, but gradually even government sectors also offered cloud-based services. There is a great demand in cloud computing technology for the big data processing from biosensors (Refer—Fig. 3 for details).

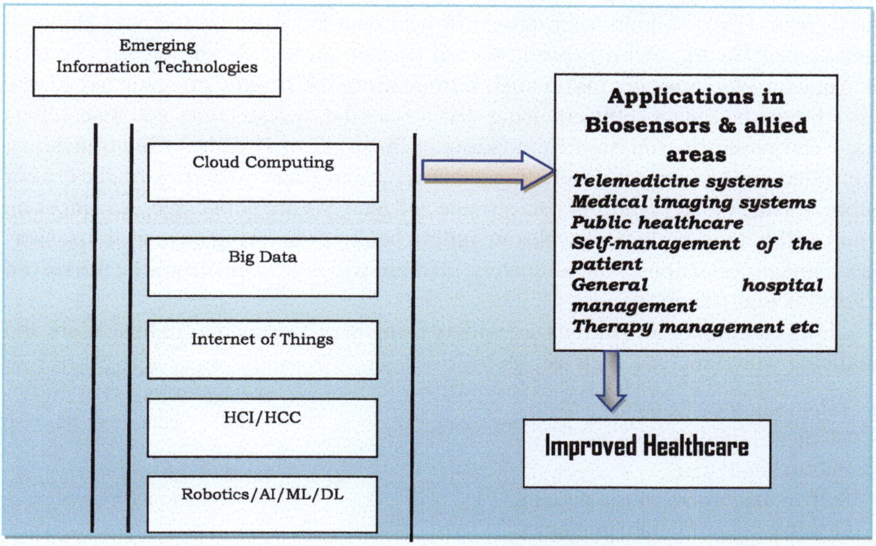

Fig. 3 Cloud and other emerging technological applications in biosensors and allied areas

It is worthy to mention that the networking technology especially the wireless sensors basically used in healthcare monitoring application including biosensor (here, the used network is personal area network). The limitations of such network (i.e., wireless network) basically collected from the following:

- Memory,
- Energy and power,
- Computational techniques,
- Communications,
- Scalability of the systems and with efficient management.

And here, big data is responsible for the collection of the facts and signals from the biosensors itself. In biosensing activities, directly and indirectly scalable high-performance IT and computing services (online/offline) are required with massive storage infrastructure. And these are required for the purpose of real-time processing including storage as well as analysis of the purpose of the biosensors data. Here, uses of complex algorithms are very important to get the extract required values from database. The telemedicine services include the interaction of the patient with the doctors or healthcare organizations; in offline or online, cloud computing technology plays a leading role. Here, the uses of storage and software services are important to note with customization and at low cost. The cloud computing is much flexible and offers a good platform for monitoring and controlling of signal in biosensing in recent past.

There are many examples in use of cloud computing in governmental organizations including healthcare settings (those are also deal with the biosensors). One such fact is from HMP-Human Microbiome Project data governed by the NIH-National Institutes of Health, and it is available and uses on the simple storage service (S3) of Amazon Web Services. With such introduction, the healthcare settings become more biomedical data set users using. Even scientist, investigators and researchers may send request/permission from National Institutes of Health for the transfer of controlled-access genomic (including phenotypic data), and in this regard, the uses of public/private cloud regarding data storage and analysis are noticeable. Advances on cloud will help in biological scales including health monitoring devices, biosensor-based image collections. And ultimately, all these will help in electronic health record platforms in different way.

Cloud computing utilizations are widely using in different areas of healthcare and medicine which include such as

- Telemedicine systems,
- Medical imaging systems,
- Public healthcare,
- Self-management of the patient,
- General hospital management,
- Therapy management.

Cloud computing is also applicable in real-time health monitoring, which helps the patients directly including in chronic conditions. In this context, health services can be managed by the cloud and virtualization systems. Cardiac arrhythmias patients ate requires continuous episode detection as well as monitoring, and in this context, cloud computing is an important name. Also in real-time electrocardiogram monitoring wireless sensors are important and applicable; and moreover in arrhythmia episode detection including in classification the virtualization techniques are important and applicable emerging.

With the help of Amazon Elastic Compute Cloud (EC2), cloud services may be combined with the mobile computing, and this can help in patient ECG, including its monitoring and further activities in recording, analyzing and displaying information in remote locations. Further, various software can be used in ECG with cloud service-SaaS, and further data can be shared for public use as well. Also, the Microsoft Azure platform has been implemented for a 12-lead ECG telemedicine services. Apart from these, in case of storing and displaying of medical images, cloud is useful. Here, images/picture may be deployed in a public cloud. Today, many companies and commercial vendors combine with the healthcare settings, viz. hospitals and healthcare for the establishment of healthy clinical systems and also in biosensor activities, viz.

- Wellness monitoring of the patient and complete healthcare units,
- Cell-based biosensors activities,
- Antibody-based biosensors analysis, imaging, etc.
- In detection technique,
- In cloud supported biosensors for the purpose of cardiovascular areas,
- Biosensors (integrated with cloud support) for the cancer applications,

- In transduction technology,
- In diabetes applications also biosensors useful and here cloud is applicable,
- Biosensor is also applicable in pharmaceutical sector, and here, cloud computing usages are important and valuable.

5.1 CloVR: An Example of Cloud in Sensor and Health Segments

The CloVR VM is an important example of cloud-based healthcare tool and software, and it decreases the barrier of entry for utilizing complex analysis protocols. It is a new application on push-button automated sequence analysis and working as a single portable VM. Further, CloVR VM provides multiple automated analyses for various activities, viz. microbial genomics with 16S, complete genome and also the metagenome sequence analysis. It uses local computer resources, and within this, the installation becomes so easy; however, here, addressing is a key challenging for the deployment of the bioinformatics workflows.

CloVR is associated with preinstalled libraries and required for the biological data analysis. Another tool, i.e., Cloud BioLinux, is another example in which resource with VMs images is offered over 100 software packages regarding bioinformatics solution. Important to note that CloVR and also BioLinux are useful in cloud IaaS environment. Here, for general big data-related problem, cloud adoption is also allowed and managed perfectly. Here, majority of companies are using machine learning as well as artificial intelligence for better services and database systems, majorly as Google BigQuery. For complex data management and large biomedical data sets, such databases can reduce management costs, ease database adoption and facilitate analysis. Apache Hadoop software library is useful.

6 Internet of Things (IoT): Overview, Functions and Framework

IoT or Internet of Things is an advanced IT component with various types of objects, and it is applicable in industrial machines to healthcare, etc. In the 1990s, Kevin Ashton (an Entrepreneur) has coined the term Internet of Things (IoT); gradually, improvement of Internet of Things (IoT) is noticeable. The Internet of Things (IoT) is applicable in different types of objects and connected with the Internet, and here, different built-in sensors gathered data, and in this context, connected network plays a leading role. Automatic adjustment is done by the sensors with various attributes, viz. heating and lighting. Internet of Things is an emerging and future technology to make efficient and smarter life. Here, each IoT device holds an IP address for collecting and transforming data with healthy intelligence systems and less manual support. It becomes more popular every day in diverse field, viz. business, industries and

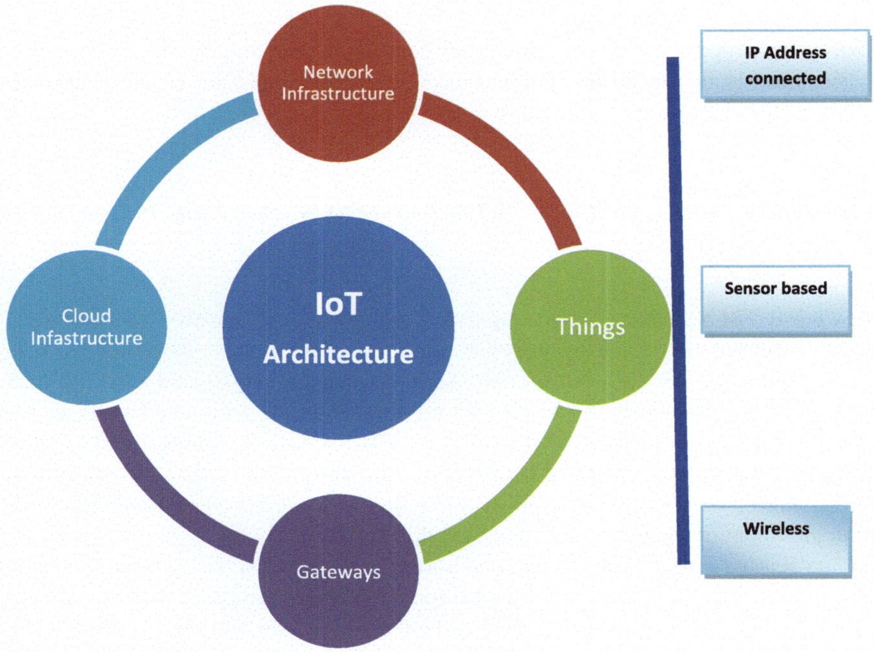

Fig. 4 IoT architecture and connected objects at a glance

commerce, education, research and training, government as well as in administration, health and medical systems, transportation, manufacturing industries, etc., and this trend is growing every day. Numerous areas exist and emerging where IoT systems are adopting. Here, the concept of wireless Internet became important. Figure 4 is depicted in the aspects of IoT in detail.

In IoT, various built-in sensors are dedicated to communicating as well as collection of data, and in this, connected network plays a leading role. For the development of smarter life and digital society building, IoT plays an important role using IP address, wireless Internet and embedded sensors which are connected and used properly. Due to its benefits, IoT is applicable in day-to-day professional and personal activities [7, 21]. Some of the important feature and fact of IoT are noted as follows

- Within IT, the emerging Internet of Things considered as smarter information and technological solutions depended on Internet and allied technologies.
- Cloud computing, data analytics, robotics, machine learning and deep learning are very important and valuable and treated as a valuable technology.
- In IoT, smart electric grids are very important and well connected with renewable resources.
- Smart electric grids may be connected with renewable resources, and it has improved system on smaller usage increments.

- Machine-based monitoring sensors gives the input which includes the repairing equipments as well as regional needs, which is most important in different context.
- The infrastructure of smart such as running waste management, efficiency, designing and development is possible with IoT-based systems.
- Home-based security may be possible with the IoT such as remote controlled-based weather controlling, down our home and open windows, based on our needs.

7 IoT, Biosensors and Healthcare

The technology and healthcare industries are very much connected these days due to various opportunities from the technology, and this trend is growing rapidly. New opportunities are also increasing due to the arrival of newer technologies, viz. cloud computing, Internet of Things (IoT), and big data. In addition, worldwide adaptations of wearable biosensors are also noticeable, and as a result, individualized eHealth and mHealth technologies have been developed. The technological input in healthcare led the highly available systems, easily accessible, personalize, cost-effectively and scalable services [3, 25].

However, there are many challenges, and issues are there in front of offering these emerging services and can create problem in reliable, adaptive, efficient healthcare systems. With adaptation of proper strategies, policies such problems can be solved, and in this context, a better relationship and collaboration among the technological stakeholders, physicians, healthcare professionals, etc. are highly required.

IoT technologies help in promoting smarter and healthy sensors for improved healthcare and individualized tele-health services as well. The Internet of Things helps in wearable and body sensors, including the pervasive healthcare systems. And additionally, cloud computing and data analytics play a leading role in complete development in healthcare segment.

Medical applications which integrate with the Internet of Things can also bear the ubiquitous services from the smartphone. Here, advanced medical diagnostics can use the biosensors for the biological data possession. Biosensors have great potential in IoT. The commercial aspects are also very important in development of IoT-based systems and that can lead the digital economy as well and therefore may improve the healthcare in different way. Both in portable and in wearable network devices, biosensors including the healthcare professionals can get wider benefits. The following aspects and initiative may be considered as valuable and important, viz.

- Biological data acquisition by biosensors,
- Point of care diagnostics,
- Medical applications based on IoT, etc.

Biosensor technologies play a leading role in development of integration of emerging information technologies with sensor systems. And this continues due to the efforts in the academia-techno industries healthcare settings. The new development of IoT will have a great transformation in the societal as well as business sectors, and according to the Gartner research, it is expected to grow the business in next 5–10 years rapidly.

Further, IoT will have the impact in medical and healthcare with various emerging and futuristics wearable devices including the home monitoring healthcare systems. IoT will also help in betterment of clinical instruments for the healthcare professionals and core physicians. As far as biosensor is concerned, the modern IoT-based biosensors even can add various kinds of tools for the chronic disease and healthcare management and importantly in the population health and wellness in better way. With the help of IoT systems in the healthcare, the continuous monitoring care of the healthcare units may be reduced radically. According to the study, it is noted that remote monitoring-based medical and healthcare sector is expected to grow $980 million by 2020. As far as 2014 data is concerned, it is about $400 million [8, 18].

The aged as well as young patients today require technology-based healthcare systems, and in this context, IoT and other emerging technologies are very much important. Cloud-Big Data-IoT-based systems no doubt helpful in the context of chronic diseases. And IoT (and partially cloud as well) helps in following treatment in better way:

- In the treatment of diabetes and allied diseases,
- In COPD, i.e., chronic obstructive pulmonary disease,
- Treatment in cardiovascular systems (includes majorly in heart attacks and stroke),
- Arthritis and long-term orthopedic treatment,
- Cancer- and oncology-based treatments,
- Obesity care and monitoring,
- Oral health-related problems,
- Neurological diseases including in epilepsy, etc.

According to the top market research organization, *Gartner* predicted that Internet of Things is the most important and growing technology trends. On the other hand, Cisco predicted that IoT and cloud-based industries will grow rapidly, and in coming days, about 50 billion smart devices may connected, and due to the importance of Internet, it is to be called as "Internet of everything (IoE)". To get an idea about the IoE, let us know the definition given by Cisco Systems Inc. "as bringing together people, process, data, and things to make networked connections more relevant and valuable than ever before—turning information into actions that create new capabilities, richer experiences, and unprecedented economic opportunity for businesses, individuals, and countries" (Source official Web site of Cisco). The networking and emerging technologies in healthcare no doubt will create a big market, and apart from the medical and healthcare systems, such will be available in other areas like in automobile, infrastructure development, transportation, emerging education as well as ICT [7, 19]. According to the report of tech giant Cisco Systems Inc, the market for Internet of everything (IoE), which includes the IoT, big data, cloud computing, etc., is expected to grow up to $106 billion within 2020. And as a whole, it can help in better data-driven patient and clinical management, which will ultimately result in more efficient, healthy and cost-effective healthcare systems including biosensors.

8 Future of Healthcare: Techno and Business Perspective

Apart from the industries and organizations, various healthcare organizations are moving toward emerging technological applications in healthcare, hospital and biomedical engineering sections, and among the emerging information technologies, few important and popular include

- IoT,
- Cloud computing,
- Big data and analytics,
- Artificial intelligence,
- Machine learning,
- Deep learning,
- Human-computer interaction systems.

And similar are becoming pervasive in healthcare industries as well. Here, each and every technology plays a leading role for the development of the techno-enable healthcare systems, and among these technologies, IoT becomes important to bring effective and low-cost biosensors, healthy medical communication and processing systems as well. Further smartphones can also be considered as important in healthcare improvement via techno support and as a component of Internet of Things.

There are huge apps and these are not developing for different sectors but also having Internet connected systems. These apps can be used in general healthcare management as well as in core treatment in coming days which includes the help during the operation by suggestions to the medical team, remote-based treatment, prediction analysis with the help of robotics and AI technologies, etc. Such apps are rising rapidly and growing internationally; the reason of such growing includes the advertisement and revenue generation from the said apps as well by different means.

As a whole, the complete market may reach up to $14 trillion, and healthcare-based technology companies can hold more development in future apart from the technological giant such as Siemens, Ericsson and Bosch. The development of pharmaceuticals companies is also important and to be noted in this context [5, 22].

Even governments of different countries are also doing well in establishing proper policies, strategies in respect of building techno-enable healthcare systems. In the developing countries, also this trend is noticeable. Certain challenges exist in healthcare improvement and integration of technologies in the public health and clinical heaths, and with proper integration of technologies and management strategies, this can be removed or waived. The global biosensors market size was valued at USD19.6 billion in 2019 and is expected to witness a CAGR of 7.9% during the forecast period. Demand for biosensors is increasing due to various applications of biosensors in the medical field, rising diabetic population, high demand for miniature diagnostic devices and rapid technological advancements. Early and precise disease diagnosis is essential for successful prognosis of the disease and survival of patient. In the recent years, the demand for disposable, cost-efficient and user-friendly devices with fast response time has extensively increased [6, 23]. These devices owing to their potential to fulfill these criteria through an interdisciplinary combination of approaches from medical science, chemistry and nanotechnology have paved their way rapidly in the medical field.

Biosensors which are most commonly used in blood metabolites which include the lactate, glucose, urea, etc. are using both optical and electrochemical modes of transduction. And these biosensors are commercially developed in highest manner in the healthcare industry. Governmental initiatives regarding genomics and proteomics help in rising and in drive of international biosensor market. In many countries, the government and institutes are taking initiatives on biosensors-related development activities, for example, NICTA, Australia, i.e., National Information and Communication Technology Australia is actively engaged in research and development activities including the data mining, networks, big data, cloud and embedded systems for the advancement of the biosensors and healthcare segment. As far as India is concerned, many leading IT companies, viz. IBM, Infosys, TCS, Capgemini, are investing in the biosensors market. Indian government is doing well in modernizing biosensing and biotechnology field. Similarly, Chinese Government employ China Grid under the leadership of Chinese Academy of Sciences and Ministry of Education and the China National Grid for the initiatives in the biosensors market [2, 5].

As of 2019, North America ranked in highest size in biosensor market. As far as North America is concerned, the POC diagnostics are the fastest-growing player in the field of biosensors, and the expansion is attributed in healthy and addressable patient population in the region. American Medical Association and American Medical Group Association expressed that above of the 50% of Americans is having chronic (one or more) diseases until 2012, and further, they state that it may reach to ~157 billion within 2022. In this regard, government and various organizations are doing well in establishing healthcare systems perfectly with biosensors market as well.

The tremendous growth in chronic and infectious diseases led the more development in biosensors including technological integration within this. Some of the leading players in biosensors and integrating technologies are also growing, and some of these are depicted in Table 2.

9 Concluding Remarks

The biosensor market is changing based on demand and technological interaction, and it may classify mainly into these—*thermal*, *electrochemical*, *piezoelectric* and*optical.* Electrochemical is holding the largest market as far as 2019 is concerned. The wide linear response range and low detection limits are the main

Table 2 Some of the leading companies reputed in biosensors in international market

Company/Organizations	Country
Abbott Point of Care Inc.	USA
Sysmex Corporation	Japan
Universal Biosensors Inc.	Australia
Medtronic Inc.	USA
Pharmaco-Kinesis Corporation	USA
Biacore	Sweden
Biosensors International	USA
DuPont	USA
Bio-Rad laboratories Inc.	USA

cause in increase of electrochemical biosensors worldwide. Additionally, some other advantages, viz. reproducibility as well as optimum stability, are considered as important for the growth in electrochemical biosensors. Electrochemical is thus leading to high consumption as well as greater market growth and penetration, internationally. The optical biosensor also witnesses the fastest growth due to versatile analytical coverage and due to its applicability in structural studies, receptor-cell interactions, kinetic, equilibrium analysis, etc. The embedded devices are also rising rapidly internationally due to its importance in the environmental monitoring, biological defense, food and nutrition, etc. and further are useful for detection in the broad spectrum of biological entities and applicable in medical laboratories, food biological analysis including in microbial detection. The technologies related to the information and computing play a leading role in enhancing and developing biosensors in different way. The emerging technologies, viz. cloud computing, IoT, robotics, etc., play a leading role in the development and modernization in biosensors activities including its operation, development, etc. Manpower development is an important issue in biosensor development and overall healthcare improvement. In this regard, research and development units, universities, training centers, industrial units, etc. have to play with more input with proper strategies and supports.

References

1. Aileni, R.M., Pasca, S., Valderrama, C.: Cloud computing for big data from biomedical sensors monitoring, storage and analyze. In: 2015 Conference Grid, Cloud & High Performance Computing in Science (ROLCG), pp. 1–4. IEEE (2015, Oct)
2. Aly, H., Elmogy, M., Barakat, S.: Big data on internet of things: applications, architecture, technologies, techniques, and future directions. Int. J. Comput. Sci. Eng. **4**, 300–313 (2015)
3. Antony, N., Unnikrishnan, L., Mohanty, S., Nayak, S.K.: The imperative role of polymers in enzymatic cholesterol biosensors-an overview. Polym. Plast. Technol. Mater. **58**(16), 1713–1741 (2019)

4. Baccarelli, E., Cordeschi, N., Mei, A., Panella, M., Shojafar, M., Stefa, J.: Energy-efficient dynamic traffic offloading and reconfiguration of networked data centers for big data stream mobile computing: review, challenges, and a case study. IEEE Network**30**(2), 54–61 (2016)
5. Caputo, A., Marzi, G., Pellegrini, M.: The internet of things in manufacturing innovation processes: development and application of a conceptual framework. Bus. Process Manage. J. **22**(2), 383–402 (2016)
6. Hassanalieragh, M., Page, A., Soyata, T., Sharma, G., Aktas, M., Mateos, G., Kantarci, B., Andreescu, S.: Health monitoring and management using internet-of-things (IoT) sensing with cloud-based processing: opportunities and challenges. In: 2015 IEEE International Conference on Services Computing, pp. 285–292. IEEE (2015, June)
7. He, D., Ye, R., Chan, S., Guizani, M., Xu, Y.: Privacy in the internet of things for smart healthcare. IEEE Commun. Mag. **56**(4), 38–44 (2018)
8. Kassal, P., Steinberg, M.D., Steinberg, I.M.: Wireless chemical sensors and biosensors: a review. Sens. Actuators B Chem. **266**, 228–245 (2018)
9. Kim, T.W., Park, K.H., Yi, S.H., Kim, H.C.: A big data framework for u-healthcare systems utilizing vital signs. In: 2014 International Symposium on Computer, Consumer and Control, pp. 494–497. IEEE (2014, June)
10. Jagadeeswari, V., Subramaniyaswamy, V., Logesh, R., Vijayakumar, V.J.H.I.S.: A study on medical internet of things and big data in personalized healthcare system. Health Inf. Sci. Syst. **6**(1), 14 (2018)
11. Li, S., Da Xu, L., Zhao, S.: The internet of things: a survey. Inf. Syst. Front. **17**(2), 243–259 (2015)
12. Lin, K., Xia, F., Wang, W., Tian, D., Song, J.: System design for big data application in emotion-aware healthcare. IEEE Access **4**, 6901–6909 (2016)
13. Mayer, M., Baeumner, A.J.: A megatrend challenging analytical chemistry: biosensor and chemosensor concepts ready for the internet of things. Chem. Rev. **119**(13), 7996–8027 (2019)
14. Miraz, M.H., Ali, M., Excell, P.S., Picking, R.: A review on internet of things (IoT), internet of everything (IoE) and internet of nano things (IoNT). In: 2015 Internet Technologies and Applications (ITA), pp. 219–224. IEEE (2015, Sept)
15. Mohammed, E.A., Far, B.H., Naugler, C.: Applications of the MapReduce programming framework to clinical big data analysis: current landscape and future trends. BioData Min. **7**(1), 22 (2014)
16. Nemane, S.T., Gholve, S.B., Bhusnure, O.G., Mule, S.T., Ingle, P.V.: Biosensors: an emerging technology in pharmaceutical industry. J. Drug Deliv. Ther. **9**(4), 643–647 (2019)
17. Patil, S.B., Annese, V.F., Cumming, D.R.: Commercial aspects of biosensors for diagnostics and environmental monitoring. In: Advances in Nanosensors for Biological and Environmental Analysis, pp. 133–142. Elsevier, Amsterdam (2019)
18. Prajapati, D.G., Kandasubramanian, B.: Progress in the development of intrinsically conducting polymer composites as biosensors. Macromol. Chem. Phys. **220**(10), 1800561 (2019)
19. Pramanik, P.K.D., Upadhyaya, B.K., Pal, S., Pal, T.: Internet of things, smart sensors, and pervasive systems: enabling connected and pervasive healthcare. In: Healthcare Data Analytics and Management, pp. 1–58. Academic Press, Cambridge (2019)
20. Ray, P.P.: Home health hub internet of things (H^3IoT): an architectural framework for monitoring health of elderly people. In: 2014 International Conference on Science Engineering and Management Research (ICSEMR), pp. 1–3. IEEE (2014, Nov)
21. Schultz, T.: Turning healthcare challenges into big data opportunities: a use-case review across the pharmaceutical development lifecycle. Bull. Am. Soc. Inf. Sci. Technol. **39**(5), 34–40 (2013)
22. Shukla, I., Suneetha, V.: Biosensors: growth and market scenario. Res. J. Pharm. Technol. **10**(10), 3573–3579 (2017)
23. Shahini, A., Hajizadegan, M., Sakhdari, M., Cheng, M.M., Chen, P.Y., Huang, H.H.: Self-powered and transparent all-graphene biosensor. In: 2016 IEEE Sensors, pp. 1–3. IEEE (2016, Oct)

24. Singh, S., Patidar, S., Hada, S., Goyal, S.: Applications of enzyme biosensors in diabetes. Int. J. Tech. Res. Appl. **4**(1), 1 (2018)
25. Swan, M.: The quantified self: fundamental disruption in big data science and biological discovery. Big Data **1**(2), 85–99 (2013)
26. Vashistha, R., Dangi, A.K., Kumar, A., Chhabra, D., Shukla, P.: Futuristic biosensors for cardiac health care: an artificial intelligence approach. 3 Biotech, **8**(8), 358 (2018)
27. Windmiller, J.R., Wang, J.: Wearable electrochemical sensors and biosensors: a review. Electroanalysis **25**(1), 29–46 (2013)
28. Yin, M.J., Gu, B., An, Q.F., Yang, C., Guan, Y.L., Yong, K.T.: Recent development of fiber-optic chemical sensors and biosensors: mechanisms, materials, micro/nano-fabrications and applications. Coord. Chem. Rev. **376**, 348–392 (2018)
29. Zaripova, V.M., Petrova, I.Y.: Information technology of concept design of biosensors. Indian J. Sci. Technol. **9**(1), 1–11 (2016)
30. Zhang, Y., Qiu, M., Tsai, C.W., Hassan, M.M., Alamri, A.: Health-CPS: healthcare cyber-physical system assisted by cloud and big data. IEEE Syst. J. **11**(1), 88–95 (2015)
31. Zhao, G., Guo, Y., Sun, X., Wang, X.: A system for pesticide residues detection and agricultural products traceability based on acetylcholinesterase biosensor and internet of things. Int. J. Electrochem. Sci. **10**(4), 3387–3399 (2015)

Platelets in Diagnostic

Ranjini Chowdhury and Suryyani Deb

Abstract The multifunctional nature of the platelets made them as a significant disease biomarker and a potential drug target in several pathophysiological conditions. Platelet counts alongside its function, dictate disease pathophysiology in different disease aetiology (e.g., cardiovascular diseases, malignancies, diabetes, etc.). Technologies for estimating platelet number or determine function have traversed a long way, through the era of manual platelet tests to an era of automated and bedside platelet analysis systems. This chapter comprises of the state-of-the-art and emerging technologies in platelet diagnosis. A special emphasis has been given on how different technologies can be beneficial for the diseases diagnosis using platelet counts and functions as biomarker.

Keywords Platelet count · Platelet function · Platelet aggregation · Haemostasis

1 Introduction

Although platelets were identified by Bizzozero in the late nineteenth century, but it took more than a decade to elucidate its comprehensive function inside the body. Platelet accounts for its pivotal role in regulating haemostasis [1]; alongside its significance in immunoregulation and inflammation is well established [2]. Owing to its dynamicity in function, platelets remain as an attractive target for both disease diagnosis and prognosis. Further to elaborate on this, if the normal functioning of the platelets become altered, due to any imbalance between pro-aggregatory stimuli and anti-aggregatory substances, it may contribute to several disease pathophysiology. Alongside its function, platelet count stands out to be a major important parameter in a clinical point of view. Normal range of platelets in a healthy individual

R. Chowdhury · S. Deb (✉)
Sangutech Enterprises Private Limited, University of Calcutta, Technology Campus JD-2, Sector 3, Salt Lake City, Kolkata 700098, India
e-mail: suryyani@gmail.com

S. Deb
Department of Biotechnology, Maulana Abul Kalam Azad University of Technology, NH 12, Haringhata 741249, India

G. Dutta et al. (eds.), *Modern Techniques in Biosensors*, Studies in Systems, Decision and Control 327, https://doi.org/10.1007/978-981-15-9612-4_13

is 150,000 to 450,000 platelets per microliter of blood. Conditions arises when the cell count decreases below this normal range termed as thrombocytopenia [3], with a clinical expression varying from asymptomatic to severe bleeding, and when the count exceeds it normal value termed as thrombocytosis [4] with the enhanced risk of abnormal blood clots.

An alteration in platelet function is often usually assisted with prolonged or excessive bleeding, sometimes lowering of platelet count also suggesting of diseases like, grey platelet syndrome [5], Bernard Soulier syndrome [6], Platelet-type von Willebrand disease (PT-vWD) [7].

Years after the discovery of platelets, its role in the implications of cardiovascular diseases, a prime case of morbidity and mortality in the world were identified [8]. Moreover, conditions of atherothrombosis and thromboembolic events and coronary artery disease are also found to be an outcome of hyperactive platelets with hyperaggregative property [9, 10]. Pivotal role of platelets in cardiovascular disease makes scientists interested in the development of antiplatelet drugs (aspirin, clopidogrel, etc.), those inhibits platelet function. Currently antiplatelet therapy is the most potent treatment in cardiovascular diseases [11].

Recent researches have reported instances of altered platelet activity in global burden diseases like diabetes and cancer [12]. In diabetes, altered platelet behaviour of hyperaggregation is prominent; platelets are also found to promote cancer by facilitating tumorogenesis, angiogenesis, metastasis and safeguarding cancer cell from immune system [13, 14]. Evidences of altered platelet activity associated with cytostatic drug treatment in cancer (as in case of Chronic Myeloid Leukemia) further depicts the effect of cytostatic drug therapy on platelet function among cancer patients [15].

Thrombocytopenia has been found to be associated with a wide spectrum of diseases (more than 200 diseases) including hematologic malignancies, infectious diseases (dengue, malaria), thrombotic microangiopathies and autoimmune disorders and side effects of medications [16–19]. As per WHO's guidelines, low platelet count or thrombocytopenia acts as a potential biomarker in the disease assessment of such vector borne diseases like malaria and dengue [19–21].

In ailments of the liver, there are reports of both impaired platelet function and decreased count [22]. Events of bleeding diathesis due to abnormalities of primary haemostasis are found in renal diseases like uremia; however, despite decreased platelet function, these patients also have a high prevalence of cardiovascular and thrombotic complications [23]. Platelets have also been evaluated as potential marker for neurodegenerative diseases like Alzhimer's disease, Parkinson's disease. Platelet can store a huge amount of amyloid precursor protein (APP) and Aβ prptide, that may accumulate in the brain and its vasculature through the blood brain barrier in such diseased conditions. In such conditions, the platelets also have increased membrane fluidity and altered secretion of its metabolites [24–26].

Therefore, abnormal platelet activity both in terms of quantity (platelet count) and quality (platelet function) are associated with several diseases and pathophysiological conditions (Fig. 1). This gives a clear idea depicting the significance of platelets in diseases and pathophysiological conditions that lead to the emergence and evolution

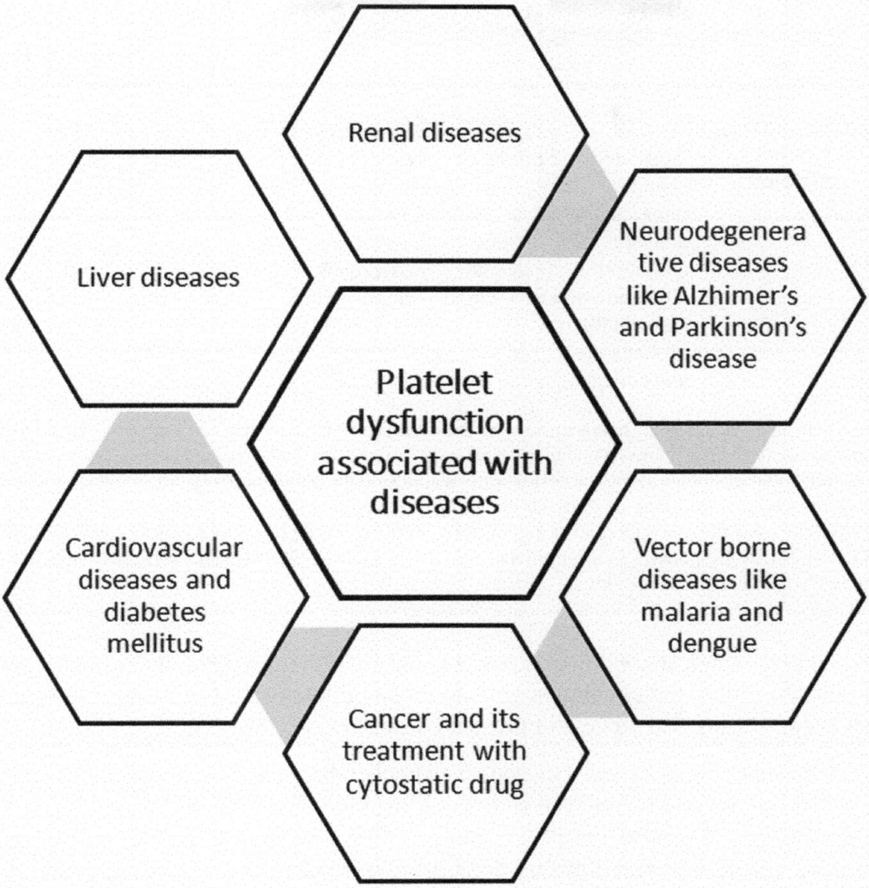

Fig. 1 Platelet dysfunction and thrombocytopenia associated with diseases

of multiple systems for platelet count and function analysis to improve health care quality and ensure increased life expectancy.

1.1 Platelet Function and Its Significance

Clear understanding of platelet function in haemostasis (prevents bleeding) can only help to understand the role of platelet in pathophysiological condition or disease; therefore, in this section, the basic of platelet function is depicted precisely. Blood platelets are produced by the process of fragmentation from bone marrow cell megakaryocytes. Platelets are small anuclear cells of size 1–2 μm. These tiny blood cells circulate in the body with a lifespan of approximately 8–10 days [27]. Under

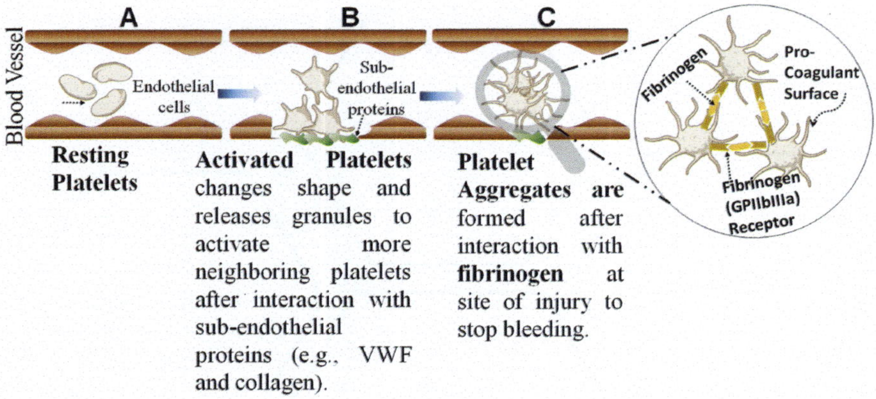

Fig. 2 Role of platelets in haemostasis (A). Normal blood flow through blood vessel when the platelets are in resting phase (B). Under vessel wall damage, the platelets gets exposed to collagen and vWF and the adheres along the vessel wall on adhesion the platelets change its morphology and becomes activated releasing its granular contents, followed by local recruitment of platelets at the site of injury. (C) Platelet aggregation mediated by fibrinogen bridging between activated GPIIbIIIa receptors forms the platelet plug to prevent blood loss and the procoagulant surface of activated platelet helps to stabilize the clot

normal physiological conditions when a blood vessel is damaged, the prime target of platelets within the circulation is to arrest blood loss [28]. The course of events that occurs during this process of prevention of blood loss are (Fig. 2):

1. **Adhesion**: Platelets get exposed to the subendothelium of the injured blood vessels it adheres to collagen and von Willebrand factor (vWF).
2. **Activation**: On adhesion to collagen or vWF the platelets become activated. They change it morphology to support the release of its granular (alpha granules; dense granules; lysosome, etc.) contents (ATP, serotonin, fibrinogen, VWF, etc.). These released materials in turn activates other neighbouring circulating resting platelets.
3. **Aggregation**: Activated platelets express fibrinogen binding receptors GPIIbIIIa and thereby interact with fibrinogen. This fibrinogen—platelet interaction helps to form aggregate at the site of injury resulting in platelet plug formation.
4. **Coagulation**: Without help of fibrin network, the platelet plug rapidly dissolves. To stabilize the clot, procoagulant surface of activated platelets and platelet derived microparticles helps to assembly of plasma coagulation factors and subsequently increase thrombin formation which in turns significantly strengthen and stabilize the clot.

2 Platforms for Diagnosis of Platelet Count and Function

The emergence of platelet counting began with the manual method of counting platelets under microscope [29]. Lack of sensitivity and reproducibility of the obtained data urged the development of more advanced, sensitive and reproducible method for counting platelets. The new upgraded methods developed for counting platelets sensed platelets depending on aperture impedance, optical scattering and fluorescence and are mostly automated (refer Table 1). The enhanced technologies of automated counters have improvised the assessment of platelet count for disease diagnosis and treatment.

In this context, it is important to note that alike count the proper assessment of platelet function is also of greater significance. It is already mentioned that lowering of platelet numbers (i.e., thrombocytopenia) can affect platelet function and promote bleeding tendencies. Interestingly, in many pathological conditions, where though the platelets count falls well within the normal range yet shows impaired haemostasis (bleeding or blood clot formations). The reason behind this impairment could be different for different pathophysiological conditions, e.g.

- **Effect of drugs or nanoparticles exposure**:
  E.g., inhibitory effect of aspirin, clopidogrel (antiplatelet drugs) [11], inhibitory effect of cytostatic drugs (tyrosine kinase inhibitors) [15], aggregatory effect of metallic nanoparticles (gold nanoparticles) [34].
- **Impairment of aggregating factor or receptors with in platelet**:
  E.g., impairment of fibrinogen receptor GPIIbIIIa in Glanzmann thrombasthenia [35]; deficiency of glycoprotein Ib receptor (GPIb) for Bernard-Soulier syndrome [6]; impairment of alpha granule release in case of grey platelet syndrome [5] and gain-of-function in GP1BA receptor in case of platelet-type von Willebrand disease [7].
- **Exposure to aggregation promoting substances**:
  E.g., exposure of endothelial matrix due to atherosclerotic plaque rupture which subsequently activates platelets, form thrombus and block blood vessels.

Impaired platelet function has been found to be associated with several diseases like cardiovascular disease, renal diseases, liver diseases, diabetes and also in cancer along with its side effect due to medications [36]. Cardiovascular diseases and diabetes are generally found to be associated with hyperactive platelet functions [8, 12], whereas in cases of liver disease not only function but count of platelets get decreased [22]. In renal diseases like uremia, patients show bleeding diathesis due to abnormalities of primary haemostasis; however, despite decreased platelet function, these patients also have a high prevalence of cardiovascular and thrombotic complications [23].

The beginning of platelet function diagnosis started with the introduction of evaluation of bleeding time (BT) by duke procedure that assess the plug forming capacity of the platelets [37]. However, due to lack of reproducibility of the test and poor sensitivity towards mild defects, the use of this procedure slowly declined.

Table 1 List of devices and technologies for counting platelets from blood samples

Sensing method	Name of the instrument	Mode of sensing	Application
Impedance, optical, and immunological	CELL-DYN 4000 (Abbott Diagnostics)	Optical analysis of blood sample using scatter angles of 7° (intermediate angle) and 90° (wide angle scatter); followed by simultaneous determination of the impedance platelet count. Algorithms employed to exclude non-platelet particles to rule out false positive count [30]	Precise platelet counting with accuracy
	CELL-DYN Sapphire (Abbott Diagnostics)		
Impedance and optical fluorescence	XE-2100, XE-5000 (Sysmex Corporation)	Staining of RNA/DNA of reticulated cells and platelet membranes and granules by polymethine dye and simultaneous counting of all the cells passed through the light beam of a semiconductor diode laser. Difference in intensity of fluorescence separates platelets from other cells. For the XN series instrument, fluorocell PLT dye is used which specifically stains platelets only [31]	Precise platelet counting with accuracy. Potent to consider the large sized platelets/thrombocytes (in case of macrothrombocytopenia) during platelet count
	XN Series (Sysmex Corporation)		

(continued)

Table 1 (continued)

Sensing method	Name of the instrument	Mode of sensing	Application
Optical	ADVIA 120 (Siemens)	Two-dimensional platelet analysis; volume and refractive index of effectively sphered individual platelets are simultaneously determined on a cell-by-cell basis by measuring two angles of laser light scatter at 2°–3° and at 5°–15°. The two scatter measurements are converted to volume (platelet size) and refractive index (platelet density) values using the Mie theory of light scattering for homogenous spheres [32]	Two-dimensional platelet count helps in accurate platelet counting even in thrombocytopenic samples
	ADVIA 2120 (Siemens)		
Impedance	LH 750, DxH 800, AC.T, Dx 300 (Beckman Coulter)	In the impedance-based method whole blood samples passed through aperture (sensing zone) is suspended in electrolyte solution, and the change in electric impedance is detected depending on the volume of each cell [33]	Automated platelet count method. Count sometimes overestimated than optical-based detection
	Pentra-DX 120, ABX Pentra-DF 120 (HORIBA Medical ABX)		
	CELL-DYN 1200 (Abbott Diagnostics)		
	Pentra 60, Micros 60 (HORIBA Medical)		
	Advia 70 (Siemens)		
	KX21, XS-1000i, pocH-100i (Sysmex Corporation)		

A plentiful of platelet function analysis instruments has been developed to fulfil the need of platelet function monitoring and analysis [38–52]. In 1960, the remarkable invention of light transmission aggregometery (LTA), revolutionised the diagnosis of platelet function. LTA is recognised as the gold standard method for platelet function test. This method relies on the sensing capability of light transmission through the platelet rich plasma (PRP) (PRP is obtained through centrifugation of the whole blood) based on aggregation property; formation of aggregation reduces turbidity of PRP (which is a colloidal suspension) and increases light transmission whereas low light transmission indicates reduced or no platelet aggregation [38, 39]. A modified version of the LTA is 96-well plate assay for platelet aggregation, where the sample volume required is markedly reduced and the well plate is pre-coated with agonists, therefore only measuring the absorbance gives the result [40, 41]. Development of impedance dependent identification of platelet function further simplified the testing procedure as it uses whole blood instead of PRP, also taking into account the platelet neutrophils interaction rather than PRP only [42]. This method senses the resistance between electrodes as platelets aggregate. Luciferine/luciferase dependent determination of platelet granule release (ATP release from dense granule) technique is often coupled with platelet aggregometry instrument [43]. Platelet works also based on impedance but instead of measuring aggregates, it measures number of free platelets [44]. Fully automated and point of care testing facility for antiplatelet therapy using whole blood have also evolved with the advent of VerifyNow instrument, based on the principle of agglutination of fibrinogen coated polystyrene beads in whole blood which is in proportion to the number of fibrinogen receptor GPIIbIIIa activated by a specific stimulus, may be blocked by antiplatelet drugs [45]. Flow cytometric analysis of platelet biomarkers serves a powerful sensing platform for expression of platelet surface receptors or any conformational changes in markers [46]. Another updated instrument for platelet function test is PFA analyser that measures the closure time through platelet plug formation in a cartridge-based method [47, 48] and shear-dependent platelet plug identification through global thrombosis test (GTT) [49]. A complete profile of clot formation, including the kinetics of clotting, clot strength and fibrinolysis is obtained from thromboelastography (TEG/ROTEM), a point of care system that determines the bleeding risk as well as antiplatelet drug effect [50, 51]. Thromboxane synthesis and release from activated platelets can be measured through radio or enzyme-linked immune assay [52]. A long list of tests for platelet function have developed based on the requirements of analysis and further progress in this field of sensing platelets function is going on. The details of the till date established techniques for platelet function analysis in the field of diagnosis is given in Table 2.

Table 2 List of different technologies and methods for platelet function evaluation

Sensing method	Mode of sensing	Application
Platelet aggregation		
Bleeding time	Time taken to stop bleeding on account of a small incision made into skin [37]	It is an in vivo screening test
Light transmission platelet aggregation (LTA)	It is a photo optical measurement of light transmission through the platelet suspension, as platelet aggregates the turbidity decreases with increase in light transmission presented as % of aggregation [38, 39]	Detection of platelet aggregation disorders in congenital/acquired bleeding disorders, VW disease, Bernard Soulier syndrome and during antiplatelet therapy
LTA on 96-well plate	Principal is same as LTA but in agonist pre-coated 96-well plates are used as reaction chamber [40, 41]	Sample volume require is very less amount
Impedance platelet aggregation	Electrical impedance is measured in whole blood sample, where increased impedance suggests platelet aggregation [42]	Similar application like LTA use of whole blood makes the method more convenient and more physiologically analogue
Lumiaggregometry	The ATP release from dense granule measured through the reaction between ATP and luciferin which gives chemiluminescence [43]	Diagnosis of platelet function defects like abnormalities in platelet, granular secretion and/or content, plasma membrane receptor defects, platelet functions during thrombocytopenia
Plateletworks	Platelet counting pre- and post-activation in whole blood [44]	Used in restricted operating areas such as cardiac surgery and cardiac operating room due to strict timing in testing after sample is collected

(continued)

Table 2 (continued)

Sensing method	Mode of sensing	Application
VerifyNow	Whole blood platelet aggregation assessment by optical turbidimetric method, where fibrinogen coated beads are used for platelet binding [45]	Point of care technology (POCT) for platelet aggregation assessment and antiplatelet therapy (abciximab or eptifibatide) monitoring for patients with cardiovascular diseases. It is mainly employed in emergency cardiac operating room
Platelet analysis based on flow cytometry		
Flow cytometry	laser-based detection of platelet surface receptors using fluorescent labelled antibodies [46]	Platelet count, alpha granule release, detection of different platelet specific markers Ideal for diagnosis of Bernard Soulier syndrome, Glanzmann's thrombasthenia and acquired platelet dysfunctions, i.e. Heparin induced thrombocytopenia (HIT) and storage pool disease, reticulated platelets
Platelet adhesion		
PFA-100; Innovance PFA-200	Evaluation of time of high shear whole blood flow blockage by platelet plug formation into a hole in activated surface [47, 48]	Assessment of bleeding tendency and drug effects. Screening of VW defects or Bernard–Soulier syndrome or Glanzmann's thromboasthenia from those due to antiplatelet therapy with aspirin. Risk assessment in cardiac surgery
Global thrombosis test (GTT)	Measurement of time cessation of WB flow by high shear-dependent platelet plug formation [49]	POCT determines thrombotic status and is employed in acute and critical care settings. Evaluation of platelet function and thrombolysis
Platelet function methods combined with viscoelastic test		

(continued)

Table 2 (continued)

Sensing method	Mode of sensing	Application
Thromboelastogram (TEG)	Assessment of rate of clot formation based on low shear-induction [50]	Though this is basically a rheometric assessment of clot formation but antiplatelet effect can be monitored through it
ROTEM platelet	This is the combination of thromboelastometry measurements and impedance-based platelet aggregation [51]	Assessment of global hemostasis along with diagnosis of platelet defects and monitoring of antiplatelet treatments effect
Platelet analysis based on flow cytometry		
Flow cytometry	Laser-based detection of platelet surface receptors using fluorescent labelled antibodies [46]	Platelet count, alpha granule release, detection of different platelet specific markers Ideal for diagnosis of Bernard Soulier syndrome, Glanzmann's thrombasthenia and acquired platelet dysfunctions, i.e Heparin induced thrombocytopenia (HIT) and storage pool disease, reticulated platelets
Evaluation of Thromboxane metabolites		
Radio- or enzyme-linked immune assays	Ligand-binding assays to measure thromboxane synthesis and release from activated platelets [52]	Evaluates activation state of platelets and identifies defects in thromboxane production, and monitors aspirin therapy

3 Current Emerging Technologies for Platelet Function

At present, multiple technologies have been developed for assessing the platelet count and platelet (dys)function in order to evaluate risk of haemorrhage or thrombus formation among individuals. The respective tests allow the assessment of platelet functioning, activation, secretion, contributing largely in disease monitoring and treatment. Despite the versatility and significance of these available platelet assays, multiple issues originated regarding the methodological aspects, data reproducibility and sensitivity of these tests. Many of the assays described are affected by platelet concentration, i.e. platelet count, e.g. the PFA-100, viscoelastic test TEG/ROTEM, multiplate analyzer and VerifyNow. Hematocrit (volume percentage of red blood cells in blood) levels are also found to influence the developed methods [43, 53]. The gold standard method of light transmission platelet aggregation also encounters the problem of manual sample processing by trained individuals, high sample volume and longer time requirement inefficient in measuring platelet function in thrombocytopenic samples [38].

To fill in the major gap in the existing technologies introduction of newer and advanced methods and point of care technologies (POCT) are in the queue. In this context, platelet adhesion testing microfluidics device, platelet spreading tests, using fluorescence microscopy or scanning electron microscopy platelet activation analysis by HPLC and fluorescence microscopy are under development, however, their feasible use in clinical purpose is uncertain due to requirement of huge laboratory set up and high expenses [54–56]. Therefore, handy POCT are emerging, those can be more affordable and feasible to execute in clinical set up on routine basis. Platelet cytoskeleton gives strength and stability to aggregate during consolidation of fibrin-rich blood clot. Viscoelastic approach cannot identify the specific contribution of platelets in clot formation, so microfluidic approach to study platelet adhesion and aggregation under clinically relevant flow conditions developed. The benefit of this technique is it requires very less amount of sample. To map individual platelet force, force-activatable biosensor called integrative tension sensor (ITS). ITS converts molecular tensions to fluorescent signals, enabling cellular force mapping directly by fluorescence imaging [55]. Diagnostic approach towards heparin induced thrombocytopenia (HIT Type II) is developed using quartz crystal microbalance with dissipation (QCM-D) platform [57]. Moreover, QCM-based measuring devices are thought to be a point of care alternative methods for rapid bedside testing of platelet aggregation with high sensitivity and low sample volume. To distinguish between normal resting platelets and inhibited platelet functionality, assay based on molecularly imprinted polymers (MIP) coated onto quartz crystal microbalance (QCM) has developed by the researcher [58]. For analysis of platelet derived extra cellular vesicles or microparticles (PMP), a major risk marker for arterial disorders, graphene oxide-based electrochemical biosensor is employed. Electrodes fabricated with immobilised graphene oxide layers and a specific antibody targeted against active conformation of integrin αIIbβ3 on PMP surface to sense PMP [59]. New aspects developed to measure platelet activity includes bio-sensing thiol reductase activity by a platelet targeting biosensor in a microfluidic thrombosis model that contains a disulphide linked glutathione (GSSG) mimicking peptide conjugated to a CD61 antibody that is able to bind to GPβ3 receptor on platelets surface. Peptide fluoresces upon disulfide bond breakage and the antibody localises the fluorescent signal on the platelet surface [60]. In this regard, it is noteworthy to mention our recent development on POCT of platelet function detecting technology using ultra violet ray. Our system to analyse blood platelet function is free of any chemical agonist assisted with deep learning image analysis and wide range of accessibility even from the remotest areas [61]. Many of these newer and advanced emerging technologies have already been developed and more are in the process of development to enrich the diagnostic platform for screening bleeding disorder and monitoring antiplatelet therapy.

4 Global Market Status

Because of their huge clinical applications in diagnosis, currently the demand for devices to test platelet function is very high all over the world [62]. The largest market of platelet function test (PFT) is North America whereas fastest growing market is in Asia Pacific region. The compound annual growth rate (CAGR) of PFT is 15% which is quite high [62]. Now in spite of their high demand, all the existing devices [light transmission aggregometer (LTA)], platelet function analyser 200 (PFA-200, VeryfyNow, Immuno-phenotyping, etc.) being very big in size, expensive and require trained personnel for instrument handling and data interpretation, are not readily available in all laboratories. Therefore, need of the handheld devices is prospering. According to the latest market reports, CAGR for the hand held devices will be 8% for period of 2019–2023 [63]. Therefore, technology with high sensitivity and reproducibility, affordability and free of trained personnel will be the emerging field of research and global demand for the future platelet diagnostics.

5 Future Prospect and Scopes

Translating emerged platelet function tests into daily or routine diagnosis is under process. Although there are many commercially available methods but owing to their high running cost and demand for skilful personnel and high laboratory set up, does not allow their widespread application in all laboratories and healthcare institutions, especially in remote areas. Thus, global or single step of platelet function with high sensitivity and specificity is need of the hour for routine laboratory diagnosis. Attempts are made to meet this need though various approaches like microfluidics, graphene oxide biosensor, force-activatable biosensor, quartz crystal microbalance (QCM)-based platelet sensing, label free, deep learning method-based platelet function detection, etc. Scientists are hopeful that these improvised and upgraded technologies to study platelet function are going to resolve the current problems encountered in the currently used technologies.

6 Discussion

Long after its discovery, the role of platelets in disease pathophysiology is well identified. The multifunctional nature of these tiny anucleated cells, makes them unique in sensitising different disease conditions. This pathophysiological sensitivity helps them to behave as disease biomarker and sometimes potential drug target also. As for example, marked increase in platelet reactivity supports the development of unwanted thrombus formation in the site of ruptured artherosclerotic plaque that can lead to myocardial infarction and other cardiovascular problems. The global burden diseases like cancer and diabetes also exhibit the pathophysiological features of enhanced platelet aggregation. However, it is not only the function of the platelets that dictates disease pathophysiology but also its count that have a marked influence in several disease aetiology. Conditions of thrombocytopenia, where the platelet count falls below its normal value also have high clinical significance. Bernard Soulier syndrome is an example of platelet disorder where platelet function and number both gets affected. In infectious diseases like malaria and dengue, quantification of platelets count has shown to be a major severity marker. Thrombocytopenia has been found to have significant clinical manifestations in several haematological malignancies and side effects on medications. In renal diseases, liver diseases alteration of both platelets' count and function have been encountered with the pathological features of bleeding diathesis. Therefore, it is obvious that altered platelet number or function have significant contributions in disease pathophysiology serving as a potent marker and further emphasising the importance of diagnosis of both platelet numbers and function.

Beginning with the method of manual count, the platelet count technology has traversed a long way to the generation of automated counting systems. The analytical procedures used for the counting relies on optical, impedance, fluorescence or immune-based principles. Similarly, for the diagnosis of platelet function dedicated technologies have been developed with specific instruments for determining specific platelet functions viz aggregation, adhesion, antiplatelet therapy, viscoelastic property, metabolite release and receptor analysis. Despite of development of such technologies, there lies a major gap for them to be used in diagnostic purposes, limiting their usage in clinical practice. To seal this gap, newer approaches of point of care technologies, microfluidics, graphene oxide biosensor, force-activatable biosensor, quartz crystal microbalance (QCM) deployed tests, label free, deep learning mediated estimation of platelet function have evolved prospering the field of platelet diagnostics. Further improvisations and upgradation in technologies for platelet diagnosis are going on to resolve the current problems encountered in the existing technologies.

References

1. Ribatti, D., Crivellato, E.: Giulio Bizzozero and the discovery of platelets. Leuk. Res. **31**, 1339–1341 (2007)
2. Garattini, S., de Gaetano, G., Samanin, R., Bernasconi, S., Roncaglioni, M.: Effects of trazodone on serotonin in the brain and platelets of the rat. Biochem. Pharmacol. **25**, 13–16 (1976)
3. Gauer, R.L., Braun, M.M.: Thrombocytopenia. Am. Family Phys. **85**, 612–22 (2012)
4. Bleeker, J., Hogan, W.: Thrombocytosis: diagnostic evaluation, thrombotic risk stratification, and risk-based management strategies. Thrombosis **2011**, 1–16 (2011)
5. Gunay-Aygun, M., Zivony-Elboum, Y., Gumruk, F., Geiger, D., Cetin, M., Khayat, M., Kleta, R., Kfir, N., Anikster, Y., Chezar, J., Arcos-Burgos, M., Shalata, A., Stanescu, H., Manaster, J., Arat, M., Edwards, H., Freiberg, A., Hart, P., Riney, L., Patzel, K., Tanpaiboon, P., Markello, T., Huizing, M., Maric, I., Horne, M., Kehrel, B., Jurk, K., Hansen, N., Cherukuri, P., Jones, M., Cruz, P., Mullikin, J., Nurden, A., White, J., Gahl, W., Falik-Zaccai, T.: Gray platelet syndrome: natural history of a large patient cohort and locus assignment to chromosome 3p. Blood **116**, 4990–5001 (2010)
6. Berndt, M., Andrews, R.: Bernard-Soulier syndrome. Haematologica **96**, 355–359 (2011)
7. Bury, L., Malara, A., Momi, S., Petito, E., Balduini, A., Gresele, P.: Mechanisms of thrombocytopenia in platelet-type von Willebrand disease. Haematologica **104**, 1473–1481 (2019)
8. Sharma, G., Berger, J.: Platelet activity and cardiovascular risk in apparently healthy individuals: a review of the data. J. Thromb. Thrombolysis **32**, 201–208 (2011)
9. Davì, G., Patrono, C.: Platelet activation and atherothrombosis. N. Engl. J. Med. **357**, 2482–2494 (2007)
10. Vorchheimer, D., Becker, R.: Platelets in atherothrombosis. Mayo Clin. Proc. **81**, 59–68 (2006)
11. Lahiri, P., Roy, S., Sardar, P., Deb, S., Chakrabarti, P., Guha, P., Guha, S., Chaudhuri, U., Dasgupta, A.: Platelet responsiveness to yohimbine hydrochloride and MRS2179 in the context of the interaction between collagen and epinephrine in acute coronary syndrome. Blood Cells Mol. Dis. **43**, 105–110 (2009)
12. Hughes, A., McVerry, B., Wilkinson, L., Goldstone, A., Lewis, D., Bloom, A.: Diabetes, a hypercoagulable state? Haemostatic variables in newly diagnosed type 2 diabetic patients. Acta Haematol. **69**, 254–259 (1983)
13. Bambace, N., Holmes, C.: The platelet contribution to cancer progression. J. Thromb. Haemost. **9**, 237–249 (2011)
14. Jain, S., Harris, J., Ware, J.: Platelets. Arterioscler. Thromb. Vasc. Biol. **30**, 2362–2367 (2010)
15. Deb, S., Boknäs, N., Sjöström, C., Tharmakulanathan, A., Lotfi, K., Ramström, S.: Varying effects of tyrosine kinase inhibitors on platelet function—a need for individualized CML treatment to minimize the risk for hemostatic and thrombotic complications? Cancer Med. **9**, 313–323 (2019)
16. Izak, M., Bussel, J.: Management of thrombocytopenia. F1000Prime Rep. **6** (2014)
17. Azeredo, E., Monteiro, R., de-Oliveira Pinto, L.: Thrombocytopenia in dengue: interrelationship between virus and the imbalance between coagulation and fibrinolysis and inflammatory mediators. Mediat. Inflamm. **2015**, 1–16 (2015)
18. Schexneider, K.: Fibrin sealants in surgical or traumatic hemorrhage. Curr. Opin. Hematol. **11**, 323–326 (2004)
19. Gupta, N., Jain, U., Sahare, K., Bansal, S.: Study of thrombocytopenia in patients of malaria. Trop. Parasitol. **3**, 58 (2013)
20. Cecilia, D.: Current status of dengue and chikungunya in India. WHO South-East Asia J. Public Health **3**, 22 (2014
21. Muley, A., Lakhani, J., Bhirud, S., Patel, A.: Thrombocytopenia in plasmodium vivax Malaria: how significant? J. Trop. Med. **2014**, 1–4 (2014)
22. Witters, P., Freson, K., Verslype, C., Peerlinck, K., Hoylaerts, M., Nevens, F., Van Geet, C., Cassiman, D.: Review article: blood platelet number and function in chronic liver disease and cirrhosis. Aliment. Pharmacol. Ther. **27**, 1017–1029 (2008)

23. Boccardo, P., Remuzzi, G., Galbusera, M.: Platelet dysfunction in renal failure. Semin. Thromb. Hemost. **30**, 579–589 (2004)
24. Catricala, S., Torti, M., Ricevuti, G.: Alzheimer disease and platelets: how's that relevant. Immun. Ageing **9**, 20 (2012)
25. Zubenko, G., Cohen, B., Reynolds, C.F., III, Boller, F., Malinakova, I., Keefe, N.: Platelet membrane fluidity in Alzheimer's disease and major depression. Am. J. Psychiatry **144**, 860–868 (1987)
26. Prodan, C., Ross, E., Stoner, J., Cowan, L., Vincent, A., Dale, G.: Coated-platelet levels and progression from mild cognitive impairment to Alzheimer disease. Neurology **76**, 247–252 (2011)
27. Hou, Y., Carrim, N., Wang, Y., Gallant, R., Marshall, A., Ni, H.: Platelets in hemostasis and thrombosis: novel mechanisms of fibrinogen-independent platelet aggregation and fibronectin-mediated protein wave of hemostasis. J. Biomed. Res. (2015)
28. Holinstat, M.: Normal platelet function. Cancer Metastasis Rev. **36**, 195–198 (2017)
29. Brecher, G., Schneiderman, M., Cronkite, E.: The reproducibility and constancy of the platelet count. Am. J. Clin. Pathol. **23**, 15–26 (1953)
30. Ault, K.: Editorial. Lab. Hematol. **11**, 235–235 (2005)
31. Briggs, C., Harrison, P., Grant, D., Staves, J., Machin, S.: New quantitative parameters on a recently introduced automated blood cell counter—the XE 2100TM. Clin. Lab. Haematol. **22**, 345–350 (2000)
32. Kunicka, J., Fischer, G., Murphy, J., Zelmanovic, D.: Improved platelet counting using two-dimensional laser light scatter. Am. J. Clin. Pathol. **114**, 283–289 (2000)
33. Coulter, W.H.: Inventor. Means for counting particles suspended in a fluid. United States patent US 2,656,508 (1953)
34. Bandyopadhyay, S., Azharuddin, M., Dasgupta, A., Ganguli, B., SenRoy, S., Patra, H., Deb, S.: Probing ADP induced aggregation kinetics during platelet-nanoparticle interactions: functional dynamics analysis to rationalize safety and benefits. Front. Bioeng. Biotechnol. **7**, (2019)
35. Solh, T., Botsford, A., Solh, M.: Glanzmann's thrombasthenia: pathogenesis, diagnosis, and current and emerging treatment options. J. Blood Med. **6**, 219–227 (2015)
36. Ghoshal, K., Bhattacharyya, M.: Overview of platelet physiology: its hemostatic and nonhemostatic role in disease pathogenesis. Sci. World J. **2014**, 1–16 (2014)
37. Duke, W.: The relation of blood platelets to hemorrhagic disease. J. Am. Med. Assoc. **55**, 1185 (1910)
38. Born, G.: Aggregation of blood platelets by adenosine diphosphate and its reversal. Nature **194**, 927–929 (1962)
39. O'Brien, J.: Platelet aggregation: Part I Some effects of the adenosine phosphates, thrombin, and cocaine upon platelet adhesiveness. J. Clin. Pathol. **15**, 446–452 (1962)
40. Sun, B., Tandon, N., Yamamoto, N., Yoshitake, M., Kambayashi, J.: Luminometric assay of platelet activation in 96-well microplate. Biotechniques **31**, 1174–1181 (2001)
41. Armstrong, P., Dhanji, A., Truss, N., Zain, Z., Tucker, A., Mitchell, J., Warner, T.: Utility of 96-well plate aggregometry and measurement of thrombi adhesion to determine aspirin and clopidogrel effectiveness. Thromb. Haemost. **102**, 772–778 (2009)
42. Cardinal, D., Flower, R.: The electronic aggregometer: a novel device for assessing platelet behavior in blood. J. Pharmacol. Methods **3**, 135–158 (1980)
43. Pai, M., Wang, G., Moffat, K., Liu, Y., Seecharan, J., Webert, K., Heddle, N., Hayward, C.: Diagnostic usefulness of a lumi-aggregometer adenosine triphosphate release assay for the assessment of platelet function disorders. Am. J. Clin. Pathol. **136**, 350–358 (2011)
44. Campbell, J., Ridgway, H., Carville, D.: Plateletworks®. Mol. Diagn. Ther. **12**, 253–258 (2008)
45. van Werkum, J., Harmsze, A., Elsenberg, E., Bouman, H., ten Berg, J., Hackeng, C.: The use of the VerifyNowsystem to monitor antiplatelet therapy: a review of the current evidence. Platelets **19**, 479–488 (2008)
46. Linden, M., Frelinger, A., Barnard, M., Przyklenk, K., Furman, M., Michelson, A.: Application of flow cytometry to platelet disorders. Semin. Thromb. Hemost. **30**, 501–511 (2004)
47. Favaloro, E.: Clinical utility of the PFA-100. Semin. Thromb. Hemost. **34**, 709–733 (2008)

48. Favaloro, E.: Clinical application of the PFA-100®. Curr. Opin. Hematol. **9**, 407–415 (2002)
49. Rand, M., Leung, R., Packham, M.: Platelet function assays. Transfus. Apheres. Sci. **28**, 307–317 (2003)
50. Whiting, D., DiNardo, J.: TEG and ROTEM: technology and clinical applications. Am. J. Hematol. **89**, 228–232 (2014)
51. Luddington, R.: Thrombelastography/thromboelastometry. Clin. Lab. Haematol. **27**, 81–90 (2005)
52. Muir, A., McMullin, M., Patterson, C., McKeown, P.: Assessment of aspirin resistance varies on a temporal basis in patients with ischaemic heart disease. Heart **95**, 1225–1229 (2008)
53. Harrison, P.: Advances in platelet counting. Br. J. Haematol. **111**, 733–744 (2000)
54. Nagy, M., Heemskerk, J., Swieringa, F.: Use of microfluidics to assess the platelet-based control of coagulation. Platelets **28**, 441–448 (2017)
55. Wang, Y., LeVine, D., Gannon, M., Zhao, Y., Sarkar, A., Hoch, B., Wang, X.: Force-activatable biosensor enables single platelet force mapping directly by fluorescence imaging. Biosens. Bioelectron. **100**, 192–200 (2018)
56. Ting, L., Feghhi, S., Taparia, N., Smith, A., Karchin, A., Lim, E., John, A., Wang, X., Rue, T., White, N., Sniadecki, N.: Contractile forces in platelet aggregates under microfluidic shear gradients reflect platelet inhibition and bleeding risk. Nat. Commun. **10** (2019)
57. Hussain, M., Gehring, F., Sinn, S., Northoff, H.: A straightforward detection of HIT Type II via QCM-D. UK J. Pharm. Biosci. **3**, 18 (2015)
58. Strallhofer, A., Jung, S., Jungbauer, C., Lieberzeit, P.: Development of a novel platelets functional assay using QCM. Proceedings **1**, 514 (2017)
59. Kailashiya, J., Singh, N., Singh, S., Agrawal, V., Dash, D.: Graphene oxide-based biosensor for detection of platelet-derived microparticles: a potential tool for thrombus risk identification. Biosens. Bioelectron. **65**, 274–280 (2015)
60. Zhu, S., Welsh, J., Brass, L., Diamond, S.: Platelet-targeting thiol reduction sensor detects thiol isomerase activity on activated platelets in mouse and human blood under flow. J. Thromb. Haemost. **14**, 1070–1081 (2016)
61. Deb, S., Chowdhury, R., Sadhu, A., Chowdhury, A.R., Dasgupta, A.K., Dhara, A.K., Roy, K., Bhattacharyya, M., Chakrabarti, A.: A system and method for detecting platelet function using UV light and deep learning analysis of microscopic images. (Indian Patent and PCT, Application number 201931048635 [TEMP/E-1/51484/2019-KOL] date 2019/11/27)
62. Platelet Aggregation Devices Market. Growth, Trends, and Forecast (2020–2025), https://mordorintelligence.com/industry-reports/platelet-aggregation-devices-market
63. Global Rugged Handheld Devices Market 2019–2023. 8% CAGR Projection Through 2023. Technavio, https://www.businesswire.com/news/home/20191213005108/en/Global-Rugged-Handheld-Devices-Market-2019-2023-8

Mimicking Human Kidney: Research Towards Better Solutions for Kidney Failure

Jasti Sateesh, Koushik Guha, Arindam Dutta, Pratim Sengupta, Ajay Agarwal, and K. Srinivasa Rao

Abstract The human body is a highly sophisticated machine that is not fully understood to date. The body function is highly dependent on the versatile functions performed by different organs, which have their diverse arrangement of cells. Every organ (perhaps every body part) is prone to damage and malfunction because of the topological conditions, lifestyle, habitats, and food chemicals. Helping humans to lead an easy, luxurious, and healthy life are the primary goal of any technology, especially extending human life is the ultimate target of medical instrumentation. The importance of developing innovative techniques for disease diagnosis and treatment are burning topics since the start of the medical and pharmaceutical industry. Kidney failure is one of the common health issues worldwide, which is a slow poison affecting 10% of the world population. The importance of replacing the kidney is essential to extend a patient life. This review focuses on Organ-on-Chip technology with a major focus on Kidney-on-Chip (KOC). The evolution of techniques to diagnose and treat organ failure is elaborately presented. Major emphasis solely on the development of kidney failure causes, diagnostic techniques, replacement techniques are reported with a timeline of developments. The major functions of the kidney that have been achieved artificially so far are reviewed to the deepest level. The future directions in this field are predicted and presented. MEMS and microfluidics allow the design and manufacturing of devices at microscale without compromising the actual

J. Sateesh · K. Guha (✉)
Department of Electronics and Communication Engineering, National MEMS Design Centre, National Institute of Technology Silchar, Silchar, Assam 788010, India
e-mail: koushik@ece.nits.ac.in

A. Dutta
RG Stone Urology & Laparoscopic Hospital, Kolkata, West Bengal, India

P. Sengupta
Belle Vue Clinic, Kolkata, West Bengal, India

A. Agarwal
Smart Sensors Area and Head, Nano Biosensors Group, CSIR-CEERI, Pilani, Rajasthan, India

K. Srinivasa Rao
Department of Electronics and Communication Engineering, MEMS Research Centre, KLEF, Guntur, Andhra Pradesh, India

G. Dutta et al. (eds.), *Modern Techniques in Biosensors*, Studies in Systems, Decision and Control 327, https://doi.org/10.1007/978-981-15-9612-4_14

functionalities, especially in terms of disease diagnosis and treatment. Microfluidics technology revolutionized the development of artificial organ industry, the chances of realizing an organ substantially improved.

Keywords Organ-on-Chip · Kidney-on-Chip · Microfluidics · Artificial kidney

1 Introduction

Ever since human civilization started, physical body care has become vital for easy living. Since then, the research on the body has started, thereafter huge inventions have been achieved. Presently, all parts of the body down to micrometer level have been revealed and are being controlled. The human body tends to affect by different diseases, of them some are life-threatening. Because of the regional and communal habits, new diseases are effecting and turning out into pandemics (Plague, Cholera, Tuberculosis, HIV, Ebola, COVID, etc.). Any disease-causing virus specifically affects a single organ and makes it fail to function. If proper medical intervention is not applied, the disease becomes fatal and leads to death. Hence, the diagnosis tools are very important in terms of disease treatment, there have been different techniques to do this job from the initial days. Presently, plenty of fast diagnosing techniques (ECG, EEG, CT scan, Ultrasound, MRI scan, etc.) are available and serving patient needs. The time frame for the development of diagnostic tools is decreasing with the advent of new technologies. As discussed earlier, the need to replace failed organs has become vital in the present day scenario. According to World Health Organization transplantation reports (Gobal observatory on donation and transplantation), 139,024 organs replaced in 2017, the number increased 7.5% compared to 2015 transplants. Of these transplants, 36% of living kidney transplants and 19% of living liver transplants.

2 Replacement Techniques

Organ-on-Chip: Bringing modernization to the human lifestyle through advancements in technology is primarily important. Electronic gadgets with minimum size and maximum functionalities are the present trend. Extending electronic device services to serve interdisciplinary applications is more promising. Analyzing the pathophysiology and replacing the failed organs is a dream for many researchers. Organ-on-Chip is a device, possessing a cell culture platform developed with microfluidic technology. The OoC has received a lot of attention in the twenty-first century delivering devices such as Kidney-on-Chip [1, 2], intestine-on-chip [2, 3], lung-on-chip [4], blood vessel-on-chip [5], cancer-on-chip [6], and bone marrow-on-chip [7]. The revolution in OoC technology took pace after the official announcement by Barak Obama, the former president of the United States of America in 2011. It was supported by 3 organizations of the US government, the NIH, FDA, and the US Department of Defence [8, 9].

3 The Flow of Organ-On-Chip

3D molding technology has enabled the fabrication of microfluidic devices with less effort. The use of 3D cell culture for biomimicking organs is gaining interest in the present scenario [10]. The physically modified microfluidic channels for making crucial subparts of Organ-on-Chip are underway. It is possible to build personalized microfluidic chips for pathophysiology assessment and treatment of an individual can be predicted in the coming future through the extension of Organ-on-Chip to Body-on-Chip applications. The possibility for the development of personalized bio-chips for replacement of organs function has started and is in progress. A person with organ failure can be treated with this technology more accurately, and it is possible to replace the failed organ which is compatible to solely to his body [11]. A monolayer cell sheet lined with endothelial cells has been developed [12]. An interesting review focusing on the microfluidics used in cell-based microchip systems is presented in [13]. For these kind of applications, it is significant to analyze the cell–cell interaction, the research of cells-on-chip for point of care devices is swiftly progressing [14] and will be a matured technology in coming future. The general flow of making OoC is presented (Fig. 1).

Drug development is a vast process involving five steps, of them four are dedicated for screening the effectiveness and safety of a drug. In the preclinical trials, the in vitro systems that are presently in use are comparatively slow which directly or indirectly

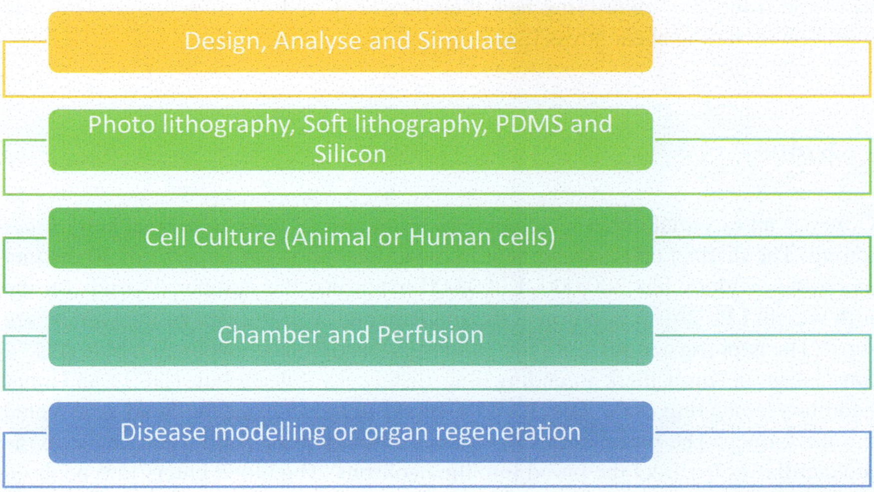

Fig. 1 The flow for Organ-on-Chip

making delays in drug marketization. Estimating the intermolecular mechanisms of new drug and its toxicity can aid in fast production. Microfluidics in drug screening through organ interaction is developed and reported [15], a review on organ-on-chip technology dedicated for drug discovery is reported in [16]. There is a microfluidic chip for drug toxicity testing developed with multiplexed microchannels, which maintains hepatocytes [17]. The usefulness of Organ-on-Chips is for drug discovery, and disease modeling has been reviewed with emphasis on the current challenges and future aspects [18].

The body parts tend to fail very often, perhaps because of lifestyle, diet, heredity, accidentally, etc., some causes can't be avoided though the person is health conscious. The necessity to replace the failed body part is extremely necessary. There are replacement techniques available for upper and lower limbs. In the same way, there are replacement techniques available for organs as well, such as therapeutic transplantation for heart, lung, liver, and kidney.

Before plunging into the kidney regeneration, it is crucial to know the organs and some other applications developed using microfluidic technologies. A detailed review of microtechnologies used for regenerating liver functioning [19]. The detection of cancer tumor cells is becoming significantly crucial, early stage detection can save lives, but the existing tests are less accurate and time-consuming. Hence, the development of cancer biomarkers using microfluidics has become a viable solution. The selective separation of circulating tumor cells in the blood is presented [20]. Importance of extension of applications of Organ-on Chip in cancer treatment is reported [21]. A microfluidics chip for regenerating lung function is developed. It is revealed that cyclic mechanical strain stimulates toxic and inflammatory responses in the lungs [4]. Microfluidic chips for investigation of axonal biology in central nervous systems is proposed and analyzed [22], for a jamming event in sickle cell disease. It provides a quantitative understanding of the rate-limiting process and opens the frontier in identifying new therapies [23], and for studying growth cone migration including shear stress [24].

4 Kidney

Kidney is an essential organ which plays several crucial tasks in normal body functioning. The distinct cell arrangement in the kidney allows it to perform 30 distinct operations. Kidney has 300,000–1,000,000 nephrons; the number depends on the birth weight [25, 26]. A nephron is the fundamental working part of the kidney, see Fig. 2. The nephron comprises of the glomerulus, proximal convoluted tubule (PCT), thin descending limb, thick ascending limb, distal convoluted tubule (DCT), and inner medullary collecting duct. The 20–25% blood pumped from the heart is directed into the kidney. The blood enters the kidney through veins, and these veins divide into capillaries to direct the blood into the nephrons. The blood enters into nephron through afferent arteriole and leaves through the efferent arteriole. Glomerulus filtration is done with the sophisticated arrangement of capillaries. RBC's and WBC's

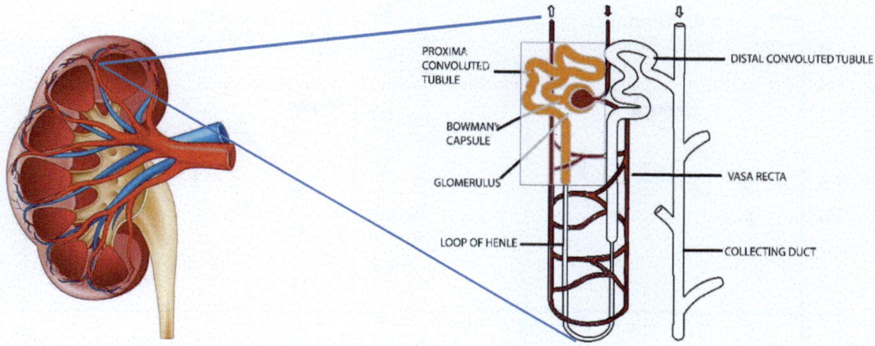

Fig. 2 Kidney and nephron. Reproduced with permission from Ref. [28]

are filtered in the glomerulus, and the rest of the blood contents are delivered into Bowman's capsule (a cup-like structure). The lumen fluid in the Boman's capsule is directed into PCT where 60% of the solutes and proteins are reabsorbed back into the bloodstream. All membranes in the body are highly selective to allow the flow of certain fluids and liquids mainly red cells. Proximal tubule cells are more permeable to water because of the presence of Aquaporins (AQP-1-10). An hourglass model to show the Aquaporin-1 (AQP-1) action is presented [27].

The rest of the tubule network will be re-absorbing the remaining 30% of the lumen fluid. There exist different types of transports namely cellular transport, extracellular transport, direct transport, facilitated transport in the nephron tubule network which allows the excretion, secretion, and reabsorption of solutes. Cellular metabolism is reflected by the transport of ions and organic solutes across the cell membrane in the human body and with the extracellular environment. Apart from above, the fluid shear stress finds a bright place in the functionality of kidney [29–31]. Like endothelial cells lining, the vasculature renal tubular epithelial cells likely experience at least three types of mechanical forces in response to variations in urinary flow rate: hydrostatic pressure, circumferential stretch, and fluid flow-induced shear or drag forces [32, 33]. Hydraulic pressure can induce tubular epithelial-myofibroblast transdifferentiation (TEMT) in a pressure magnitude- and duration-dependent manner [34].

The shear stress in medullary governs the production of nitric oxide (NO). The nitric oxide in inner medullary collecting duct (IMCD) regulates the reabsorption of sodium and water. The increase in fluid shear stress increases the production NO. By experimenting, it is noted that the expression of three NO isoforms (NOS-1, NOS-2, and NOS-3 (they produce NO when pressurized [35]). The effect of low and stretch on calcium levels in principle and intercalated cells is presented in [36]. They demonstrated that an increase in flow rate produces a three-fold increase in Ca^{2+} response within 10 s [36]. Fluid shear stress (FSS) can potentially affect and modulate re-absorption and excretion of Na^{+}, K^{+}, and HCO_3 and Cl^{-} [37–41]. The paper that deals with the flow-dependent reabsorption of Na^{+} [42]. An increase in luminal diameter will affect the flow velocity in microvilli's. The urine is generated

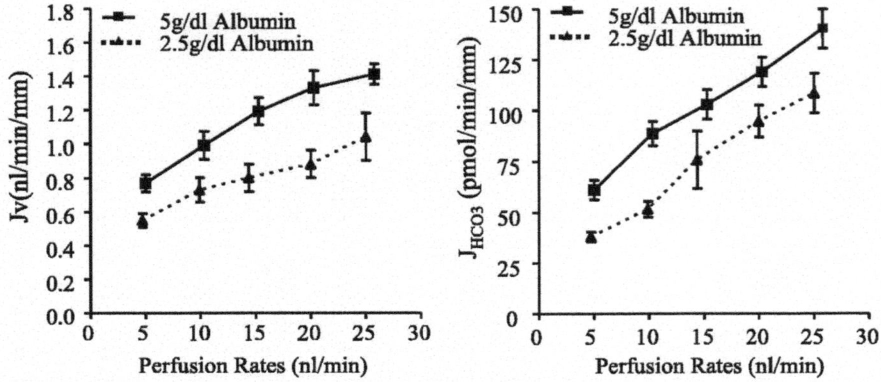

Fig. 3 Effect of different bath albumin concentrations on J_v of Na^+ and J_{HCO3} in isolated proximal tubules of mouse kidney. Data show that an increased albumin concentration from 2.5 to 5 g/dl enhanced J_v and J_{HCO3} at all flow rates and was additive with the flow-dependent changes in tubular transport. Reproduced with permission from Ref. [39]

by kidney tubules through most of the electrolytes that are transported back to the plasma, eventually leaving the waste solutes in the luminal fluid. For identification of volumetric changes inside the tubules, insulin is added to the solution of luminal perfusate. The equations defining the drag force is derived and validated. For this, different mice kidney were utilized [42] (Fig. 3).

Apart from filtering blood, the kidney also performs solute clearance, excretion, secretion, blood homeostasis maintenance, regulation of blood pressure, regulation of protein level in blood. The diet and heredity affect drugs (especially pain killers) kidney adversely. The decline of kidney function can be estimated by the number of working nephrons in it. Kidney failure or kidney disease follows the process: acute kidney injury (AKI), chronic kidney disease (CKD), and end-stage kidney disease (ESRD). The ESRD patient is in the requirement of replacement of the organ. If the CKD is assessed in the earlier stages, the kidney disease is curable by maintaining a proper diet and suitable medical intervention.

The renal replacement therapy that is presently in use is the dialysis technique. The dialysis has invented long ago in 1960, where filtration of blood is done by passing it into tiny tubules that are surrounded by pure water. The dialysis generally of two types: hemodialysis and peritoneal dialysis. In hemodialysis, a fistula is made to draw blood from the patient body. The blood that is drawn from the body is allowed to pass into a dialysis machine which is in a nearby dialysis center. The machine is composed of dialysate (a special fluid to treat blood), pressure pumps, flow pumps, and several tiny tubes. The impure blood will enter the machine, and purified blood will be drawn out which will be redirected into the patient body. In peritoneal dialysis, a catheter will be implanted into the patient's body in the abdominal cavity. The filtration will be done inside the body. From the starting stages, dialysis did not see many technical changes, even though neighboring machines to treat different diseases got revolutionized with technology improvement. Recently, a wearable dialysis bag is developed and is being utilized, but in terms of cost, it is huge.

There are majorly two types of dialysis techniques: hemodialysis and peritoneal dialysis. In hemodialysis, the blood is drawn out of patients to feed into a bulkier machine where the blood cleaning is performed. Once the purification of blood is completed, the purified blood is again injected back into the patient's body. Whereas in peritoneal dialysis, a catheter is implanted into the patient's abdomen, and the blood cleaning happens in that catheter itself (the catheter is supplied with dialysate). Dialysis is a time taking process, averaging 4–6 h per treatment. On top of that, it is very costly, painful, not user-friendly, need of a trained person and less availability. The important aspect to look at is it provides only 20% of kidney operation. It is true that dialysis is nowhere providing the patient body requirements. Kidney transplantation is another technique to fight the disease. The process is isolating a healthy kidney from a donor and implant it in the patient body. The implanted kidney will be placed in parallel working with the failed kidney. Initially, transplantation has a very less success rate, but presently with the advent of medical tools, it is possible to improve the success rate to 90%. The scarcity of donating organs is a major limiting factor for the kidney transplantation. 5 patients out of 100 are actually receiving the transplant. There is no technology for replacing the kidney failure, but Kidney-on-Chip.

The statistics of kidney disease are devastating, 735,000 deaths are happening annually, kidney failure statistics in India [43], see Fig. 3. Kidney disease associated with diabetes and/or blood pressure is a big problem which gradually leads to death of a person, in this case, the fatality rate is very high. The curing of this is associated with costlier interventions, providing less relief to the patient. There is a burning need to develop new technology to overcome this ever-increasing problem one of the better solutions will be Kidney-on-Chip technology. This has gained interest presently and is being rigorously researched.

The Kidney-on-Chip (KOC) development can be divided into two categories (1) Disease diagnosis, (2) Kidney replacement.

5 Disease Diagnosis

As explained earlier, if kidney disease is diagnosed in the early stages of AKI or CKD, certainly the cure is possible. The kidney may be injured in different ways, of them, drug-induced kidney damage is very fatal. It is generally seen in pharmacy therapy, the dosage of the drug decides the amount of damage. Extensive efforts have been conducted to accurately predict nephrotoxicity. Animal models are much trusted for their fast response. But, the animal models are not suitable as they cannot predict the problem accurately. 2D and in vitro models for testing and predicting have gained interest in later stages. The cells that are cultured in such techniques are carried out in highly sophisticated labs. The cultured cells are placed in bioengineered containers that specially developed to partially replicate the human body (replicating human

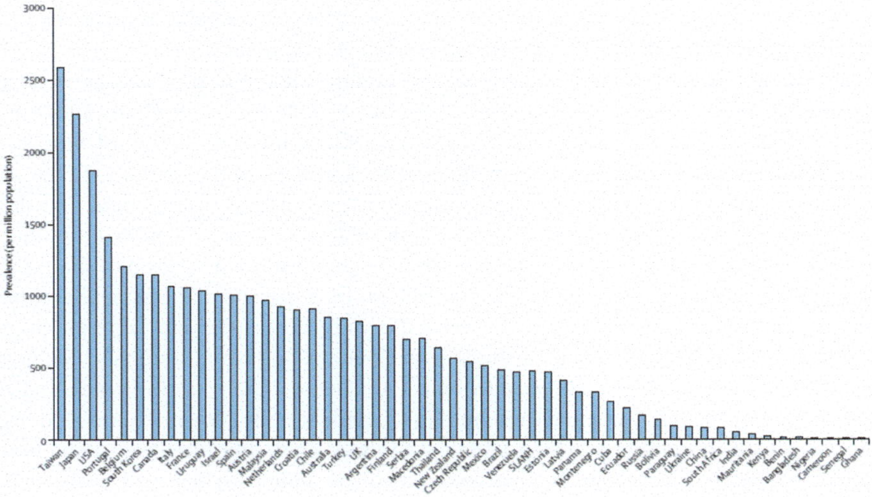

Fig. 4 Annual incidence of end-stage kidney disease in different countries. Reproduced with permission from Ref. [43]

body environment). These techniques are extremely complicated and are associated with burdensome process. The ultimate response will not be satisfying. In this context, KOC will perform better. There are crucial developments that contributed to the diagnosis of kidney damage through KOC. Prediction of drug-induced nephrotoxicity using microfluidics technology presented for the first time. Cultured proximal tubule cell in a microfluidics device on top of a porous membrane. The cells were exposed to fluid shear stress to estimate the proximal tubule functioning [44]. Furthermore, calcium phosphate stone formation in the nephron is examined and analyzed by utilizing microfluidic channels [45].

6 Artificial Kidney

In the development of artificial kidney biologists, medical practitioners, bio-engineers, and microfluidic researchers are collaborating to realize more realistic techniques for serving the kidney failure. The design of KOC can be divided into two stages (1) Realization ultrafilter, (2) Realization of bio-reactor as depicted in Fig. 5. This method is hypothesized based on the analogy that glomerulus acts as ultrafilter. And Tubule network which reacts with lumen fluid for solute clearance, reabsorption, excretion, and secretion act as bio-reactor. However, these two should be integrated in a proper way that both produce no complication upon association.

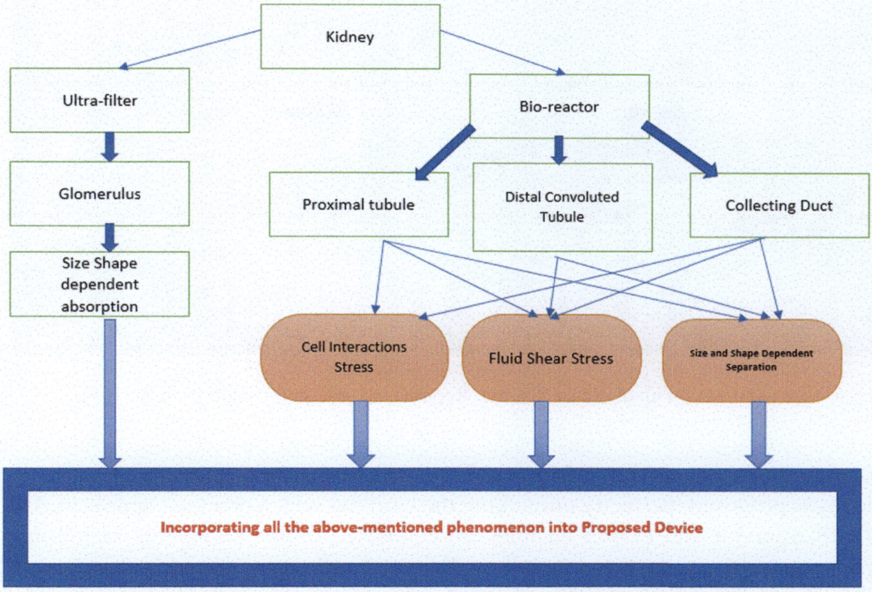

Fig. 5 The block diagram of Kidney-on-Chip demonstrating the possible nodes that must be incorporated by designer

7 Ultrafilter

Blood enters nephron firstly into the glomerulus, where most of the solutes get filtered. The glomerulus is a sponge-like structure, where the pores uniformly distributed in equidistance with equal pore size. Other than RBC's, WBC's and big molecules, all the left out ingredients are filtered out in glomerulus. The finer arrangement of the pores and collecting chamber of filtered fluid (Bowman's capsule) provides ultrafiltration. To realize this function, Si-nano pore technology has been developed (Fig. 6).

8 Bio-reactor

Once the blood enters the glomerulus, it gets filtered and the filtrate will be collected in Bowman's capsule. The lumen fluid (plasma) will be directed into the tubule network. While passing through this tubule network, the whole lumen fluid will be reabsorbed back into the blood. On average kidneys filters 180 L/day, but the urine output is 2–2.5 L/day the rest is reabsorbed back. It is predicted that the complete process can be achieved on a chip with the help of microfluidics technology.

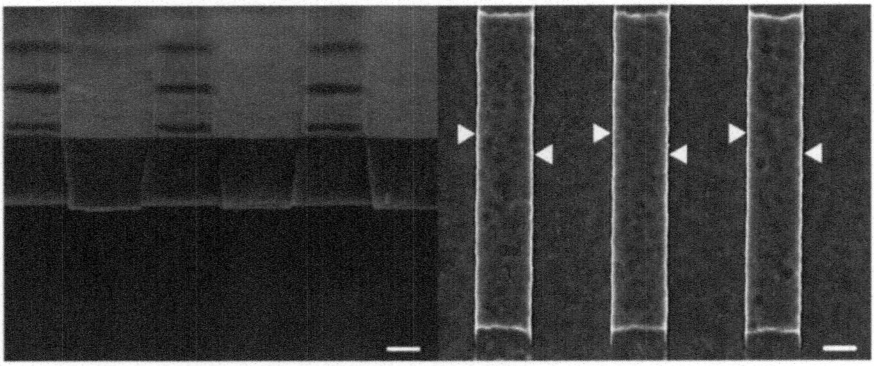

Fig. 6 The nanopores with diameter 5–8 nm. Reproduced with permission from Ref. [46]

A detailed explanation of the microfluidic chip for analysis of the working of the kidney is proposed, perfectly mimicking the working with a practical approach. The PDMS reservoir and PDMS channel are connected by the paralyzing membrane. Where the reservoir is composed of medium and cells are cultured in the channel. After three days of confluency, the cells are exposed to a fluid shear stress of 1 dyn/cm^2. Cell viability analysis is also conducted [1]. Following this, blood cleaning is achieved using optical trapping technique. The trapping forces are formed by the mutual working of the gradient field and scattering photons [47]. The Ca^{2+} increase with an increase in flow rate is discussed in [48]. Autosomal recessive polycystic kidney disease (ARPKD) is characterized using mechano-induced Ca^{2+} signaling [48]. The paper dealing with the flow-dependent reabsorption of Na^+. An increase in luminal diameter will affect the flow velocity in microvilli's. The urine is generated by kidney tubules through most of the electrolytes that are transported back to the plasma eventually leaving the waste solutes in the luminal fluid. For identification of volumetric changes, insulin is added to the solution of luminal perfusate. The equations defining the drag force is derived and validated. For this, different mice kidney cells were utilized [42].

In practical, it is not possible to make a full artificial kidney on a single chip at one go, as the functions to recreate are extremely complex and big in number. The research started to venture towards regenerating different parts of nephron individually. The major parts of nephron like glomerulus, proximal tubule, distal tubule, and collecting duct were artificially generated. Li Wang et al. developed a model of glomerulus-on-chip for studying diabetic nephropathy. A less complex model with three channels namely glomerulus inlet, gel inlet, and collecting unit. The fluid that enters the glomerulus inlet will pass through a gel containing a channel to reach the collecting unit [49]. Likewise, Mengying Zhou et al. developed a disease model for hypertensive nephropathy [50] (Fig. 7).

There are attempts to mimic proximal tubule functions by culturing cells on membranes (a note about the membranes used in the KOC is presented). Kidney toxicity estimation was achieved by mimicking only PCT function which expresses

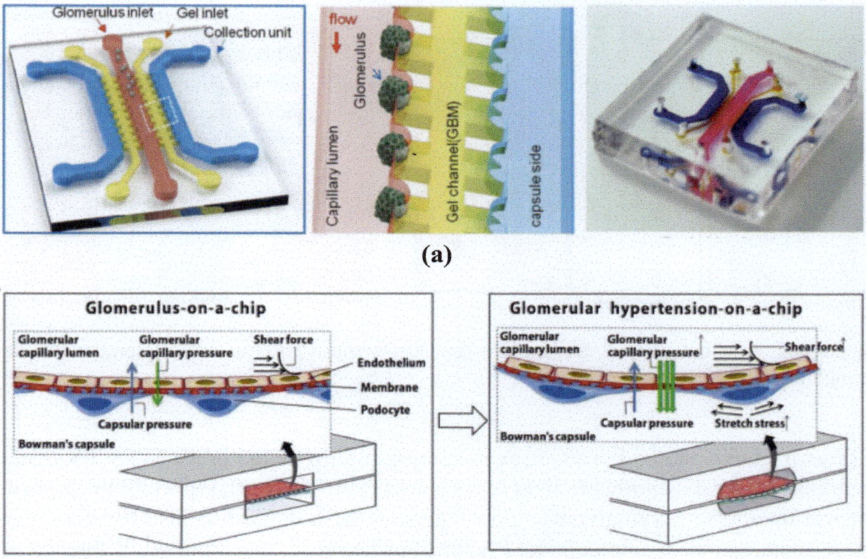

(a)

(b)

Fig. 7 **a** Model for predicting diabetic nephropathy. Reproduced with permission from Ref. [49]. **b** Model for estimating hypertensive nephropathy. Reproduced with permission from Ref. [50]

marker proteins, secretion, and reabsorption functions [51, 52]. The drug transport interactions were analyzed by the 3D microfluidic model [53]. The realization extracellular matrix of the PCT to mimick the reabsorption of glucose and album is presented [54], for more clear insight the reader may refer to [55].

As discussed earlier reabsorption in kidney tubules is an essential function to retain required solutes and electrolytes. The phenomenon behind the reabsorption is not fully established yet. Transport in cells can occur in many different ways such as diffusion, endocytosis, basolateral, apical, kiss and run, cellular, paracellular, hyper osmatic, molecular, etc. It is highly uncertain to estimate which transport actually works at a given particular instance. Reabsorption is majorly dependent on particle size, shape, charge, and fluid shear stress. In the same manner, our previous works includes regeneration of proximal tubule functions, more specifically the reabsorption function. Our design has comprised a total of three channels: a main tubule and two blood tubules. The blood tubules are placed on both sides of the main tubule and all are connected through transport channels as shown Fig. 8. The transporting channels allow the flow of solutes into the blood tubule selectively depending on the size. It is observed that the FSS and hydraulic pressure makes considerable effects on the reabsorption [28, 56, 57]. Collecting duct is the extreme part of the nephron, the secretion, and reabsorption of electrolytes, and fluid balance of final lumen fluid is carried out in here. The fluid that exits the collecting duct can be considered as urine. The possibility of disease diagnosis through analysis of collecting duct working is

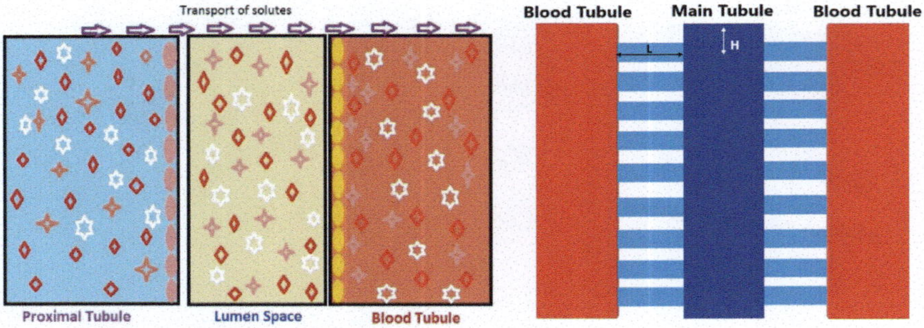

Fig. 8 The model developed for the size-dependent reabsorption of PCT. Reproduced with permission from Ref. [28]

easy as it carries the end product. A collecting-o-chip model considering FSS, and the transepithelial osmotic gradient is presented [58]. The work reports actin depolymerization and re-polymerization depending on the luminal FSS and regulation of the reorganization of intercellular F-actin. Furthermore, the immense influence of FSS on the AQP2 trafficking and cytoskeleton reorganization is observed.

9 Fabrication

Another significant aspect in any microfluidic device production is its fabrication. Fabrication my include any of these four arrangements: membranes, microchannels, gell based, and tubule and vessel embedded. Nowadays, membrane-based biodevices are gaining interest. To crucially categorize, there are two types of membranes: apical and basolateral. The cell membrane functions include molecular recognition, enzymetric catalysis, cellular adhesion, and membrane fusion. In actual sense, the membranes are made of thermoplastic or elastometric materials to support the heavy flow. The selection of materials for a device holds a huge impact on device acceptance and reliability, especially in the medical device industry it is more critical [59]. The selection of biocompatible materials for in-vivo devices is compulsory. Presently, the trusted materials are polydimithyl siloxane (PDMS) [60] and treated silicon [61]. PDMS is a polymeric material that has advantages like transparent, high bio-compatibility with body fluidics, and easily moldable. The fabrication process of PDMS is very simple and involves less cost; it is reported that reliability is also moderate. The problem comes with PDMS is, it is not possible to fabricate high aspect ratio structures owing to its moldable nature. On the other hand, silicon has a matured fabrication process. It is possible to make any kind of structure, but the need for surface modification is troubling the researchers. Recently, people are also

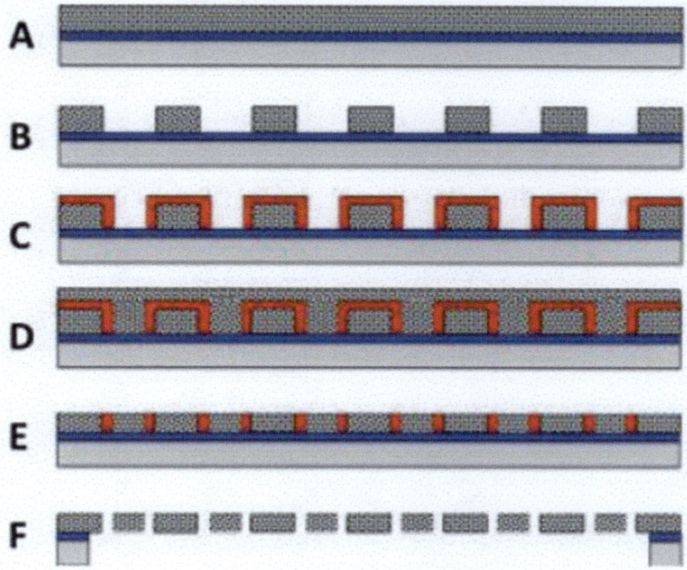

Fig. 9 Fabrication steps of silicon nano pore membranes, **a** Thin layers of silicon dioxide (blue) and polycrystalline silicon ("polysilicon," stippled gray) are deposited on a silicon wafer (gray). **b** The polysilicon layer is patterned into parallel bars using standard photolithography. **c** A 5–7 nm-thick conformal layer of silicon oxide (red) is grown on the polysilicon. **d** A second layer of polysilicon is deposited (stippled gray). **e** The surface is planarized and thinned to expose the 5–7 nm veins of silicon oxide (red). **f** The oxide is etched away to create 5–7 nm pores, and windows are etched into the wafer to expose the backside of the thin-film membrane. Reproduced with permission from Ref. [46]

using material mixtures such as polyether sulfon and PVP mixed with N-methyl-2-pyrrolidone [62]. Soft lithography is gaining interest nowadays owing to its advantages over conventional fabrication techniques. The state of art is possible to fabricate structures of atomic-scale sizes [63–67]. It is up to the designer to select the material for fabrication based on his/her device model complexity. The fabrication and material selection for microfluidics are discussed. The focus on the mechanical operation principles of microfluidics is presented [68] (Figs. 9 and 10).

10 Cell Culture

The functioning versatility of any organ is because of the behavior of cells continuously working inside it. In all the reported works reviewed above utilize living cells extracted from either animals or humans. The Organs-on-Chip differs from 2D models in the aspect of more precise results for this reason. Cell culturing is a huge process involving highly complicated machinery, intensive care, and trained persons

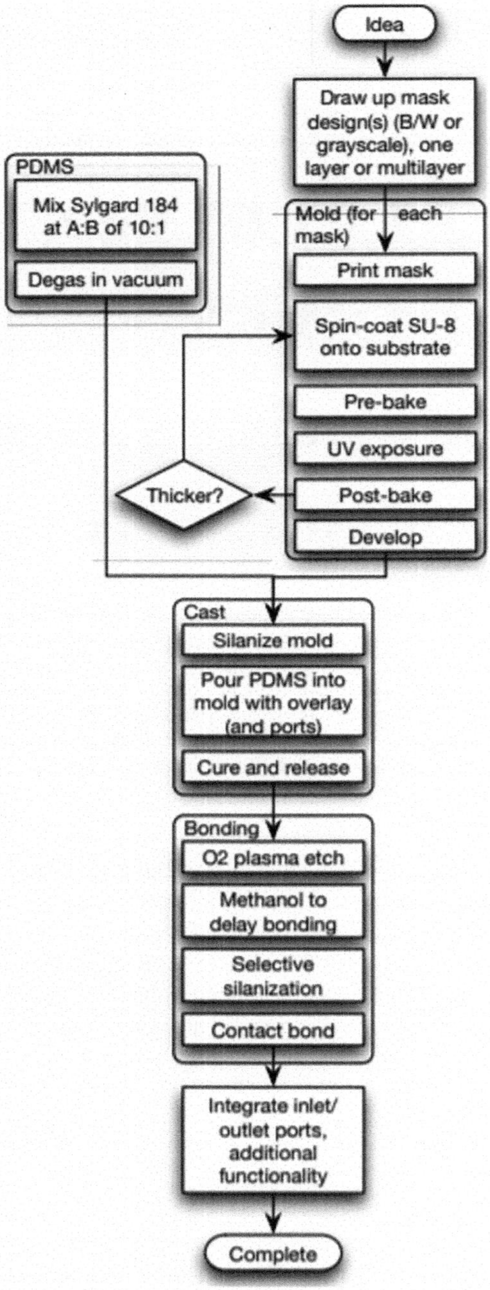

Fig. 10 Process of fabrication of PDMS microfluidic devices. Reproduced with permission from Ref. [60]

to carry the work. The kind of cells used in kidney regeneration is human primary renal proximal tubule cells (HPTC's), adult renal stem/progenator cells (ARPCs), glomerular epithelial cells (GECs), opossum kidney cells (OK), and rat kidney cells [69–75]. However, cells require intensive attention to harvest and culture. The process consumes a huge amount of work time, the chemical composition for cells is very peculiar, the behavior of cells changes with time and kind of chemicals used. The involvement of big labs and expensive equipment turns the cell culture platforms costly and unaffordable upon marketization. The life of devices equipped with cell culture are considerably low making it viable only for the diagnostic purpose (use and throw). The device that is implanted inside a body should be long-lasting, but cell cultured devices raise complications in this context. In a practical way, there is no replacement found till now for the cell culture, the practicality, accuracy, and response of these systems are very high, upcoming researchers should pave new ways to eradicate these issues.

While culturing the cells, every designer wishes to achieve all the below-mentioned targets. The precise model should have all these characteristics. Cell viability, reliability, availability, activity, reliability at different flowrate (for hypertension and diabetic patients), all the functions of the kidney, no coagulation of blood and immunostaining. To date, there is no model to satisfy the above-mentioned requirements, hence showing a large room for upcoming researchers in this field.

11 Conclusions

Any organ in the body (any cell) is hard to replace, taking good care of the body is crucial for healthy living. The habits are primary enemies for the body that are making it to malfunction. For the dysfunction of any organ, replacement might be a compulsory solution. The research on regenerating all organs is going on very swiftly. Artificial heart (pacemaker), artificial blood vessel is matured now, but the significance to develop the rest of organs is also essential. The research is being conducted to develop kidney, liver, lungs, bran, etc. Most importantly, artificial kidney research has seen a lot of advancements. The fluidic effects like FSS, hydrodynamic pressure, flow velocity in kidney tubules that are crucial for kidney function have reveled to a more precise level. There are reported works that developed diagnosis devices for renal dysfunction which can provide high accuracy and quick response [76]. In-vitro models realizing actual kidney function using microfluidics technology has been attempted and are partially successful. The development of every specific kidney function is underway (Like reabsorption, excretion, secretion, etc.) But, still there is no commercialized artificial kidney that can perform all the required functions (at least 50% of all kidney functions). The dream of replacing a failed kidney with fully developed kidney-on-chip is yet to be fulfilled.

12 Future Scope

The technology is progressing, the possibilities to mimic an entire organ will increase in the coming future. Apparently, the development of body-on-chip will become an achievable target. We can predict, there will be personlized diagnosis and treatment. Kidney-on-Chip has a lot of areas to explore for new researchers. The still usage of living cells for diagnosis and replacement is a major concern to look at. It is highly important to make everything artificial to fulfill the coined term, artificial kidney. The replacement of cells is not yet found, it may be a good option to artificially regenerate a single cell anatomy and functions. Furthermore, the usage of different separation techniques like electrophoresis and di-electrophoresis is limited because of the involvement of electrical potential. A new researcher can find a way to use these techniques by utilizing very little voltage (which does not affect other body functions and alter blood characteristics). The exploration of magnetophorosis and accoustophorosis for particle separation may provide some interesting outcomes. The coagulation of blood inside the device once it is transplanted is another mammoth limiting factor for social approval and marketization of artificial kidney. The continuous supply of anti-coagulants for a transplanted device is not possible. As discussed in the objectives of the cell culture section, all the cell-related objectives do require intensive research for device commercialization.

References

1. Jang, K.J., Suh, K.Y.: A multi-layer microfluidic device for efficient Culture and analysis of renal tubular cells. Lab Chip **10**, 36–42 (2010)
2. Maschmeyer, I., Lorenz, A.K., Schimek, K., Hasenberg, T., Ramme, A.P., Hubner, J., Lindner, M., Drewell, C., Bauer, S., Thomas, A., Sambo, N.S., Sonntag, F., Lauster, R., Marx, U.: A four-organ-chip for interconnected long-term co-culture of human intestine, liver, skin and kidney equivalents. Lab Chip **15**, 2688–2699 (2015)
3. Kimura, H., Yamamoto, T., Sakai, H., Sakai, Y., Fujii, T.: An integrated microfluidic system for long-term perfusion culture and on-line monitoring of intestinal tissue models. Lab Chip **8**, 741–746 (2008)
4. Huh, D., Matthews, B.D., Mammoto, A., Montoya-Zavala, M., Hsin, H.Y., Ingber, D.E.: Reconstituting organ-level lung functions on a chip. Science **328**, 1662–1668 (2010)
5. Tsai, M., Kita, A., Leach, J., Rounsevell, R., Huang, J.N., Moake, J., Ware, R.E., Fletcher, D.A., Lam, W.A.: In vitro modeling of the microvascular occlusion and thrombosis that occur in hematologic diseases using microfluidic technology. J. Clin. Invest. **122**, 408–418 (2012)
6. Walsh, C.L., Babin, B.M., Kasinskas, R.W., Foster, J.A., McGarry, M.J., Forbes, N.S.: A multipurpose microfluidic device designed to mimic micro-environment gradients and develop targeted cancer therapeutics. Lab Chip **9**, 545–554 (2009)
7. Torisawa, Y.S., Spina, C.S., Mammoto, T., Mammoto, A., Weaver, J.C., Tat, T., Collins, J.J., Ingber, D.E.: Bone marrow-on-a-chip replicates hematopoi-Eticniche physiology in-vitro. Nat. Methods **11**, 663–669 (2014)
8. Zhang, C., Zhao, Z., Abdul Rahim, N.A., van Noort, D., Yu, H.: Lab Chip **9**(22), 3185–3192 (2009)
9. Fitzpatrick S, Sprando R (2019) Advancing regulatory science through innovation: in vitro microphysiological systems. Cell. Mol. Gastroenterol. Hepatol. **7**(1), 239

10. Liu, Y., Gill, E., Huang, Y.Y.S.: Microfluidic on-chip biomimicry for 3D cell culture: a ft-for-purpose investigation from the end user standpoint. Future Sci. OA **3**(2), FSO173 (2017)
11. Williamson, A., Singh, S., Fernekorn, U., Schober, A.: The future of the patient-specific body-on-a-chip. Lab Chip **13**, 3471–3480 (2013). https://doi.org/10.1039/c3lc50237fhttps://doi.org/10.1039/c3lc50237f
12. Kim, K., Ohashi, K., Utoh, R., Kano, K., Okano, T.: Preserved liver-specific functions of hepatocytes in 3D co-culture with endothelial cell sheets. Biomaterials **33**, 1406–1413 (2012). https://doi.org/10.1016/j.biomaterials.2011.10.084https://doi.org/10.1016/j.biomaterials.2011.10.084
13. Xu, Y., Jang, K., Yamashita, T., Tanaka, Y., Mawatari, K., Kitamori, T.: Microchip-based cellular biochemical systems for practical applications and fundamental research: from microfluidics to nanofluidics. Anal. Bioanal. Chem. **402**, 99–107 (2012). https://doi.org/10.1007/s00216-011-5296-5https://doi.org/10.1007/s00216-011-5296-5
14. El-Ali, J., Sorger, P.K., Jensen, K.F.: Cells on chips. Nature **442** (2006). https://doi.org/10.1038/nature05063
15. Bhushan, A., Martucci, N.J., Usta, O.B., Yarmush, M.L.: New technologies in drug metabolism and toxicity screening: organ-to-organ interaction. Expert Opin. Drug Metab. Toxicol. **12**, 475–477 (2016). https://doi.org/10.1517/17425255.2016.1162292 (PMC free article)
16. Esch, E.W., Bahinski, A., Huh, D.: Organs-on-chips at the frontiers of drug discovery. Nat. Rev. Drug Discov. **14**, 248–260 (2015). https://doi.org/10.1038/nrd4539[PMCfreearticle]https://doi.org/10.1038/nrd4539[PMCfreearticle]
17. Toh, Y.C., Lim, T.C., Tai, D., Xiao, G., van Noort, D., Yu, H.: A microfluidic 3D hepatocyte chip for drug toxicity testing. Lab Chip **9**, 2026–2035 (2009). https://doi.org/10.1039/b900912dhttps://doi.org/10.1039/b900912d
18. Ghaemmaghami, A.M., Hancock, M.J., Harrington, H., Kaji, H., Khademhosseini, A.: Biomimetic tissues on a chip for drug discovery. Drug Discov. Today **17**, 173–181 (2012). https://doi.org/10.1016/j.drudis.2011.10.029https://doi.org/10.1016/j.drudis.2011.10.029
19. No, D.Y., Lee, K.H., Lee, J., Lee, S.H.: 3D liver models on a microplatform: well-defined culture, engineering of liver tissue and liver-on-a-chip. Lab Chip **15**, 3822–3837 (2015). https://doi.org/10.1039/C5LC00611Bhttps://doi.org/10.1039/C5LC00611B
20. Nagrath, S., Sequist, L.V., Maheswaran, S., Bell, D.W., Irimia, D., Ulkus, L., Smith, M.R., Kwak, E.L., Digumarthy, S., Muzikansky, A., Ryan, P., Balis, U.J., Tompkins, R.G., Haber, D.A., Toner, M.: Isolation of rare circulating tumour cells in cancer patients by microchip technology. Nature **450** (2007)
21. Young, E.W.K.: Cells, tissues, and organs on chips: challenges and opportunities for the cancer tumor microenvironment. Integr. Biol. **5**(9), 1096–1109 (2013). https://doi.org/10.1039/c3ib40076jhttps://doi.org/10.1039/c3ib40076j
22. Taylor, A.M., Blurton-Jones, M., Rhee, S.W., Cribbs, D.H., Cotman, C.W., Jeon, N.L.: A microfluidic culture platform for CNS axonal injury, regeneration and transport. Nat. Methods **2**(8), 599 (2005)
23. Higgins, J.M., Eddington, D.T., Bhatia, S.N., Mahadevan, L.: Sickle cell vasoocclusion and rescue in a microfluidic device. PNAS **104**(51), 20496–20500 (2007)
24. Joanne Wang, C., Li, X., Lin, B., Shim, S., Ming, G., Levchenko, A.: A microfluidics-based turning assay reveals complex growth cone responses to integrated gradients of substrate-bound ECM molecules and diffusible guidance cues. Lab Chip **8**, 227–237 (2008)
25. McNamara, B.J., Diouf, B., Douglas-Denton, R.N., Hughson, M.D., Hoy, W.E., Bertram, J.F.: Comparison of nephron number, glomerular volume and kidney weight in Senegalese Africans and African Americans. Nephrol. Dial Transplant. **25**, 1514–1520 (2010)
26. Manalich, A., Reyes, L., Herrera, M., Melendi, C., Fundora, I.: Relationship between weight at birth and the number and size of renal glomeruli in humans: a histomorphometric study. Kidney Int. **58**, 770–777 (2000)
27. Jung, J.S., Preston, G.M., Smith, B.L., Gugginoll, W.B., Agre, P.: Molecular structure of the water channel through aquaporin CHIP. J. Biol. Chem. **269**(20), 14648–14654 (1994)

28. Sateesh, J., Guha, K., Dutta, A., Sengupta, P., Agarwal, A., Srinivasa Rao, K.: Recreating the size dependent re-absorption function of proximal convoluted tubule towards artificial kidney applications-structural analysis and computational study. Artif. Organs (2020)
29. Baquet, A., Gaussin, V., Bollen, M., Stalmans, W., Hue, L.: Mechanism of activation of liver acetyl-CoA carboxylase by cell swelling. Eur. J. Biochem. FEBS **217**, 1083–1089 (1993)
30. Peak, M., Al-Habori, M., Agius, L.: Regulation of glycogen synthesis and glycolysis by insulin, pH and cell volume. Interactions between swelling and alkalinization in mediating the effects of insulin. Biochem. J. **282**(3), 797–805 (1992)
31. Hamill, O.P., Martinac, B.: Molecular basis of mechanotransduction in living cells. Physiol. Rev. **81**, 685–740 (2001)
32. Davies, P.F.: Flow-mediated endothelial mechanotransduction. Physiol. Rev. **75**, 519–560 (1995)
33. Traub, O., Berk, B.C.: Laminar shear stress: mechanisms by which endothelial cells transduce an atheroprotective force. Arterioscler. Thromb. Vasc. Biol. **18**, 677–685 (1998)
34. Li, F., Xie, X., Fan, J., Li, Z., Wu, J., Zheng, R.: Hydraulic pressure inducing renal tubular epithelial-myofibroblast transdifferentiation in vitro. J. Zhejiang Univ. Sci. B **10**(9), 659–667 (2009)
35. Cai, Z., Xin, J., Pollock, D.M., Pollock, J.S.: Shear stress-mediated NO production in inner medullary collecting duct cells. Am. J. Physiol. Renal. Physiol. **279**, F270–F274 (2000)
36. Liu, W., Xu, S., Woda, C., Kim, P., Weinbaum, S., Satlin, L.M.: Effect of flow and stretch on the $[Ca^{2+}]_i$ response of principal and intercalated cells in cortical collecting duct. Am. J. Physiol. Renal. Physiol. **285**, F998–F1012 (2003)
37. Schnermann, J., Wahl, M., Liebau, G., Fischbach, H.: Balance between tubular flow rate and net fluid reabsorption in the proximal convolution of the rat kidney. I. Dependency of reabsorptive net fluid flux upon proximal tubular surface area at spontaneous variations of filtration rate. Pflugers Arch. **304**, 90–103 (1968)
38. Giebisch, G., Windhager, E.E.: Characterization of renal tubular transport of sodium chloride and water as studied in single nephrons. Am. J. Med. **34**, 1–6 (1963)
39. Du, Z., et al.: Axial flow modulates proximal tubule NHE3 and H-ATPase activities by changing microvillus bending moments. Am. J. Physiol. Renal. Physiol. **290**, F289–F296 (2006)
40. Malnic, G., Berliner, R.W., Giebisch, G.: Flow dependence of K+ secretion in cortical distal tubules of the rat. Am. J. Physiol. **256**, F932–F941 (1989)
41. Satlin, L.M., Sheng, S., Woda, C.B., Kleyman, T.R.: Epithelial Na(+) channels are regulated by flow. Am. J. Physiol. Renal. Physiol. **280**, F1010–F1018 (2001)
42. Du, Z., Duan, Y., Yan, Q.S., Weinstein, A.M., Weinbaum, S., Wang, T.: Mechanosensory function of microvilli of the kidney proximal tubule. PNAS **101**(35), 13068–13073 (2004)
43. Jha, V., Garcia-Garcia, G., Iseki, K., Li, Z., Naicker, S., Plattner, B., Saran, R., Wang, A.M., Yang, C.W.: Chronic kidney disease: global dimension and perspectives. Lancet **382**(9888), 260–272 (2013)
44. Jang, K.-J., Mehr, A.P., Hamilton, G.A., McPartlin, L.A., Chung, S., Suh, K.-Y., Ingber, D.E.: Human kidney proximal tubule-on-a-chip for drug transport and nephrotoxicity assessment. Integr. Biol. **5**, 1119 (2013)
45. Weia, Z., Amponsah, P.K., Al-Shatti, M., Nie, Z., Bandyopadhyay, B.: Engineering of polarized tubular structures in a microfluidic device to study calcium phosphate stone formation. Lab Chip **12**(20), 4037–4040 (2012)
46. Kensinger, C., et al.: First implantation of silicon nanopore membrane hemofilters. ASAIO J. (Am. Soc. Artif. Inter. Org. 1992) **62**(4), 491–495 (2016). https://doi.org/10.1097/MAT.0000000000000367
47. Suwanpayak, N., Jalil, M.A., Aziz, M.S., Ismail, F.D., Ali, J., Yupapin, P.P.: Blood cleaner on-chip design for artificial human kidney manipulation. Int. J. Nanomed. **6**, 957–964 (2011)
48. Liu, W., Murcia, N.S., Duan, Yi., Weinbaum, S., Yoder, B.K., Schwiebert, E., Satlin, L.M.: Mechanoregulation of intracellular Ca2 concentration is attenuated in collecting duct of monocilium-impaired orpk mice. Am. J. Physiol. Renal. Physiol. **289**, F978–F988 (2005)

49. Wang, L., Tao, T., Su, W., Yu, H., Yu, Y., Qin, J.: A disease model of diabetic nephropathy in a glomerulus-on-a-chip microdevice. Lab Chip **17**(10), 1749–1760 (2017)
50. Zhou, M., et al.: Development of a functional glomerulus at the organ level on a chip to mimic hypertensive nephropathy. Sci. Rep. **6**, 31771 (2016). https://doi.org/10.1038/srep31771
51. Weber, E.J., Chapron, A., Chapron, B.D., Voellinger, J.L., Lidberg, K.A., Yeung, C.K., Wang, Z., Yamaura, Y., Hailey, D.W., Neumann, T., Shen, D.D., Thummel, K.E., Muczynski, K.A., Himmelfarb, J., Kelly, E.J.: Development of a microphysiological model of human kidney proximal tubule function. Kidney Int. **90**(3), 627–637 (2016). ISSN 0085-2538. https://doi.org/10.1016/j.kint.2016.06.011
52. Jang, K.-J., Mehr, A.P., Hamilton, G.A., McPartlin, L.A., Chung, S., Suh, K.-Y., Ingber, D.E.: Human kidney proximal tubule-on-a-chip for drug transport and nephrotoxicity assessment. Integr. Biol. **5**(9), 1119–1129 (2013). https://doi.org/10.1039/c3ib40049b
53. Vriend, J., Nieskens, T.T.G., Vormann, M.K., et al.: Screening of drug-transporter interactions in a 3D microfluidic renal proximal tubule on a chip. AAPS J **20**, 87 (2018). https://doi.org/10.1208/s12248-018-0247-0https://doi.org/10.1208/s12248-018-0247-0
54. Lin, N.Y.C., Homan, K.A., Robinson, S.S., Kolesky, D.B., Duarte, N., Moisan, A., Lewis, J.A.: Renal reabsorption in 3D vascularized proximal tubule models. Proc. Natl. Acad. Sci. **116**(12), 5399–5404 (2019). https://doi.org/10.1073/pnas.1815208116
55. Weber, E.J., Chapron, A., Chapron, B.D., Voellinger, J.L., Lidberg, K.A., Yeung, C.K., Wang, Z., et al.: Development of a microphysiological model of human kidney proximal tubule function. Kidney Int. **90**(3), 627–637 (2016)
56. Guha, K., Sateesh, J., Dutta, A., et al.: Mimicking kidney re-absorption using microfluidics by considering hydrostatic pressure inside kidney tubules: structural and analytical study. Microsyst. Technol. (2019). https://doi.org/10.1007/s00542-019-04720-9https://doi.org/10.1007/s00542-019-04720-9
57. Sateesh, J., Guha, K., Dutta, A., et al.: Regenerating re-absorption function of proximal convoluted tubule using microfluidics for kidney-on-chip applications. SN Appl. Sci. **2**, 39 (2020). https://doi.org/10.1007/s42452-019-1840-2https://doi.org/10.1007/s42452-019-1840-2
58. Jang, K.-J., Cho, H.S., Kang, D.H., Bae, W.G., Kwon, T.-H., Suh, K.-Y.: Fluid-shear-stress-induced translocation of aquaporin-2 and reorganization of actin cytoskeleton in renal tubular epithelial cells. Integr. Biol. **3**(2), 134–141 (2011)
59. Shum, H.C., Kim, J.-W., Weitz, D.A.: Microfluidic fabrication of monodisperse biocompatible and biodegradable polymersomes with controlled permeability. J. Am. Chem. Soc. **130**(29), 9543–9549 (2008)
60. Friend, J., Yeo, L.: Fabrication of microfluidic devices using polydimethylsiloxane. Biomicrofluidics **4**(2), 026502 (2010)
61. Ren, K., Zhou, J., Hongkai, Wu.: Materials for microfluidic chip fabrication. Acc. Chem. Res. **46**(11), 2396–2406 (2013)
62. Lai, T.T., Xie, D., Zhou, C.H., Cai, G.X.: Copper-catalyzed inter/intramolecular N-alkenylation of benzimidazoles via tandem processes involving selectively mild iodination of sp3 C-H bond at α-position of ester. J. Org. Chem. **81**(19), 8806–8815 (2016)
63. Xia, Y., Whitesides, G.M.: Soft lithography. Annu. Rev. Mater. Sci. **28**(1), 153–184 (1998)
64. Whitesides, G.M., Ostuni, E., Takayama, S., Jiang, X., Ingber, D.E.: Soft lithography in biology and biochemistry. Annu. Rev. Biomed. Eng. **3**(1), 335–373 (2001)
65. Kane, R.S., Takayama, S., Ostuni, E., Ingber, D.E., Whitesides, G.M.: Patterning proteins and cells using soft lithography. Biomaterials **20**(23–24), 2363–2376 (1999)
66. Rogers, J.A., Nuzzo, R.G.: Recent progress in soft lithography. Mater. Today **8**(2), 50–56 (2005)
67. Paul, M.T.Y., Kim, D., Saha, M.S., Stumper, J., Gates, B.D.: Patterning catalyst layers with microscale features by soft lithography techniques for proton exchange membrane fuel cells. ACS Appl. Energy Mater. (2020)
68. Kim, S.M., Leeb, S.H., Suh, K.Y.: Cell research with physically modified microfluidic channels: a review. Lab Chip **8**, 1015–1023, 1015 (2008)

69. Striker, G.E., Striker, L.J.: Glomerular cell culture. Lab. Investig. J. Tech. Methods Pathol. **53**(2), 122–131 (1985)
70. Tibbitt, M.W., Anseth, K.S.: Hydrogels as extracellular matrix mimics for 3D cell culture. Biotechnol. Bioeng. **103**(4), 655–663 (2009)
71. Di Carlo, D., Wu, L.Y., Lee, L.P.: Dynamic single cell culture array. Lab on a Chip **6**(11), 1445–1449 (2006)
72. Hung, P.J., Lee, P.J., Sabounchi, P., Lin, R., Lee, L.P.: Continuous perfusion microfluidic cell culture array for high-throughput cell-based assays. Biotechnol. Bioeng. **89**(1), 1–8 (2005)
73. Nagy, A., Rossant, J., Nagy, R., Abramow-Newerly, W., Roder, J.C.: Derivation of completely cell culture-derived mice from early-passage embryonic stem cells. Proc. Natl. Acad. Sci. **90**(18), 8424–8428 (1993)
74. Edmondson, R., Broglie, J.J., Adcock, A.F., Yang, L.: Three-dimensional cell culture systems and their applications in drug discovery and cell-based biosensors. Assay Drug Dev. Technol. **12**(4), 207–218 (2014)
75. Mehling, M., Tay, S.: Microfluidic cell culture. Curr. Opin. Biotechnol. **25**, 95–102 (2014)
76. Sung, J.H., Kam, C., Shuler, M.L.: A microfluidic device for a pharmacokinetic pharmacodynamics (PK-PD) model on a chip. Lab Chip **10**, 446–455 (2010)

The manufacturer's authorised representative in the EU is Springer Nature Customer Service Centre GmbH, Europaplatz 3, 69115 Heidelberg, Germany. If you have any concerns regarding our products, please contact ProductSafety@springernature.com

Printed and bound by CPI Group (UK) Ltd, Croydon, CR0 4YY
30/09/2026
02225247-0005